Solar Cooling

S. C. Kaushik · S. K. Tyagi · V. Baiju

Solar Cooling

Basics and Advances

CP

Springer

S. C. Kaushik
Former Emeritus Professor
Indian Institute of Technology Delhi
New Delhi, India

S. K. Tyagi
Associate Professor
Indian Institute of Technology Delhi
New Delhi, India

V. Baiju
Assistant Professor
TKM College of Engineering Kollam
Kerala, India

ISBN 978-3-031-42409-0 ISBN 978-3-031-42410-6 (eBook)
https://doi.org/10.1007/978-3-031-42410-6

Jointly published with Capital Publishing Company
The print edition is not for sale in Afghanistan, Bangladesh, Bhutan, India, the Maldives, Nepal, Pakistan and Sri Lanka. Customers from Afghanistan, Bangladesh, Bhutan, India, the Maldives, Nepal, Pakistan and Sri Lanka please order the print book from: Capital Publishing Company.

This Springer imprint is published by the registered company Springer Nature Switzerland AG
The registered company address is: Gewerbestrasse 11, 6330 Cham, Switzerland

Paper in this product is recyclable.

Foreword

It gives me immense pleasure to congratulate Dr. S. C. Kaushik and his learned colleagues in writing a comprehensive book on solar cooling system which is also the need of the time. The book, 'Solar Cooling: Basics and Advances' is based on extensive research done by the authors and is presented systematically and coherently for the sake of reader's convenience. The book would be of great use for the solar scientists, engineers, students and researchers in the field of Applied Thermal Engineering and HVACR System. Some novel concepts of Solar Cooling have been analyzed and presented which are not common. I have great appreciation of the authors for their pain staking efforts and once again congratulate them for the book.

Prof. M. S. Sodha
Padmashree and SS Bhatnagar
Awardee; Former Vice Chancellor,
Lucknow University; DAAV Indore
and Barkatullah University Bhopal;
Honorary Visiting Professor at CES,
IIT Delhi

Preface

Solar energy is one of the most abundant sources of energy which is not only freely available but also free from harmful gaseous pollutants and can be used for a number of useful applications and cooling is one of them. The production of cooling using solar energy is known as solar refrigeration and airconditioning in which solar photovoltaic cells could be used to generate electricity to run the vapour compression refrigeration (VCR) system and/or solar thermal collector to generate heat to run sorption refrigeration systems and thermoelectric/thermoacoustic cooling systems.

Although there are number of books of solar energy and several books on refrigeration and air conditioning published separately, so far, but none on solar cooling systems. This is because this topic has been confined only to presentations and discussions in various conferences, workshops and research publications in Journals of repute. The authors have realized this gap in the available literature and proposed a unique book on basics and advances in solar cooling systems which is of great interest to solar scientists and practicing engineers, especially, the young researchers in the field of solar energy and thermal engineering.

This book is mainly based on the research work carried out by the authors covering the basics of solar energy, solar energy options and potential for cooling, solar energy collector, solar energy storage, solar operated vapour compression, vapour absorption and vapour adsorption systems, desiccant cooling systems, thermoelectric cooling systems, alternative cooling systems, hybrid cooling systems, and economic consideration in solar cooling systems. Further, the relevant research papers, review articles, and technical reports available in the literature were also consulted for the completeness of the book.

The content of this book has been divided into 12 chapters starting from basics of solar energy to economics of solar cooling. The first chapter presents the basic concept of thermodynamics and thermodynamics systems viz., heat engine and refrigeration cycle and psychrometry while the second chapter deals with the solar energy options and potential for cooling. Chapter 3 presents the solar collector options and different types of solar collections and solar energy storage options and types of energy storages are presented in Chap. 4.

In Chap. 5, the vapour compression refrigeration and their solar cooling options is presented while vapour absorption and vapour adsorption cooling systems also known as heat-driven cooling system are presented in Chaps. 6 and 7 respectively. The desiccant cooling systems are given in Chap. 8 while the thermoelectric cooling system is presented in Chap. 9. On the other hand alternate cooling system options and hybrid cooling systems are presented in Chaps. 10 and 11 respectively. Finally, the economic considerations in solar cooling systems are given in Chap. 12 while the relevant references are given at the end of the book.

The book would be of great interest to scientists and practicing engineers for the design of solar cooling systems.

New Delhi, India S. C. Kaushik
New Delhi, India S. K. Tyagi
Kerala, India V. Baiju

Acknowledgements

The authors gratefully acknowledge the support from the following:

- Director/Dy. Directors, Indian Institute of Technology, Delhi for providing the necessary facilities and financial assistance in writing this book at Department of Energy Science and Engineering (formerly, Centre for Energy Studies).
- Professor M. S. Sodha, former Vice Chancellor of DAVV, Indore, Lucknow University and Bhopal University. Former Dy. Director/Acting Director, IIT Delhi and Former Head & Honorary Visiting Professor, IIT Delhi for his overall guidance and writing the Foreword for this book in time.
- Head, Department of Energy Science and Engineering for providing overall support and facilities.
- Research Associates viz. Drs. Pramod Kumar, P. S. Bilga, A. Arora, V. V. Tyagi, A. Mahesh, V. Sivareddy, Rajesh & Ranjana, K. Manjunath, M. Dixit, S. Manikandan, R. Lamba, S. Verma, O. P. Sharma, A. K. Pandey and S. S. Bhatti for their moral cooperation.
- Research students viz., G. Raveesh (especially), Himanshu, A. Verma, S. K. Singh, P. R. Chauhan, A. Kumar, Riya Sharma, and project fellows for their help and cooperation to this end.
- Energy experts in the scientific community and various organizations whose work have been consulted, referred and used in writing this book.
- Last but not the least the family members (wife and children) for their high patience, cooperation and everlasting moral support.
- And above all, the Almighty God for giving intellect and strength to write this book with wisdom.

S. C. Kaushik
S. K. Tyagi
V. Baiju

About This Book

This book presents the concept of solar cooling and applications of solar energy for the operation of different types of cooling systems such as:

- Vapour compression (refrigeration) system
- Vapour absorption system
- Vapour adsorption system
- Desiccant cooling system
- Thermoelectric cooling systems
- Alternative cooling systems
- Hybrid cooling systems

Further, this book provides the technical and economic consideration for solar cooling options and finally, equips the readers with key techniques associated with various solar cooling options and their potential for real-life applications.

This book will appeal to the broad readership particularly Energy Engineers viz. graduate students, engineers, and scientists working in the area of solar refrigeration systems.

Contents

About the Authors

Prof. S. C. Kaushik former Emeritus Professor and Head, Centre for Energy Studies, IIT Delhi received his Ph.D. in Plasma Science from IIT Delhi after his distinguished First Position in Master's degree in electronics. His research fields include the activities in Plasma Science and Thermal Science and Engineering; Energy Conservation and Heat Recovery, Solar Refrigeration and Air-conditioning, Solar Architecture, and Thermal Storage and Power Generation. He has made significant contributions in these fields as evident by his over 600 research publications in Journals of repute at National and International levels having more than 21000 citations so far. He has guided 72 Ph.D. Theses and 120 M.Tech Projects and has authored several books. Dr. Kaushik has also been the Post-doctoral Fellow at Queensland University, Brisbane, Australia, Visiting Professor at LES-IIM-UNAM, Mexico and Madam Curie Visiting Fellow of European Commission, Paris, France.

As per, DST-Elsevier survey report-2015, Prof. Kaushik was ranked No.1 Academic Research Performer in India in Energy and he was conferred the Bharat Ratna-Dr. A. P. J. Abdul Kalam Gold Medal Research Award (2017) from GEPRA (India) and received the Outstanding Research Faculty Award (2018) from Careers360—an educational counselling academic publisher. Dr. Kaushik has also been recognized among top 1% Global Scientists from India in Energy through an Independent Study conducted by Stanford University Scientists in 2020 and Elsevier in 2021 and 2022 respectively.

Dr. S. K. Tyagi is presently working as Associate Professor at Department of Energy Science and Engineering (formerly Centre for Energy Studies), Indian Institute of Technology Delhi since April, 2017. Earlier he was working as Director/Scientist-E at Sardar Swaran Singh National Institute of Renewable Energy, Kapurthala. He has also worked as Assistant Professor at SMVDU, Katra (J&K) during Nov, 2008–March, 2011. Dr. Tyagi has worked as Invited Scientist at Korea Institute of Energy Research, South Korea, and Post-Doctoral Fellow, Hong Kong and China during 2002–2008. He has more than 25 years of research experience and expertise in the area of Renewable Energy including solar thermal and biomass related fields. He has made a significant contribution in RD&D activities and having more than 250 publications of repute. He has also guided 12 Ph.D. Theses and 18 M.Tech. and B.Tech. projects, edited four books, 12 Book Chapters and one Special Issue of Biofuels Journal and filed four Patents. Dr. Tyagi has been recognized among top 2% Global Scientists from India in Energy through an Independent Study conducted by Stanford University Scientists in 2020, Elsevier in 2021 and 2022 respectively. He has organized four Conferences and two Training Programs, while attended several others and delivered more than 30 invited talks, globally. He has taken up several R&D projects worth 15 Million INR (as PI and Co-PI) funded by different agencies including Indian Industry. He is also member of several reputed professional societies.

Dr. V. Baiju is currently working as Assistant Professor, Department of Mechanical Engineering, TKM College of Engineering, Kollam, Kerala. He did Post Doctoral programs in the area of solar cooling systems from IIT Delhi. He established an Energy Research Centre at TKM College of Engineering with indigenous systems of solar vapour adsorption, desalination, etc. He has more than fifty articles in the area of solar cooling systems, solar vapour adsorption cooling systems in particular. Currently he is supervising five research scholars and other funded projects in the area of solar cooling and applications. He has 15 years of teaching and research experience in the field of refrigeration and airconditioning.

Chapter 1
Introduction: Basic Concepts

Solar energy is one of the most abundant source of energy, which is freely available in nature. The most significant advantage of solar energy as related to other types of energy is that it is clean and can be delivered without ecological issues such as pollution, greenhouse effect and global warming, even if the associated costs are a little higher. Moreover, solar power costs have steadily declined over the last few decades. Solar energy received on the earth's surface is amounting to approximately 10,000 times the world's present yearly energy consumption. It makes solar energy as one of the promising renewable energy sources. So the utilisation of solar energy and its technologies have received much attention.

The above said attractive features of solar energy make it as a viable alternative to the traditional energy sources and spread its presence in all engineering areas. Solar energy is beneficial for power generation in remote areas; it can be utilised for electric power generations using photovoltaic cells or through thermal energy using a turbine. The major solar energy applications are found in domestic and industrial heating processes. For domestic purposes, solar energy can be used for cooking, water heating, solar lamps, etc. In chemical industry the applications are found in the distillation of saline water and other chemical compounds. Water pumping and drying of fruits are the common applications in agriculture sector. Other purpose of solar energy includes the power source for vehicles, satellites, sea vessels, aircraft, street lights, watches and calculators etc.

There are number of useful and unique applications of solar energy and cooling is one of them. The production of cooling using solar energy is known as solar refrigeration. The conventional vapour compression refrigeration (VCR) system uses a substantial amount of high-grade energy (electricity), and nearly 40% of global CO_2 emissions are produced from this power generation. At the same time, leaking of CFCs from those devices have a significant environmental impact, such as ozone depletion. By replacing the traditional VCR system with solar cooling technology, great relief can be achieved on environmental issues. Solar energy can be used to

S. C. Kaushik et al., *Solar Cooling*, https://doi.org/10.1007/978-3-031-42410-6_1

provide cooling in various ways, including PV-assisted VCR systems, solar absorption systems, solar adsorption systems, and thermoelectric cooling. Whether the cooling is solar-assisted or electrically operated, the working of the refrigeration system is nearly identical. This chapter provides a brief overview of the refrigeration system as well as basic thermodynamics applicable to the refrigeration system.

1.1 Basic Thermodynamics

Thermodynamics is the branch of science that deals with energy in the form of heat and work and their transformation. There are wide variety of applications of thermodynamics, such as in day to day life, heat transfer equipment, power plants, IC engines, turbines, compressors, heat pump and refrigeration is one of them. Refrigeration is the science that deals with the production and maintenance of the temperature of a space or a product below that of the surroundings for either comfort or preservation. In other words, refrigeration is the process of removing heat from a space at a temperature lower than its surrounding temperature. Devices that produce refrigeration are called 'refrigerators', and the cycles on which they operate are 'refrigeration cycles'. According to the Clausius statement of the second law of thermodynamics, heat flows in the decreasing direction of temperature spontaneously, that is, from higher temperature regions to lower temperature regions without external work. This heat transfer process occurs in nature without any devices. But for the reverse process, the transfer of heat from a low-temperature region to a high-temperature requires power-consuming devices called refrigerators. The working fluid used for achieving the refrigeration is called refrigerants (Kaushik 1989). There are two basic laws of thermodynamics that are important from the subject point of view: (i) first law of thermodynamics and (ii) second law of thermodynamics.

1.1.1 First Law of Thermodynamics

The first law of thermodynamics was formulated by Joule in the nineteenth century, based on the repetitive experiments. The first law defines a useful property called energy. The first law of thermodynamics is also called the law of conservation of energy. According to this law, the total energy of the system is constant; energy can be transformed from one form to another, but cannot be created or destroyed. If Q is amount of heat supplied to the system by its surroundings and W is work done by the system then, according to the first law of thermodynamics, the net heat transfer to the system is equal to the net work done by the system in a cycle.

$$\oint dQ = \oint dW \tag{1.1}$$

1.1.2 Second Law of Thermodynamics

The inadequacy of the first law to identify whether the process can take place or not is remedied by introducing the second law of thermodynamics. Satisfying the first law alone does not guarantee that the process will take place in real. In short, a process will take place if it satisfies both the first and second law of thermodynamics.

The second law of thermodynamics has two statements, as explained below.

(a) Kelvin-Planck statement
(b) Clausius statement

(a) Kelvin-Planck Statement
According to Kelvin-Planck statement, it is impossible to construct a cyclically operating device that receives heat from a single reservoir and produce a net amount of the work (Fig. 1.1).

This means that the complete conversion of heat energy into work is impossible, that is, an engine with 100% efficiency is impossible to construct. This indicates that a heat engine must exchange heat with a low-temperature sink and a high-temperature source.

(b) Clausius Statement
As explained in the previous sections, heat always flows from a body at higher temperature to another body at lower temperature (Fig. 1.2) spontaneously. According to

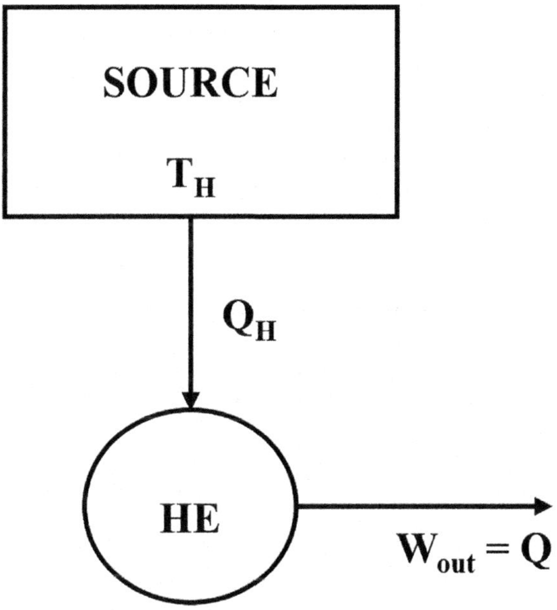

Fig. 1.1 Schematic representation of a heat engine (HE) (which is impossible according to the Kelvin-Planck statement)

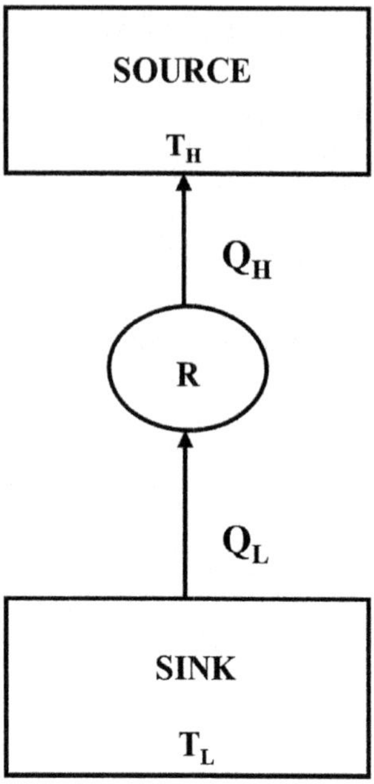

Fig. 1.2 Schematic representation of a refrigerator (R) (which is impossible according to the Clausius statement)

the Clausius statement, "It is impossible to construct a device that operates on a cycle and produces no effect other than the transfer of heat from lower temperature to higher temperature body". This means that the transfer of heat from the body at a lower temperature to a higher temperature does not take place naturally. For this to happen, the device is to be powered by an external agency. For example, house hold refrigerator cannot work if it is not powered by an external source such as an electric motor coupled to a compressor.

Both Kelvin-Planck and Clausius statements are formulated based on the experimental observations and cannot be proved until now, nobody conducted an experiment that contradicts the first and second laws of thermodynamics.

1.1.3 Thermodynamic Systems

The thermodynamic system may be defined as the space confined by definite boundaries within which the thermodynamic processes take place. Anything outside of such boundaries is referred to be surroundings, while both system and surroundings are referred to as universe. Boundaries may be fixed (storage tank for compressed air) or movable (cylinder piston arrangement for enclosing gas). Thermodynamic systems are classified into three types: (i) Open system (ii) Closed system and (iii) Isolated system.

(i) **Open system:** In open system, mass of the working substance crosses the boundary of the system during the process. Here the boundary is taken as imaginary and it allows transfer of energy as well as matter.

(ii) **Closed system:** Closed system is referred to as the system having fixed mass, and the boundary does not allow the mass transfer. However, the possibility of energy transfer is there through the boundary.

(iii) **Isolated system:** In this type of system, there is no energy or mass transfer, across the system boundary. As a result, there is no interaction between the system and the environment since it is fully isolated. The schematic of different types of thermodynamics systems are shown in Fig. 1.3.

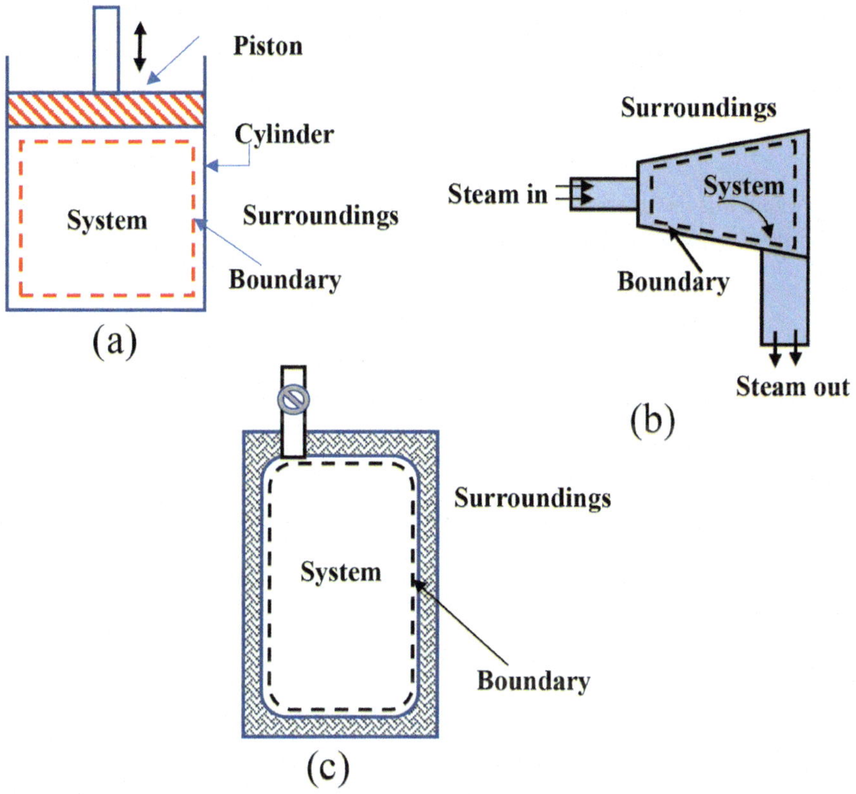

Fig. 1.3 Schematic of thermodynamic systems **a** open system **b** closed system and **c** isolated system

1.1.4 Thermodynamic Properties

Specific Heat (C)
The amount of heat required to raise the temperature of 1 kg of a substance by 1 °C is called specific heat of that substance, i.e.

$$Q = mCdT \tag{1.2}$$

where m is the mass (in kg), C the specific heat and dT the temperature difference (C: J/kg K).

Two types of specific heats are:

(i) Specific heat at constant volume (C_v)

$$C_v = \left.\frac{\partial u}{\partial T}\right)_v \tag{1.3}$$

where u is the internal energy (J/kg) and T is the absolute temperature (K).

(ii) Specific heat at constant pressure (C_p)

$$C_p = \left.\frac{\partial h}{\partial T}\right)_p \tag{1.4}$$

where h is the enthalpy.

Enthalpy (H)
Enthalpy (H) is the total heat content of the system. It is the sum of internal energy (U) and pressure volume (PV) work.

The enthalpy of a substance is a property since it consists of the sum of property and product of the two other properties, i.e.,

$$\begin{aligned} H &= U + PV \\ h &= u + pv \text{ (specific, for } m = 1\,\text{kg)} \\ dh &= du + d(pv) \\ d(pv) &= pdv + vdp \end{aligned} \tag{1.5}$$

At constant pressure, $vdp = 0$

$$\begin{aligned} dh &= du + pdV \\ pdV &= RdT \\ dh &= du + RdT \\ du &= mC_v dT \\ dh &= C_v dT + RdT \end{aligned} \tag{1.6}$$

$$dh = (C_v + R)dT$$
$$dh = C_p dT$$
$$C_p = C_v + R \tag{1.7}$$

The total enthalpy is given by

$$H = mC_p dT \tag{1.8}$$

Thermal Reservoir

Thermal reservoir is a body with ability to store or supply large quantity of heat without appreciable change in its temperature. Source is a thermal reservoir which supplies heat and sink is the reservoir that absorbs heat energy.

Heat Engine

The conversion of other forms of energy into work is ambitious, whereas, the conversion of work into other energy forms is easy. It requires a special device. A heat engine is a cyclic device that converts heat energy into mechanical work. It receives energy as heat from a high-temperature reservoir and part of the energy received is converted into useful work and remaining is rejected into the sink (Fig. 1.4).

$$W_{out} = Q_{in} - Q_{out} = Q_H - Q_L \tag{1.9}$$

where Q_H is the heat input to the heat engine, Q_L the heat rejected to the surroundings and W_{out} the work produced by the heat engine.

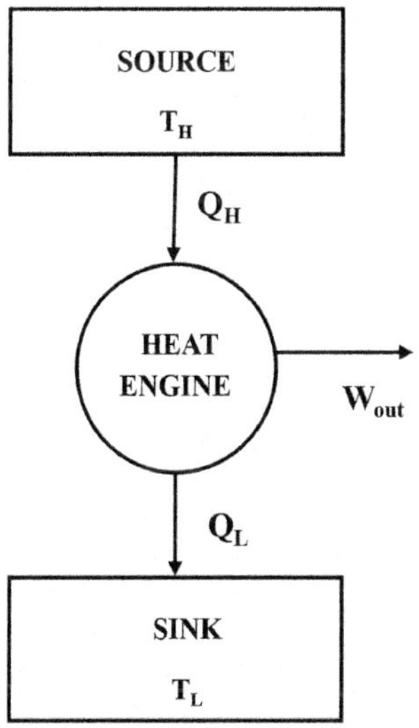

Fig. 1.4 Schematic of a heat engine (HE)

1.1.5 Thermal Efficiency

A part of net heat supplied to the system (Q_{in}) is used for converting into the useful work and the remaining (Q_{out}) is exhausted to the surroundings. Since Q_{out} is never equal to zero, the network output of the engine is always less than the amount of heat input. The fraction of heat input converted into useful work (W) is a measure of performance of a heat engine, known as the thermal efficiency and given as below:

$$\text{The efficiency, } (\eta) = \frac{\text{Work output}}{\text{Heat input}} \tag{1.10}$$

$$\eta = \frac{W_{out}}{Q_{in}} \approx 1 - Q_{out}/Q_{in} < 1 \tag{1.11}$$

1.1.6 Refrigerator and Heat Pump

In nature, heat always transfers from a greater temperature region to a smaller temperature. The reverse process, however, cannot occur naturally, that is, the transfer of heat from lower temperature to higher temperature requires some special device called as the heat pump or refrigerator.

Refrigerator

A refrigerator is a cyclic device which is used for maintaining a system at temperature lower than the surrounding temperature. Figure 1.5 shows the schematic of a refrigerator.

As can be seen from Fig. 1.5, it is a cyclic device, which absorbs heat from a low-temperature reservoir to a high-temperature reservoir when the work is performed on the body.

Let Q_L be the heat extracted from the system and Q_H be the heat rejected to the surroundings.

Then W_{in} is the work input to the device, given as (Cengel et al. 2011):

$$W_{in} = Q_H - Q_L \tag{1.12}$$

Heat Pump

A device which transfers heat from a lower temperature space/body to a higher temperature space/body with the aid of an external agency is called a heat pump. The objective of both refrigerator and a heat pump are different. The objective of the refrigerator is to maintain a space at a lower temperature than the surroundings (space cooling), while a heat pump is used to maintain a space at a higher temperature than that of surroundings (Fig. 1.6). In other words, a heat pump is employed in the cold

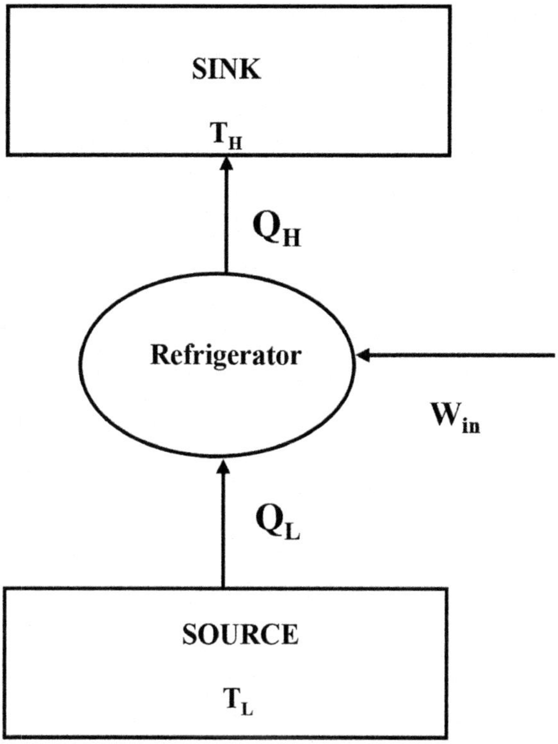

Fig. 1.5 Schematic of a refrigerator (R)

regions to heat up the space for maintaining the thermal comfort, while the refrigerator is employed at both hot and cold climate to preserve something including but not limited to medicines (vaccines), food products, fruits, vegetables, and so on.

1.1.7 Coefficient of Performance (COP)

The performance of the refrigerators and heat pump is usually expressed as the ratio of cooling or heating produced to work input known as coefficient of performance (COP) given as below;

$$\mathrm{COP} = \frac{\text{Desired effect}}{\text{Work input}} \tag{1.13}$$

For a refrigerator, the desired effect is the cooling load or refrigerating effect and the performance ($\mathrm{COP_R}$) is given by:

$$\mathrm{COP_R} = \frac{Q_L}{W} = \frac{Q_L}{Q_H - Q_L} = \frac{T_L}{T_H - T_L} \tag{1.14}$$

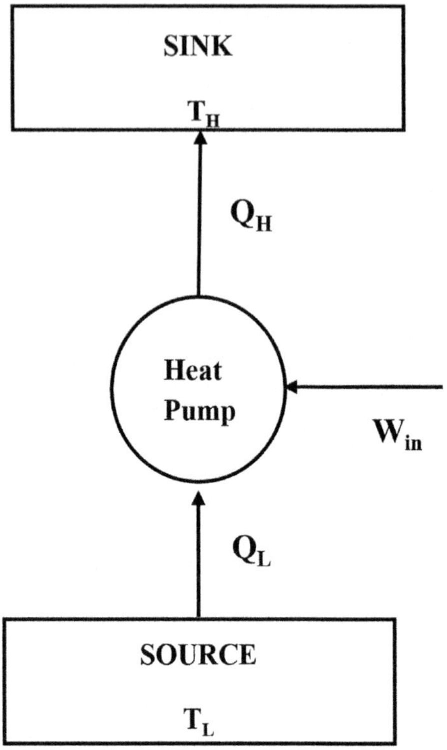

Fig. 1.6 Schematic of a heat pump (HP)

The value of COP_R can be greater than unity. This shows that the amount of heat extracted from the refrigerated space can be more than the work spent on it. The efficiency of a heat engine is always less than unity. This is one of the reasons why the performance of a refrigerator and heat pump is expressed in terms of COP rather than efficiency to avoid the oddity.

Similarly the coefficient of performance of heat pump is given by:

$$COP_{HP} = \frac{Q_H}{(Q_H - Q_L)} = \frac{T_H}{(T_H - T_L)} \tag{1.15}$$

where, the desired effect is heating load and Q_H is the heat supplied to the (heating) space and Q_L is the heat extracted from the ambient (surroundings), while W is the work supplied to the machine.

COP_{HP} is always greater than COP_R.
Using Eqs. (1.14) and (1.15), we have:

$$COP_R = \frac{Q_L}{(Q_H - Q_L)} \tag{1.16}$$

Adding 1 on both sides,

$$\begin{aligned} COP_R &= \frac{Q_L}{(Q_H - Q_L)} + 1 = \frac{(Q_L + Q_H - Q_L)}{(Q_H - Q_L)} \\ &= \frac{Q_H}{(Q_H - Q_L)} = COP_{HP} \end{aligned} \tag{1.17}$$

Therefore,

$$COP_{HP} = COP_R + 1 \tag{1.18}$$

In other words, the COP of a heat pump is always more than unity, while COP of a refrigerator may be greater than or equal to unity.

1.2 Psychrometry

Psychrometry is a branch of science, which describes the study of properties of moist air. The moist air is a combination of dry air and water vapour present in it.

1.2.1 Terms Used in Psychrometry

(a) *Dry air*: It contains N_2, O_2 etc. and does not contain water vapour in it.
(b) *Moist air*: The atmospheric air which contains water vapour and dry air.
(c) *Saturated air*: It contains the maximum amount of water vapour at a particular pressure and temperature.
(d) *Dry-bulb temperature* (t_{db}): It is the actual temperature of moist air recorded by an ordinary thermometer with a dry bulb.
(e) *Wet-bulb temperature* (t_{wb}): The wet bulb temperature represents the temperature of air recorded by a thermometer when its bulb is covered with a wetted wick (cloth) and is exposed to moving air (e.g., the temperature felt when skin is wet). The device used for measuring humidity is called a psychrometer. The psychrometer consists of two thermometers, viz. wet-bulb thermometer and dry-bulb thermometer.
(f) *Wet-bulb depression* (*WBD*): It is the difference between dry-bulb temperature and wet-bulb temperature:

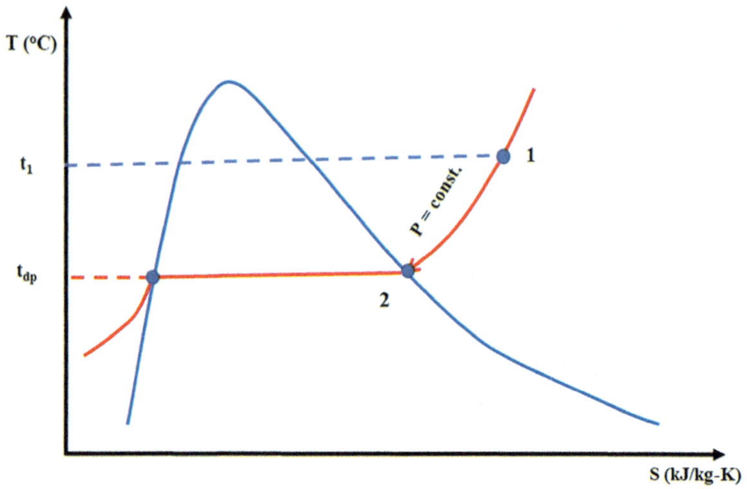

Fig. 1.7 Thermodynamic state of water vapour in moist air

$$\mathrm{WBD} = t_{\mathrm{dbt}} - t_{\mathrm{wbt}} \tag{1.19}$$

(g) *Dew point temperature* (t_{dp}): The dew point temperature is the temperature at which the water suspended in atmospheric air starts condensing when it it is cooled at constant pressure. The horizontal line drawn from the corresponding state to the saturation line represents the dew point temperature in the chart. The dew point temperature lines are shown in the Fig. 1.7 along the saturated relative humidity lines. When an unsaturated quantity of air is cooled isobarically, the temperature of the mixture will decrease to the saturation temperature, t_{dp} of the water vapour and the water. The water vapour begins to condense at this point. The temperature t_{dp} is called as the dew point temperature.

(h) *Dew point depression* (DPD): It is the difference between dry-bulb temperature and dew point temperature:

$$DPT = t_{\mathrm{db}} - t_{\mathrm{dp}} \tag{1.20}$$

(i) *Specific humidity or humidity ratio* (ω):
The ratio of mass of water vapour to the mass of dry air present in moist air is defined by the specific humidity or humidity ratio (ω).

$$\omega\ (\text{kg of water vapour/kg of dry air}) = m_{\mathrm{v}}/m_{\mathrm{a}} \tag{1.21}$$

we have,

$$PV = mRT \tag{1.22}$$

for water vapour,

$$P_v V = m_v R_v T$$

for dry air,

$$P_a V = m_a R_a T$$

$$R_v = \frac{R_u}{M_v} = \frac{8314}{18} = 462\,\text{J/kg K} (\text{M}_\text{v} = 18, \text{M}_\text{a} = 28.87)$$

$$R_a = \frac{R_u}{M_v} = \frac{8314}{28.97} = 286.9\,\text{J/kg}\, K$$

so,

$$\omega = \frac{m_v}{m_a}$$

$$\omega = \frac{(P_v V)(R_a T)}{(P_a V)(R_v T)} \tag{1.23}$$

$$\omega = \frac{(P_v R_a)}{P_a R_v} \tag{1.24}$$

i.e.,

$$\omega = 0.622 \times \frac{P_v}{P_a} \tag{1.25}$$

According to Dalton's law of partial pressure, total pressure P_t is equal to the sum of the partial pressure of dry air P_a and partial pressure of water vapour P_v.

$$P_t = P_a + P_v$$

or

$$P_a = P_t - P_v \tag{1.26}$$

hence,

$$\omega = 0.622 \times \frac{P_v}{P_t - P_v} \tag{1.27}$$

(j) *Degree of saturation* (μ): Degree of saturation is defined as the ratio of actual specific humidity to the specific humidity when the moist air is saturated at the same temperature of air considered for analysis, that is,

$$\mu = \frac{\text{Actual specific humidity}}{\text{Specific humidity at saturated condition}} = \frac{\omega}{\omega_s} \tag{1.28}$$

were ω_s is given by:

$$\omega_\text{s} = 0.622 \times \frac{P_\text{vs}}{P_\text{t} - P_\text{vs}} \tag{1.29}$$

$$\mu = \frac{P_\text{v}}{P_\text{vs}} \times \frac{(P_\text{t} - P_\text{vs})}{(P_\text{t} - P_\text{v})} \tag{1.30}$$

(k) *Relative humidity* (φ): The ratio of the mass of water vapour in a given volume of moist air to the mass of water vapour in the same volume, when the air is saturated at the given temperature is termed as the relative humidity.

$$\varphi = \frac{m_\text{v}}{m_\text{vs}} \tag{1.31}$$

we have,

$$PV = mRT$$

$$P_\text{v}V = m_\text{v}R_\text{v}T$$

(R_v is same whether it is saturated or not as $R_\text{v} = R_\text{u}/M_\text{v}$, since both R_u and M_v remains constant, R_v remains constant)

$$\varphi = \frac{m_\text{v}}{m_\text{vs}} \tag{1.32}$$

$$\varphi = \frac{P_\text{v}}{P_\text{vs}} \tag{1.33}$$

The relative humidity is also defined in terms of the partial pressure of water vapour as indicated in equation (1.34).

$$\mu = \frac{P_\text{v}}{P_\text{vs}} \times \frac{(P_\text{t} - P_\text{vs})}{(P_\text{t} - P_\text{v})} \tag{1.34}$$

becomes

$$\mu = \varphi \times \frac{P_t - P_{vs}}{P_t - P_v} \tag{1.35}$$

$$= \varphi \times \frac{P_t - P_{vs}}{P_t - \phi P_{vs}} \tag{1.36}$$

(l) *Enthalpy of moist air* (H): The total enthalpy of moist air is equal to the sum of enthalpy of dry air and enthalpy of water vapour.

$$H = H_a + H_v$$
$$H = m_a h_a + m_v h_v \tag{1.37}$$

Since the mass of dry air remains constant, enthalpy of moist air is expressed in terms of the mass of dry air:

$$\frac{H}{m_a} = h_a + \left(\frac{m_v}{m_a}\right) \times h_v$$
$$h = h_a + w h_v \tag{1.38}$$

where h_a is the enthalpy of dry air part and ωh_v the enthalpy of the water vapour part.

(m) *Adiabatic saturation process*: Consider an insulated chamber containing sufficient quantity of water in it. When the unsaturated air having specific humidity ω, dry bulb temperature t_{db1} flows over the water, water evaporates. The humidity ratio (specific humidity) of the air increases during this process. Since the latent heat of vapourisation of water is taken from incoming unsaturated air, the temperature of air decreases (Fig. 1.8).

When the container is long enough, the exit air becomes saturated without heat addition, hence, the process is called as adiabatic saturation process. The corresponding exit temperature of air is called *adiabatic wet-bulb temperature.*

Applying mass balance to the system (since mass of dry air at inlet and outlet are the same)

$$m_{a1} = m_{a2} = m_a \tag{1.39}$$

Applying mass balance for water vapour

$$m_{v1} + m_f = m_{v2},\ \omega = \frac{m_v}{m},\ \therefore\ m_{v1} = \omega_1 m_a \text{ and } m_{v2} = \omega_2 m_a$$

therefore,

$$m_a \omega_1 + m_f = m_a \omega_2 \tag{1.40}$$

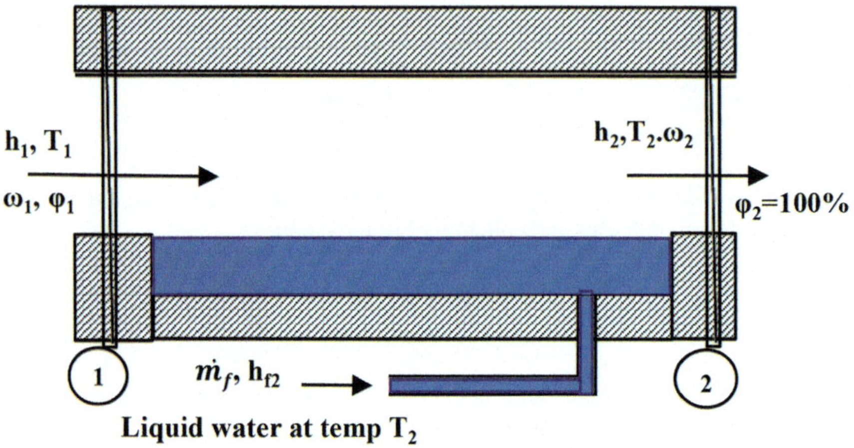

Fig. 1.8 Adiabatic saturation process (Cengel et al. 2011)

$$m_{\mathrm{f}} = m_{\mathrm{a}} \times (\omega_2 - \omega_1) \tag{1.41}$$

Applying energy balance (since the process takes place at the adiabatic condition and without any work input):

$$Q = 0\,(\text{heat}),\ W = 0\,(\text{work done})$$

Also, potential and kinetic energies are neglected
So,

$$h_1 + h_{\mathrm{f}} = h_2 \tag{1.42}$$

We have,

$$h = h_{\mathrm{a}} + \omega \times h_{\mathrm{v}}$$

$$h_1 = h_{\mathrm{a}} + \omega_1 h_{\mathrm{v1}} = C_{\mathrm{p}} t_{\mathrm{db1}} + \omega_1 h_{\mathrm{v1}}$$
$$h_2 = h_{\mathrm{a}} + \omega_2 h_{\mathrm{v2}} = C_{\mathrm{p}} t_{\mathrm{db2}} + \omega_2 h_{\mathrm{v2}}$$

Therefore,

$$C_{\mathrm{p}} t_{\mathrm{db1}} + \omega_1 h_{\mathrm{v1}} + h_{\mathrm{f}} = C_{\mathrm{p}} t_{\mathrm{db2}} + \omega_2 h_{\mathrm{v2}} \tag{1.43}$$

$$h_1 + h_{\mathrm{f}} = h_2 \tag{1.44}$$

considering mass

$$m_a h_1 + m_f h_f = m_a h_2 \tag{1.45}$$

Dividing Eq. (1.46) by m_a

$$h_1 + \left(\frac{m_f}{m_a}\right) \times h_f = h_2 \tag{1.46}$$

Substituting (Eq. 1.42) for m_f

$$h_2 = h_1 + \left(\frac{m_a}{m_a}\right) \times (\omega_2 - \omega_1) \times h_f$$

$$h_2 = h_1 + (\omega_2 - \omega_1) \times h_f \tag{1.47}$$

Substituting for h_1 and h_2

$$\begin{aligned} C_p t_{db1} + \omega_1 h_{v1} + (\omega_2 - \omega_1) \times h_f &= C_p t_{db2} + \omega_2 h_{v2} \\ \omega_1 \times (h_{v1} - h_f) &= C_p \times (t_{db2} - t_{db1}) + \omega_2 (h_{v2} - h_f) \end{aligned} \tag{1.48}$$

$$w_1 = \frac{\left(C_p \times (t_{db2} - t_{db1}) + w_2 \times (h_{v2} - h_f)\right)}{(h_{v1} - h_f)} \tag{1.49}$$

Note: After the adiabatic saturation process, the exit air is saturated.

Therefore, $t_{db2} = t_{wb2}$, the temperature at exit (temperature of saturated air) is also called adiabatic saturation temperature or thermodynamic *wet-bulb temperature* (Fig. 1.8).

1.2.2 Psychrometric Processes

Maintenance of the desired temperature and humidity level in the living space is necessary. Various thermodynamic psychometric processes that are commonly adopted are presented here (Fig. 1.9). Such processes are called psychrometric processes. The different processes are:

(i) Sensible heating
(ii) Sensible cooling
(iii) Cooling and dehumidification
(iv) Heating and humidification
(v) Adiabatic mixing of the air stream

(i) *Sensible heating*: During this process, the temperature of incoming air increases from t_1 to t_2. The specific humidity of the moist air remains at constant. The sensible heat, DBT and WBT of the air increases while latent heat of air and

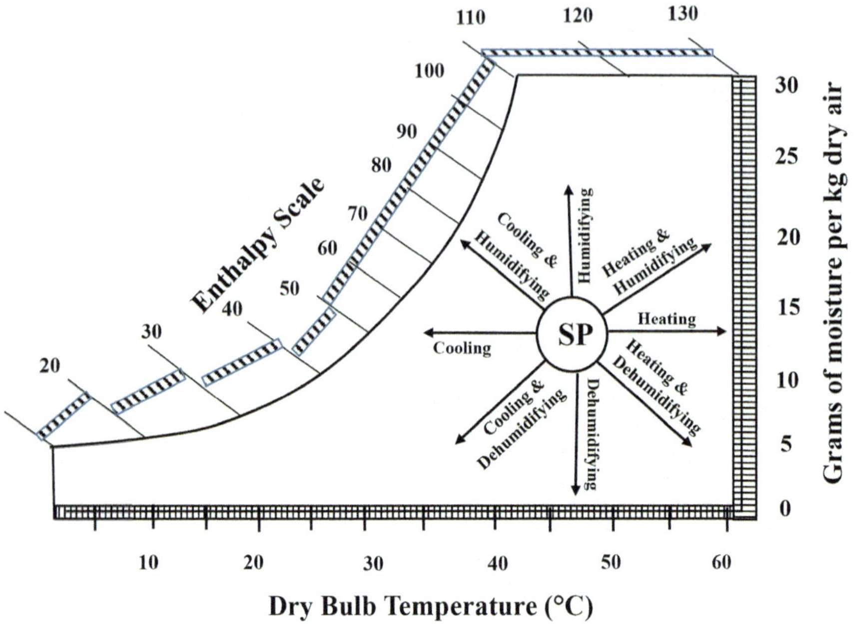

Fig. 1.9 Psychrometric chart showing various processes

the DPT of air remain constant. The coil is heated by using the refrigerant hot water, steam or an electric heater integrated to the equipment (Fig. 1.10a).

When the moist air of temperature t_{db1} is passed through a heating coil (temperature of heating surface > temperature of the air), the air will be heated up and the temperature will increase from t_{db1} to t_{db2} keeping specific humidity ($\omega_1 = \omega_2$) constant. During this process, there is no addition or removal of water vapour from the air. Therefore, m_v is constant. $\omega = m_v/m_a$ since m_v and m_a are constant, ω will remain constant. The horizontal line parallel to the x axis on the psychrometric chart represents the sensible heating process (see Fig. 1.10b). The sensible heating line is also a constant DPT line. The relative humidity decreases with sensible heating, since the dry-bulb temperature increases the capacity to hold water vapour, but the mass of water vapour remains constant, hence, φ decreases.

(ii) *Sensible cooling*: In this process, the moist air is allowed to passes over a surface, which is at temperature lower than the surroundings. The temperature of the surface is higher than dew point temperature (t_{dp}) of air. During this process the temperature of incoming air decreases from dry-bulb temperature,

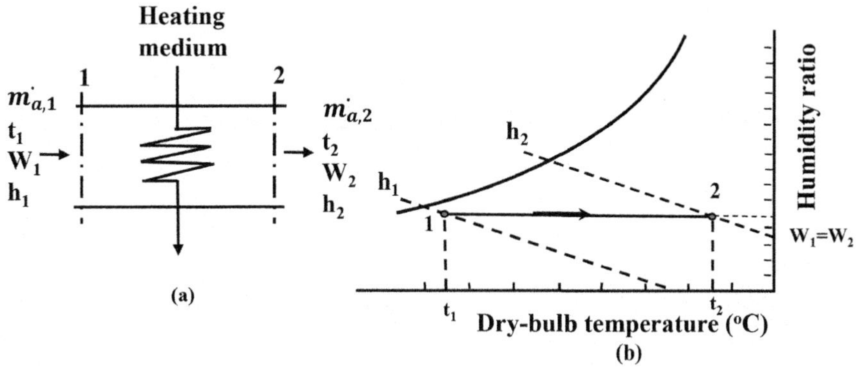

Fig. 1.10 **a** Schematic of sensible heating process. **b** Sensible heating process shown in psychrometric chart

t_{db1} to t_{db2} at a constant specific humidity. The relative humidity, φ of air increases from φ_1 to φ_2 (Fig. 1.11).

(iii) *Cooling and dehumidification*: When the moist air is allowed to pass over the cooling coil which is maintained at a temperature less than dew point (t_{dp}) and less than dry-bulb (t_{db}) temperature of air, the air will be subjected to sensible cooling and dehumidification. For dehumidification of air, it is necessary to cool the air below its dew point temperature. When air reaches dew point temperature a part of moist air will be subjected to condensation and the cool saturated air at state point 3 is mixed with fresh air to obtain the required dew point temperature and relative humidity (Fig. 1.12).

(iv) *Heating and humidification*: In this process, the moist air is first passed through a heating coil kept at a temperature greater than the temperature of incoming air ($t > t_{db}$). This increases the temperature of the incoming air from t_{db1} to t_{db2}. Then it is passed through a humidifier where it is subjected to the humidification

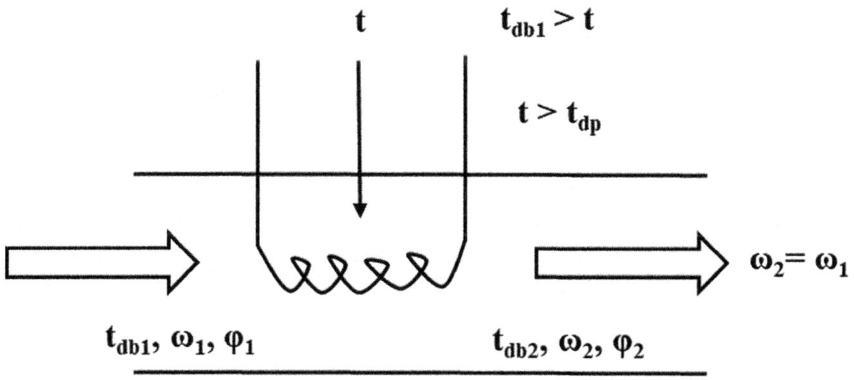

Fig. 1.11 Schematic of the sensible cooling process

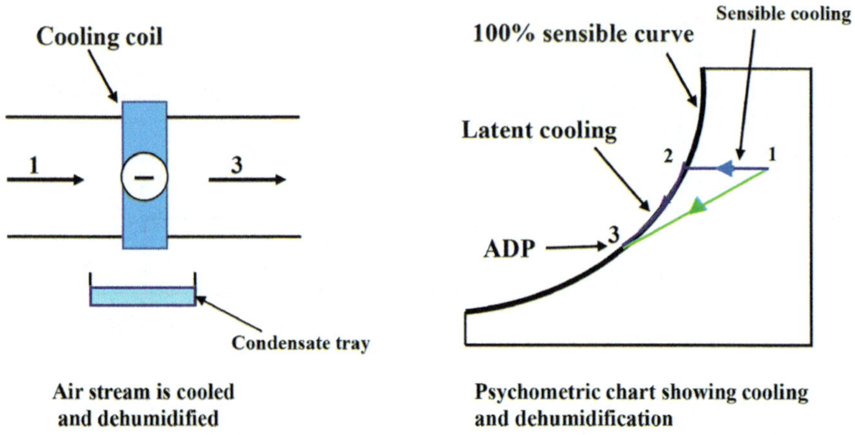

Fig. 1.12 Cooling and dehumidification

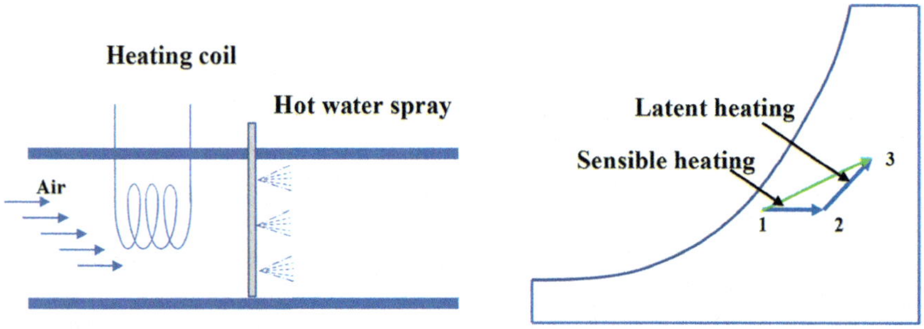

Fig. 1.13 Heating and humidification. **a** Schematic of the process and **b** in psyhromeric chart

process. In the humidifier, the hot water which is maintained at a temperature higher than the dry-bulb temperature of the air is sprayed over the incoming air (Fig. 1.13) to increase the humidity. Figure 1.13a represents the schematic of the process and its representation in psychrometric chart is shown in Fig. 1.13b.

(v) *Adiabatic mixing of air stream*: Many airconditioning devices needs the mixing of air streams. The applications such as airconditioning in large buildings, hospitals, production industries etc. require the mixing of conditioned air with the outside fresh air before it is supplied to the living space. During the process, the heat transfer with surroundings is negligibly small, and hence the process is considered as adiabatic (Fig. 1.14). In the adiabatic mixing process, there involves no work interactions, kinetic and potential effects are neglected.

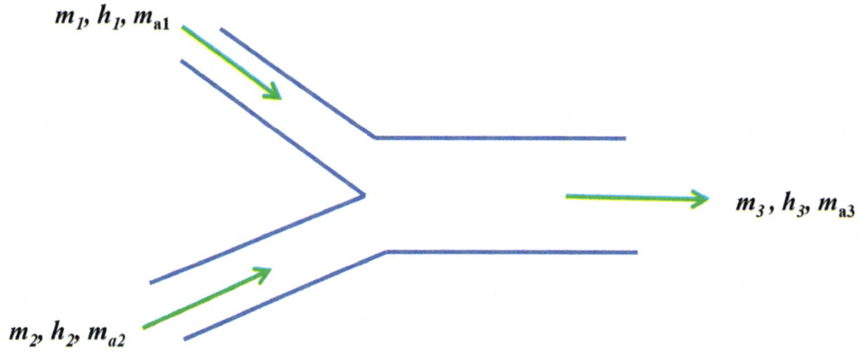

Fig. 1.14 Adiabatic mixing

Consider mixing of two streams of moist air having specific humidity ω_1 and ω_2, the mass of water vapour m_{v1} and m_{v2}, mass of the dry air m_{a1} and m_{a2}. Assume ω_3, m_{v3} and m_{a3} be the specific humidity, the mass of water vapour and mass of dry air of the leaving stream (Fig. 1.14).

$$\text{Mass balance: } m_{a1} + m_{a2} = m_{a3} \tag{1.50}$$

$$\text{Mass of the water vapour: } \omega_1 m_{a1} + \omega_2 m_{a2} = \omega_3 m_{a3} \tag{1.51}$$

$$\text{Energy balance: } m_{a1} h_1 + m_{a2} h_2 = m_{a3} h_3 \tag{1.52}$$

Substituting (1.50) in (1.51) and (1.52) eliminating m_{a3}, the equation becomes

$$\frac{m_{a1}}{m_{a2}} = \frac{\omega_2 - \omega_3}{\omega_3 - \omega_1} = \frac{h_2 - h_3}{h_3 - h_1} \tag{1.53}$$

1.3 Classification of Refrigeration

Refrigeration systems are classified into three types:

1. Traditional (non-cyclic) refrigeration systems → Ice refrigeration systems
2. Cyclic refrigeration systems →
   - Vapour compression refrigeration systems
   - Air cycle refrigeration systems
   - Vapour absorption refrigeration systems
   - Vapour adsorption refrigeration systems
3. Non-conventional →
   - Thermo electric refrigeration systems
   - Magnetic refrigeration systems
   - Pulse tube refrigeration systems
   - Stirling refrigeration systems
   - Vortex tube refrigeration systems

The most commonly used method of cooling is the vapour-compression cycles because it is fairly easy to construct a cooling device employing this method and the cost is low. Airconditioning devices also work based on the vapour compression technology. Vapour compression refrigeration works as a reverse heat engine and hence, heat energy is taken from a cold reservoir and given to a hot reservoir. According to the second law of thermodynamics, heat energy does not spontaneously transfer from low temperature body to a high temperature body. To have heat transfer in that reverse direction (and not from the high temperature to lower temperature reservoir, as the system is naturally inclined to do), it is necessary to do work on the system. Therefore, the vapour compression refrigeration systems are work consuming devices. The science of refrigeration utilises some physical property of the matter for producing cold. Various traditional methods of cooling are described in brief as follows:

1. *Ice Refrigeration*: The most traditional method of cooling is the ice refrigeration systems. Ice collected from the surface of ponds and lakes in winter stored in the insulated ice house for use year-round. Temperatures inside the refrigerator cabin were between 5 and 10 °C and could be controlled by the flow of air over the ice. For lower temperature below 0 °C, salt mixtures along with the ice were used and brines were used for making indirect contact refrigeration. Dry ice (solid CO_2) is also commonly used for frozen food refrigeration.
2. *Evaporative refrigeration*: This is based on the principle that heat is absorbed when liquid evaporates. A large part of this heat necessary to cause evaporation comes from the liquid itself. A simple example is the desert water bag. Another example is the method of making artificial snow when the temperature of the surrounding air is 0 °C or below. In case of low humidity, artificial snow can also be made even with 2 °C air temperature due to the rapid evaporation and evaporative cooling caused by low relative humidity.

3. *Vapour refrigeration*: The vapour refrigeration and cooling devices are works based on the reversed Carnot cycle and employs vapours of various refrigerants as working fluid instead of air. The heat carried away by the refrigerant vapour produces cooling effect, hence the refrigeration effect is higher than for air refrigeration system. The necessary components of the system are, a compressor, a condenser, a throttling device and an evaporator. The compressor may be mechanical compressor as for vapour compression refrigeration system or thermocompressor as for vapour absorption refrigeration system.

1.4 Energy Conservation in RAC Systems

The energy conservation in cooling systems plays a vital role. The conservation means the reduction in energy consumption (Sur and Das 2010). Some of the suggestions for conserving energy and reduce global warming and save money in the long run are:

(i) Use maximum daylight.
(ii) Conduct an energy audit to understand the problem and to take the remedial steps.
(iii) Use energy-efficient appliances and lighting systems.
(iv) Drive less, walk more and carpooling: This will help to reduce the carbon footprint.
(v) Plant shady landscaping outside the home will help to reduce the entry of immense heat during hot and sunny days and chilly winds during the winter season.
(vi) Energy-efficient windows.
(vii) Switch off appliances when not in use.
(viii) Lower the room temperature.
(ix) The efficient architecture of the building such as providing natural lighting and proper natural ventilation etc. reduce energy consumption.

Energy required to drive the cooling systems plays a significant role in the world energy scenario. According to the International Institute of Refrigeration (IIR) about 15% of the all-electric energy produced worldwide is used to run refrigeration and airconditioning devices (Sarbu and Adam 2011). According to the statistics survey (Mazzaferro et al. 2020), the demand for airconditioners worldwide has a fundamental tendency of steady increase and the global growth rate is about 17%.

According to BEE, about 32% total energy produced in India is consumed for cooling purposes. According to the various surveys, about 35,000 MW of electricity is used for the cooling applications. Vapour compression cooling systems are universally accepted for the refrigeration and airconditioning purposes. The vapour compression cooling systems requires mechanical energy as the driving source and is responsible for the emission of CO_2 and the other greenhouse gases such as CFCs and HFCs, which are considered major ozone-depleting gases. The productions of oil and natural gas have become scarcer and their reserves are likely to be consumed by the year 2040.

The production of coal is also not so promising and its maximum will touch somewhere around in 2045 worldwide. With the advent of the energy crisis and the realisation of the depleting nature of fossil fuels, the search for the use of alternative energy source becomes essential. Due to increased use of energy, nowadays it is important to use alternative energy resources as efficiently as possible. The main renewable energy resources include biomass, solar energy, wind energy, etc. Out of these, the solar energy is the most promising among all the renewable source of energy since it is abundantly available and is an inexhaustible global source of energy having no pollution or hazards associated with it. The cooling load is generally high when solar radiation is high. Together with existing technologies, solar energy can be converted to both electricity and heat, either of these can be used to power refrigeration systems.

Refrigeration and airconditioning are highly energy-consuming processes. Most parts of India get sunlight throughout the year. The demand for energy is increasing from day-to-day and due to the energy crisis by the increased use of fossil fuels, the development of ways of utilisation of renewable energy resources becomes very urgent (Jouhara and Sayegh 2018). Indeed, the prospects of climate change and, eventually, fossil fuel depletion, trigger a growing interest in renewable energies in general and solar energy in particular. According to recent surveys conducted in rural areas, where electricity is not available, 30–40% of agricultural products are being destroyed due to lack of proper post- harvest cooling facilities. Hence, solar cooling options not only mitigate energy shortage and environmental pollution but also contribute to the reduction of food spoilage.

Some measures to be taken for conserving energy and improving the energy efficiency in the refrigeration and airconditioning systems are:

(i) *Compressors*: VCR systems are driven by conventional high-grade energy operated mechanical compressors. In a typical refrigeration system, 40–60% of the total energy supplied is utilised by compressors, 10–25% by condenser fans, 10–20% evaporator fans and 5–10% defrosts systems. Hence, the compressor is the key element in the refrigeration system to take care of the energy conservation. Use of high-efficiency variable speed compressors will provide large energy savings in VCR systems. The solar Rankine cycle VCR or PV integrated VCR systems are also options to reduce the consumption of conventional energy sources. The other way is to replace the mechanical compressors by thermal compressors as in the case of vapour absorption (VAB) and vapour adsorption (VAD) refrigeration systems. These are purely run by low-grade energy such as heat. The advantages of the VAB and VAD systems are that they can be effectively operated by solar energy, engine exhaust gases, waste heat sources etc.

(ii) *Condensers and evaporators*: The maximum efficiency in the operation of airconditioners is ultimately obtained by the combined performance of all the major components, namely, condenser and evaporator (along with compressor).

So, thermal design of condenser and evaporator along with economic analysis is an important aspect in higher performance of HVAC systems (Lv et al. 2020). Condenser rejects a huge quantity of heat as latent heat of condensation when the refrigerant converted its vapour phase to the liquid phase. Hence, there is potential for the conservation of rejected heat for further use. The heat transfer rate of condenser and evaporator coils also plays an important role in the performance of refrigeration systems. The optimum air flow through refrigerated space reduces fan energy. Condenser and cooling tower fans can also be optimised. To reduce the fan power, use variable speed control fans in the system.

(iii) *Expansion devices*: Use of the electronic expansion valves, which can provide better super heat control and system control is recommended.

Some of the other low cost, simple ways of making the air conditioning equipment's in energy-efficient ways are:

(i) temperature control
(ii) use of variable speed drives
(iii) use of free cooling coils.

Temperature control: The operation of the airconditioner is so adjusted with the temperature of the cabin by electronically controlling the speed of fan and compressor. The system automatically modifies the refrigeration cycle in response to variations in load, weather conditions, and running time to minimise the system's energy use. Temperature sensors placed at the evaporator and condenser side helps the control unit to evaluate the speed of the fan and compressor.

Variable speed drive: Generally, airconditioning systems are designed for full-load conditions, although most of the time, the cooling load is less than this number and full motor capacity is not required. Constant speed motors running at a higher speed than necessary to produce more capacity, consumes a significant amount of energy under these conditions. Variable speed drives can overcome this problem by adjusting the speed based on the load value, resulting in improved product environment.

Free cooling coils: In free cooling colder ambient air is used to perform cooling rather than the refrigeration cycle. This system uses the cool outdoor air as a free cooling source when the temperature outside is lower than the temperature inside. It is a cost-effective way to reduce energy consumption while also lowering the systems environmental impact.

These factors are to be taken into account for the conservation of energy in different kind of refrigeration and airconditioning devices.

References

Cengel, Y. A., Boles, M. A. & Kanoğlu, M. (2011). *Thermodynamics: An engineering approach* (Vol. 5, p. 445). New York: McGraw-hill.

Jouhara, H. & Sayegh, M. A. (2018). Energy efficient thermal systems and processes. *Thermal Science and Engineering Progress, 7*, e1–e2.

Kaushik, S. C. (1989). Solar Refrigeration and Space Conditioning, Divya Jyoti Prakashan House.

Lv, Y., Chen, W. & Cheng, J. (2020). Effects of urbanization on energy efficiency in China: New evidence from short run and long run efficiency models. *Energy Policy, 147*, 111858.

Mazzaferro, L., Machado, R. M., Melo, A. P. & Lamberts, R. (2020). Do we need building performance data to propose a climatic zoning for building energy efficiency regulations? *Energy and Buildings, 225*, 110303.

Sarbu, I. & Adam, M. (2011). Applications of solar energy for domestic hot-water and buildings heating/cooling. *International Journal of Energy, 2*(5), 34–42.

Sur, A. & Das, R. K. (2010). Review on solar adsorption refrigeration cycle. *International Journal of Mechanical Engineering and Technology, 1*(1), 190–226.

Chapter 2
Solar Energy Option and Potential for Cooling

The conventional refrigeration and space conditioning devices are based on vapour compression machines and need electrical energy and/or fossil fuels for their operation. India to become carbon neutral economy by 2070. The production of coal is also not promising and its maximum utilisation will touch somewhere around 2045 on a world-wide basis (source: World Energy Outlook, 2021). In addition to supplying energy, fossil fuels are also being used extensively as fuel stock material for the manufacture of various chemicals. With the advent of the energy crisis and the realisation of the depleting nature of fossil fuels, the search for the use of alternative energy sources has become essential. Solar energy has a very high potential amongst renewable sources of energy since it is abundantly available and is an inexhaustible global source of energy having no pollution or hazards associated with it. The power from the Sun intercepted by the Earth is approximately 1.3×10^{13} MW, which is 10,000 times than the present energy consumption rate on the Earth. In principle, solar energy can supply all the present and future energy needs of the world continuingly.

2.1 Availability of Solar Energy

Sun is a normal middle-aged star. It has been on the main sequence for several billion years in hydrostatic equilibrium maintained between the collapsing gravitational force and the expanding pressure force. It is estimated to be staying on the main sequence for several billion years for radiating energy at about the same rate (Fig. 2.1).

The diameter of the Sun is about 109 times that of the Earth. The total mass of the Sun is about 330,000 times that of the Earth and consists of hydrogen gas, helium and small quantities of oxygen, carbon, neon and iron (Table. 2.1). The sources of solar radiation are the nuclear fusion reactions that are taking place in the interior core of the Sun where the temperature is of the order of 10^6 K. The hydrogen ions

S. C. Kaushik et al., *Solar Cooling*, https://doi.org/10.1007/978-3-031-42410-6_2

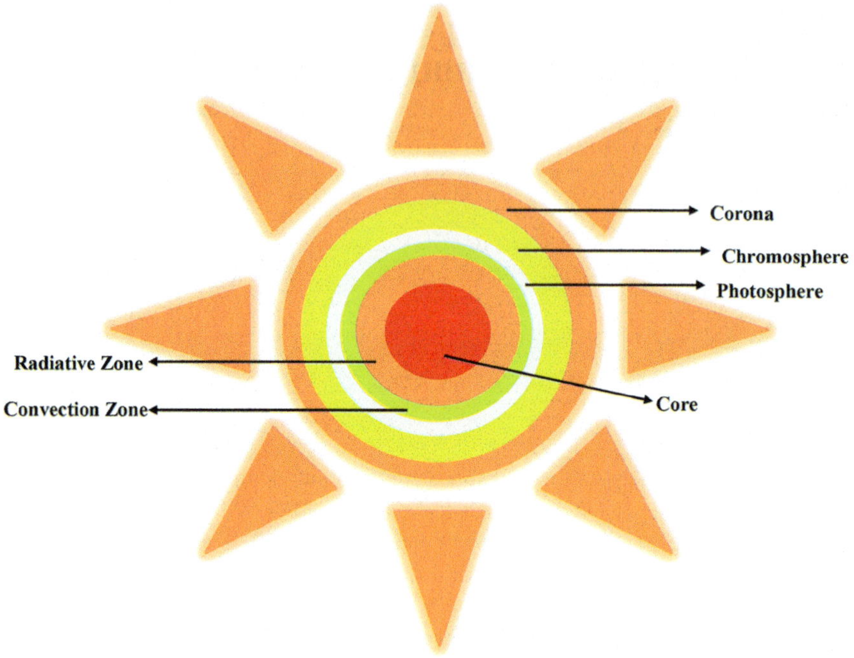

Fig. 2.1 Structure of the Sun

are transferred into helium and energy released is as follows:

$$4H^1 \rightarrow He^4 + 26.3\,\text{MeV} \tag{2.1}$$

Thus, the Sun has been emitting electromagnetic thermal radiation. Viewed from the Earth, the radiation coming from the Sun appears to be essentially equivalent to that coming from a black body at about 6000 K. The rate at which energy is received, from the Sun on a unit area perpendicular to the Sun rays at a mean distance of the Earth, is known as solar constant (its standard value is 1353 W/m^2). It is also useful to know the spectral distribution of the extraterrestrial solar radiation. The spectrum of electromagnetic radiation coming from the Sun is divided into three main regions: (i) the radiation below 3500 Å (ultraviolet), (ii) region between 3500 and 7000 Å (visible region) and (iii) above 7000 Å which is infrared. The total energy reaching the surface of the Earth in these three regions is distributed in the ratio of 2%, 51% and 47%, respectively. The utilisation of solar energy means using the solar radiation available in the visible and near-infrared regions of the solar spectrum [for detail discussion on the availability of solar radiation, see Sukhatme and Nayak (2017), Charters and Pryor (1980), Mohanty et al. (2017) and Kaushik et al. (1988)].

Solar radiation is received at the Earth's surface after being subjected to attenuation, reflection and scattering in the Earth's atmosphere. The Earth's energy budget

Table. 2.1 Characteristics of the Sun

Mean distance from the Earth	1.496×10^8 km
Visual brightness	26.74
Spectral classification	G2V
Equatorial radius	695,700 km
Circumference (equatorial)	4.379×10^6 km 109 times that of the Earth
Volume	1.41×10^{18} km^3 130,000 times that of the Earth
Mass	$(1.98855 \pm 0.00025) \times 10^{30}$ kg 333,000 times that of the Earth
Surface area	6.09×10^{12} km 12,000 times that of the Earth
The surface gravity at the equatorial	274.0 m/s^2
Escape velocity (from the surface)	618 km/s 57 times that of the Earth
Temperature	
(i) Centre	1.57×10^7 K
(ii) Photosphere	5782 K
(iii) Corona	5×10^6 K
Luminosity	3.828×10^{26} W
Age	Approximately 4.6 million years

published by NASA is shown schematically in Fig. 2.2. The radiation received without change of direction is called the beam or direct radiation, while that received after its direction has been changed by reflection etc. is called the diffuse radiation. The sum of the beam and diffuse radiation is referred as the total or global radiation.

In order to find the beam radiation falling on the surface having any orientation it is necessary to convert the value of beam flux coming from the direction of the Sun to an equivalent value corresponding to the normal direction to the surface. If θ is the angle between an incident beam of flux, I_{bn} and normal to a plane surface, then the equivalent flux falling normal to the surface is given by

$$I = I_{\text{bn}} \cos \theta \tag{2.2}$$

The angle θ is related by a general equation with other solar earth angles (Fig. 2.3) as given by (Sukhatme and Nayak 2017):

$$\begin{aligned}\cos \theta = &\cos \delta \cos \omega (\cos \phi \cos \beta + \sin \phi \sin \beta \cos A)\\ &+ \sin \delta (\sin \phi \cos \beta - \cos \phi \sin \beta \cos A)\\ &+ \cos \delta \sin \beta \sin A \sin \omega\end{aligned}$$

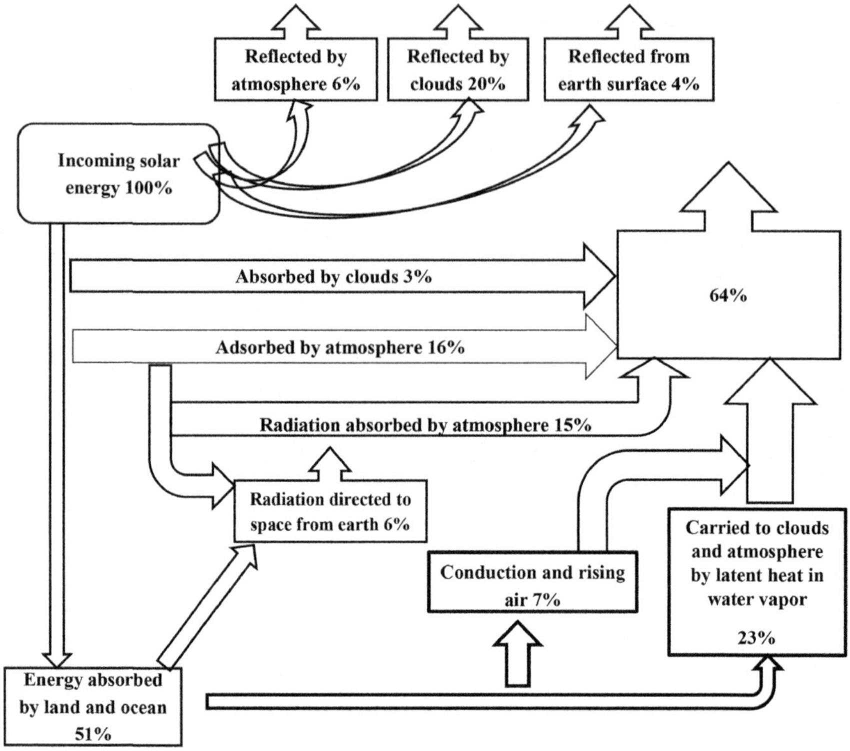

Fig. 2.2 Earth's energy budget

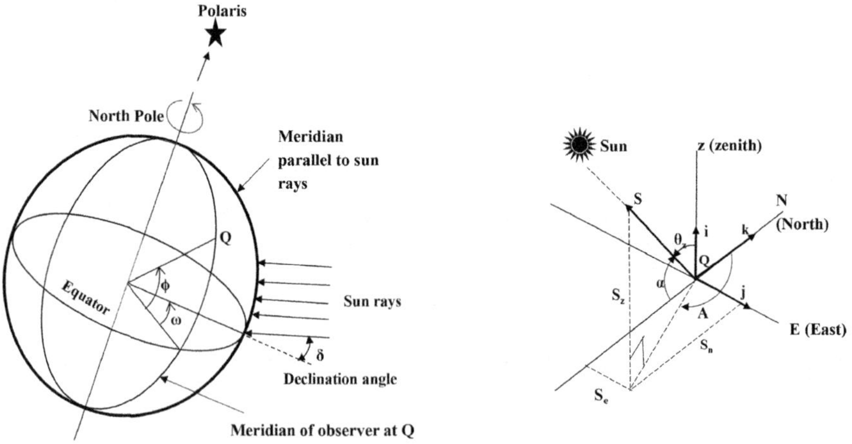

Fig. 2.3 Schematic of the Sun–Earth angles

where β is the slope made by the plane surface with the horizontal. It is taken to be positive for surface sloping towards the south and negative for surfaces sloping towards the north, ϕ is the latitude of a location, δ the declination angle, *A* the surface azimuth angle and ω the hour angle as defined elsewhere. Thus, the amount of the solar radiation to be collected on the Earth's surface depends on the location of the Earth's surface, the position of the Earth and orientation of the collection device relative to the direct beam radiation as well as the emission and absorption characteristics of the collector materials. The favourable places for collecting and exploiting solar energy are confined to areas between latitudes 35° north and south of the equator. These areas receive about 5 kWh/m^2 day on a horizontal surface. Solar research and technology aimed at finding the most efficient and inexpensive ways of capturing the dispersed solar energy and developing systems to store the collected heat for use at night or during off sunshine hours. The rate at which solar energy will become widely used depends on the research and development efforts to overcome the technical, economic and social barriers. The major technical obstacle for solar applications is the intermittent nature of the resource both on daily and annual cycle basis, which leads to storage requirements, not present in non-solar systems. The extra cost and complexity of storage associated with the solar system and the requirement of good 'solar climate' for efficient system operation are the negative aspects which attribute to the solar system. On the other hand, several advantages of solar systems over the more scarce fossil fuels include the simple technology and ease of materials used for solar collection and controls. Moreover, there is an excellent thermodynamic match between solar energy resource and many end uses, which are rare in conventional energy sources.

2.2 Indirect Solar Energy Options

The Earth receives around 1×10^{18} kWh of energy from the Sun annually. The incoming solar radiation is absorbed by different processes, viz. partly absorbed by the green plants for the growth of organic matter. A part of the radiation is being absorbed by the Earth's atmosphere, which is converted into wind energy. A part of the radiation is used for the evaporation of water, which are consequently returned to the Earth as running water energy. The remaining of the radiation goes into heating of the land and water masses.

It is thus evident that the energy flow through the Earth surface has three sources: (i) solar radiation, (ii) thermal energy conveyed to the surface of the Earth from the warmer interior by conduction of heat and by convection in hot springs and volcanoes and (iii) tidal energy derived from the combined kinetic and potential energy of the Earth–Moon–Sun system. This gives a total power influx on the Earth surface environment of 1.73×10^{17} W of which solar radiation accounts for 99.98%. About 30% of this incident solar energy is directly reflected and scattered back into space as short wavelength radiation. Another 47% is absorbed by the atmosphere, the land surface and the oceans and thus converted into heat at the ambient surface temperature. Around 23% is consumed in the evaporation, convection and precipitation in the hydrological cycle. A small fraction (0.2%) is derived by the atmospheric winds, oceanic currents and waves and an even smaller fraction are absorbed by the chlorophyll of the plant leaves, where it becomes the energy supply of the photosynthesis processes of the plants and trees. There are many ways in which solar energy can be indirectly transformed into a useful form of energy. It is normally useful to transform into a form in which it can be used for mechanical energy and electric energy etc. Some of the indirect sources for solar utilisation are described as follows:

1. Wind energy can be considered as an indirect source of solar energy. The air pressure differences cause air to flow and thus produce wind due to the change of temperatures produced when solar radiation is absorbed by different materials of the Earth's surface. As the water heats up much more slowly than dry land, the air near each type of surface will be heated at different rates. This, in turn, causes different air densities, different air pressures and consequently the movement of large volumes of air. In many areas of the world, these are heavy winds that develop into storms, cyclones and tornadoes etc. This enormous amount of energy can be converted into mechanical and electrical energy using windmills. This energy source is also a localised resource for generating electricity in remote areas. The potential areas could be promising and are likely to be made viable by the end of this century.
2. Ocean thermal energy provides another renewable energy resource as an indirect use of solar energy. A considerable amount of solar energy is absorbed in the ocean water of the tropical regions. The surface temperature of these oceans is found to be maintained at about 3 °C while the bottoms are cold at about 2 °C. This temperature gradient in the ocean near the tropical regions thus raises the possibility of utilising it to run a heat engine between the hot surface temperature and the cold bottom temperature. This method is quite promising even if the efficiency is low, as the source of energy is free. The above concept could be coupled with the *V-C* refrigeration cycle for the cooling application. Heat is provided by the warm surface water of the ocean to boil a special low boiling point organic fluid (such as propane), which then drives a turbine. The cooling water required for such a system is provided from the deeper and cooler layers of the ocean through large pipes to the surface.
3. Ocean waves generated by the interaction of winds with the surface of the ocean also provide another possible renewable energy resource. It could also be considered as an indirect solar option. The energy available from ocean waves varies with the size and frequency of the waves. On an average, it is estimated that about 10 kW are available for 1-meter amplitude of wavefront. However, due to wide fluctuations in frequency and amplitude at any location; wave energy is difficult to collect from a practical stand point. In addition, devices built for the purpose have to transfer the collected energy to the end-use and have to be strong enough to cope with adverse weather conditions. Therefore, problems such as the

development of large corrosion-resistant structures, the low efficiency due to the friction losses and the transporting of the energy produced are major obstacles.

4. Alternative methods of using solar radiation are based on photochemical conversion and photobiological processes, i.e., biomass conversion. In the former case, solar radiations are used to achieve photolysis of water yielding H_2 and O_2 in the presence of a reducing agent acting as a catalyst. The production of H_2 gas carries great promise to be used as fuel without storage problem. In the case of biological conversion, the natural process of photosynthesis is promising and solar energy is stored in organic matter called biomass. Energy can be recovered from plant materials (biomass) by direct combustion, fermentation or other microbiological actions leading to combustible liquids such as alcohol or methane gas. Plants growing in tropical countries produce such materials at a high rate. It would permit 3% of solar radiation to be recovered as heat annually. Alternatively, lower plants-algae are grown in open tanks or sewage can be used since such organisms are more efficient.
5. Hydroelectric power can also be considered as another indirect source of power available from the Sun. The processes that produce rainfall via evaporation of water from the Earth's surface and the transport of this water by winds are both powered by solar energy. Hydroelectricity is then generated by extracting some of the potential energy stored in water that has fallen as rain in mountainous regions.

The more traditional energy input options for cooling systems could be direct use of industrial waste heat, firewood, biomass/biogas and geothermal energy. Firewood has been used for producing heat, steam, as well as electricity but the quantity of energy produced from this source is expected to decrease due to scarcity of firewood.

Another good option of renewable energy is biomass energy, especially for agro wastes available in rural areas, where lots of organic materials are being wasted and extraction of biomass energy without pollution is possible. Biogas operated refrigeration systems are found to be more cost-effective but limited in use due to the non-availability of biomass in urban areas. Very recently Tyagi (2022) have developed a clean combustion device which has the emission factor comparable to that of the LPG burner using agro-residues (biomass) pellets. Therefore, heat-driven refrigeration system can be powered by such energy renewable energy systems to achieve the desired temperature.

Geothermal energy is the heat generated inside the Earth, which has been transported to the Earth surface. Because of the non-homogeneities in the Earth's crust, there are local hot spots just below the surface where the Earth temperature is much greater than the average value expected. When groundwater comes in contact with the hot rocks in some of these locations, either dry steam or wet steam and hot water are formed. A well drilled to these locations causes the dry or wet steam to emerge at the surface (Singh et al. 2020). This energy can be utilised for generating electricity or supplying low-grade heat. The geothermal heat emerges at an average rate of 82 mW/m^2 while the radial temperature gradient, which causes heat flow, is about 0.03 °C/m.

2.3 Utilisation of Solar Energy

Solar energy in the form of heat can be used for a variety of applications as follows:

(i) Heating of water for domestic and industrial purposes
(ii) Air heating/drying for agricultural and industrial products
(iii) Heating and cooling for space conditioning applications
(iv) Cooking of food and milk products
(v) Desalination of the brackish water
(vi) Mechanical shaft power production, and
(vii) Electricity production.

Energy research and development programmes in developing countries are oriented towards low-temperature applications like water heating/distillation, crop drying, greenhouses, while in developed countries, the main emphasis is on high technology applications of solar energy, viz. space heating/cooling, solar refrigeration and solar electricity production. However, there is a sharp transition on solar energy utilisation everywhere in the world and the use of solar energy for a variety of applications is most attractive because it is a freely available, abundant and infinite source of energy, although it is dilute and intermittent in nature.

Considerable attention is also being devoted to the collection and storage of solar energy on account of solar radiation being dilute and intermittent. Low-temperature heat production and short-term storage of solar energy are being considered. Long-term high-temperature storage devices will include thermochemical storage and considerable research and development work is being done.

As regards solar power production, conventionally there are two methods used to convert solar radiant energy into the useful electrical energy-photo voltaic conversion route, which converts solar radiation directly into electrical energy and thermal conversion route, which first converts solar radiation into heat and then it can be converted into mechanical and electrical energy. The conversion efficiencies associated with these routes that have been achieved, indicates that the second route is more efficient while the first one is cleaner and involves no moving part and hence, fewer maintenance problems.

It is thus evident that both direct and indirect ways for utilisation of solar energy offer a great promise for the future of mankind, especially for the developing nations who seem to have a bright share of sunshine. However, the merits of solar energy development must be considered accordingly to the following:

(i) The existence of technology or its development to transform solar energy into a useful form.
(ii) Economics of its transformation.
(iii) Effect of development on the environment and social welfare.
(iv) Efficiencies of transformation and utilisation because of the thermodynamic limitations.

Any analysis of solar energy development will be incomplete and inaccurate without looking at it from the thermodynamic point of view. The laws of thermodynamics play a key role in understanding the energetic universe not only from the physics point of view but also from the economic point of view. In solar energy development, we come across many possibilities regarding the collection, conversion and storage for the utilisation of solar energy. In all these phases of development, the overall system efficiency will be given by the product of the individual efficiencies, that is,

$$\eta_s = \eta_{\text{collection}} \times \eta_{\text{conversion}} \times \eta_{\text{storage}} \tag{2.3}$$

The problem, therefore, is of optimisation which requires system cost, efficiency. Another important aspect is the economic consideration, which may also be looked at in a way similar to technical analysis based on the laws of thermodynamics.

Although solar energy is a clean source, freely available in all parts of the world yet there are many problems associated with its use, mainly due to (i) its dilute nature and (ii) the intermittent nature. It has been found that even in the hottest region on the Earth, solar intensity rarely exceeds 1 kW/m^2, which is a low value for any technological utilisation. Consequently, large collector areas are required and the collection cost is of significant value. Some efficient and technically advanced solar collection systems are needed for making the solar system viable for practical use as discussed later.

The intermittent nature of solar energy is on a daily basis, because of the Earth's day-night cycle and also on a seasonal basis because of the Earth's orbit around the Sun, so the availability of solar radiation becomes intermittent in nature. The variation in availability occurs at a specific location because of the local weather conditions. Thus, solar energy available, during sunshine should be stored for use during off-sunshine hours. The intermittency of solar energy needs the storage systems and adds significantly to the energy cost of the solar system. Some solar energy storage methods are also discussed later. The real challenge in utilising solar energy for cooling is to develop a cheaper method of collection and storage besides more efficient cooling systems so that the large initial investments for solar cooling systems could be reduced.

2.4 Potential for Solar Cooling

Solar cooling is one of the most important aspects of solar thermal utilisation and quite similar to other solar thermal utilisation systems, the solar cooling system needs solar collection and storage units along with the cooling device for its continuous solar operation. Figure 2.4 shows that the demand for cooling is greatest at times of the maximum availability of solar radiation. The solar cooling systems are more attractive because of the following factors:

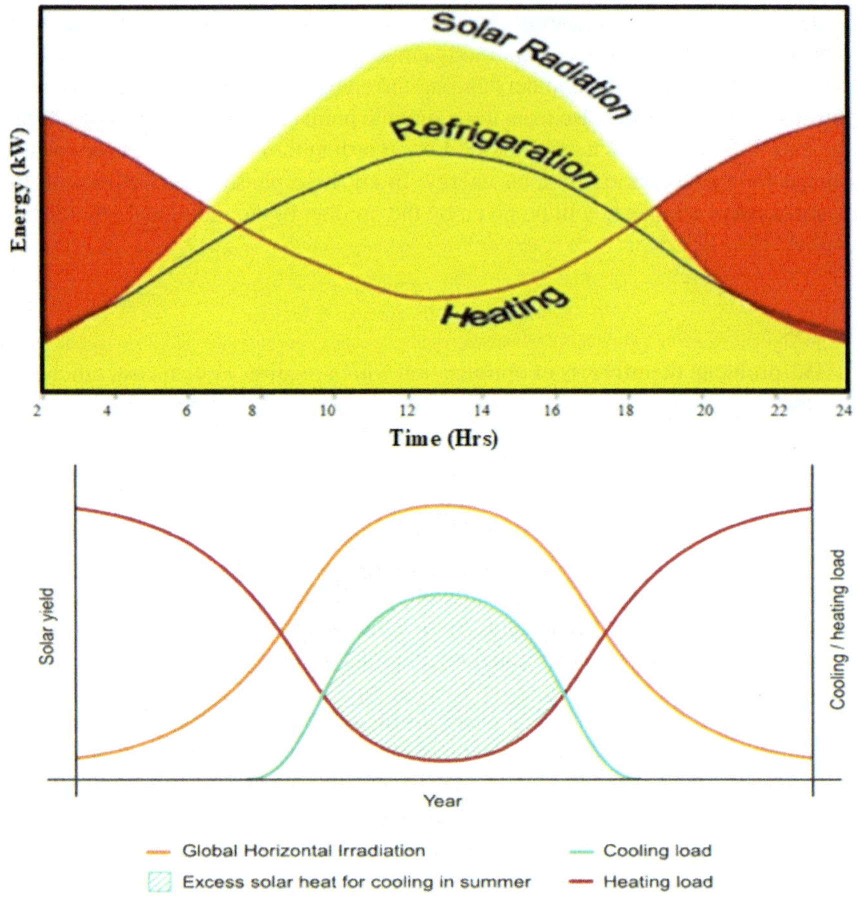

Fig. 2.4 Time variations of the **a** solar radiation and demands of refrigeration and heating **b** cooling/heating load and solar yield

- Solar energy is renewable and abundantly available in nature
- Major consumption of electricity is due to refrigeration
- The demand for cooling and availability of solar radiation are in phase
- Food preservation (agriculture, milk, fish) in remote rural areas
- Preservation of life-saving medicines in the remote areas.

In tropical countries, the solar energy operated cooling systems and cold storage facility has great demand. In the rural areas due to the shortage of power, there is an acute shortage of cold storage facilities. The surveys clearly demand the need for cold storage facilities in rural areas, where electricity is a major concern. Thousands of tonnes of perishable materials like vegetables, fruits, fish and similar other food materials are being wasted every year. The rural medical care centers face the difficulty of preservation of essential medicines for emergency use, because of the unavailability of cold storage facilities. Similarly, due to the shortage of electrical energy, in rural areas, people find it difficult to live because of the non-availability of the human comforts like air conditioning facilities during extremely hot and cold seasons. Solar cooling systems can provide these facilities since solar energy is freely available all over the country.

In a country like India, food production centres are usually situated in areas where electric supply for running large refrigeration plants is not available. Therefore, a cooling system, which does not depend completely on electricity for its operation will be the most ideal under such circumstances. The food production centre in rural and coastal regions happens to be most suitable for harnessing solar energy over vast open areas. In a tropical country like India, the importance of refrigeration can hardly be over-emphasised. Agricultural products play a significant role in the national economy. Some of the products could yield substantial foreign exchange, potatoes and onions for instance. Steps are being taken to increase the production and export of these commodities. The most important method used for preventing spoilage is preservation in cold storages. In 1947, there were only four cold storage units in the country, whereas nowadays there are thousands (Acharya and Sadath 2019). This continual increase in storage capacity is still dwarfed by the growth in the production of cold storage items (Dixit 1984).

Moreover, geographically India is situated between latitude 8° N and 36° N and longitude 62° E and 96° E. It has certain parts that experience extreme conditions of climate. A large amount of solar radiation with an average value of 600 cal/cm^2 day for the entire country falls in India. Even in the month of December, the value is around 500 cal/cm^2 day for some cities in India (Kaushik et al. 1988). During summer, the interior parts of the country become very hot with the maximum ambient air temperature reaching around 48 °C. Northern parts experience dry heat wave while the coastal regions are slightly cooler but due to extreme humidity, it becomes uncomfortable for human beings. Due to the extreme conditions in India, cooling of houses in summer and heating in winter is imperative, particularly in the interior areas and away from the coast. In addition to the improved human comfort, the space conditioning of buildings also leads to an increase in commercial production and feasibility of establishing high technology industries, which require a controlled environment. The existing cold storage and air conditioning units in India are not quite satisfactory.

It is also worthwhile to mention about the typical climatic conditions existing in India. The typical climatic zones of India are shown in Fig. 2.5. Generally, cooling is mainly desirable in two types of climates, viz. hot and dry, and hot and humid (Mullick 1976). Hot and dry climate exists in Delhi, Jodhpur and Nagpur etc. while hot and humid climate exists in coastal regions like Mumbai, Kolkata and Chennai etc. The first type of climate exists for the central and northern parts including the National Capital Region (NCR). The harsh temperature range for Delhi lies within the period from April to July while the uncomfortable high humidity is from August to September (water vapour pressure being 21–23 mm Hg), which is comparable to Chennai during April to May. The second type of climate in India exists in coastal regions (close to the sea) like Mumbai, Kolkata and Chennai etc. Chennai bears a harsh temperature (34–39 °C) in the period April to July and pinching humidity (water vapour pressure being 20–24 mm of Hg) from March to October.

This type of climate needs cooling as well as dehumidification. While Delhi's climate needs only cooling for most of the seasons except during August to September. Thus, in India, both type of places are abundant and justify the need for both cooling and dehumidification at one place or the other.

Solar cooling is, therefore, most relevant to Indian conditions because plenty of sunshine is available throughout the year and vast open area is easily accessible for harnessing the solar energy.

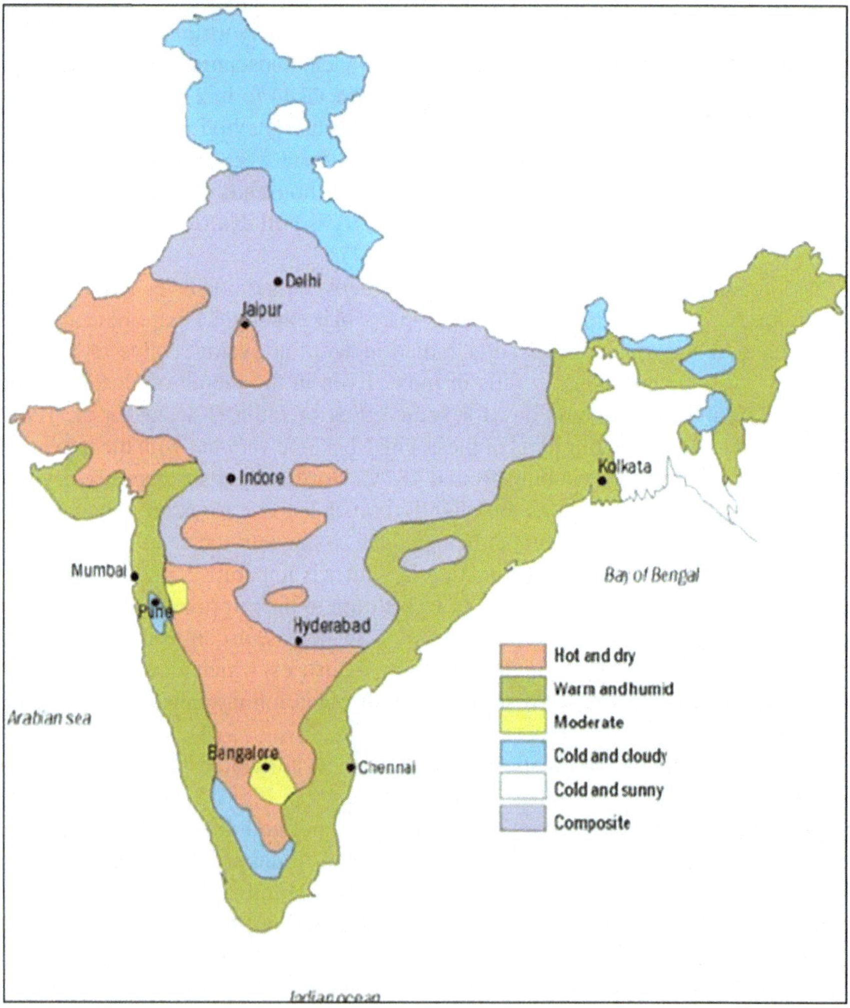

Fig. 2.5 Typical climatic zones in India (Chaney et al. 2007)

2.5 Types of Solar Cooling Systems

There is a huge scope for the utilization of solar energy and its application because the availability of solar energy is in phase with the demand for cooling, which is maximum in summer and minimum in winter. Moreover, the solar collector efficiency is relatively high in summer because of the high ambient air temperature as compared to winter. The development of solar cooling devices has been dwarfed due to the availability of relatively low temperatures from solar collectors. However, with modern collectors, there is new hope for solar cooling systems. The solar cooling is one of the most important aspects of solar energy utilisation. A solar refrigerator is a device that is operated completely by the solar energy harnessed from the Sun. For decades, peoples have used vapour compression refrigeration systems to store their medicines, food and other products. After Otto Mohr's work in 1935, people learnt that solar power can run the refrigerator just like the electricity. Otto Mohr wanted to create a refrigerator that would not require any energy other than exposure to sunlight for two hours per day. He was able to describe all of the correct components needed to build the solar refrigerator and make it work. Since that time, more engineers, scientists, and researchers have all developed prototypes based on Otto Mohr's original work. Even though the idea of a solar refrigerator was created by Otto Mohr in 1935, people are, however, more likely to choose highly efficient VCR systems, hence, the RD&D on solar thermal driven refrigeration systems noticeably reduced. Subsequently, electricity-driven vapour compression systems have played a significant role in the market. In the 1970s, the energy crisis became an issue and people started to work on PV-based systems. But at that time, the cost of photovoltaic was very high while the efficiency was low. After the energy crisis in the 1970s, solar energy became a popular energy source and in the 1990s, the research and development in the field of solar cooling became more familiar. Figure 2.6 illustrates the possible refrigeration systems and their applications. Solar refrigeration systems are relevant to India because:

(i) Throughout the year, the ambient temperatures are high. Higher ambient temperatures mean more energy is consumed in refrigeration.
(ii) Solar insolation is high in most part of the country.
(iii) In rural areas, where electricity is a major concern, solar cooling systems are best suited.

As explained in the previous section, the supply of solar energy and demand for cooling are in phase, hence, solar cooling systems are a very good option in rural areas. The solar cooling systems are most economical in the regions where the electricity is a major concern as, it can be used for the storage of agricultural products, food and medicines (e.g., vaccines). Solar cooling can be achieved by many alternative methods as shown in Tables 2.2 and 2.3. Passive solar space conditioning is based on the evaporative cooling and radiative cooling options and can be integrated with the building structure itself. Active solar cooling systems mostly used are based on the vapour refrigeration cycles, which include the vapour compression refrigeration

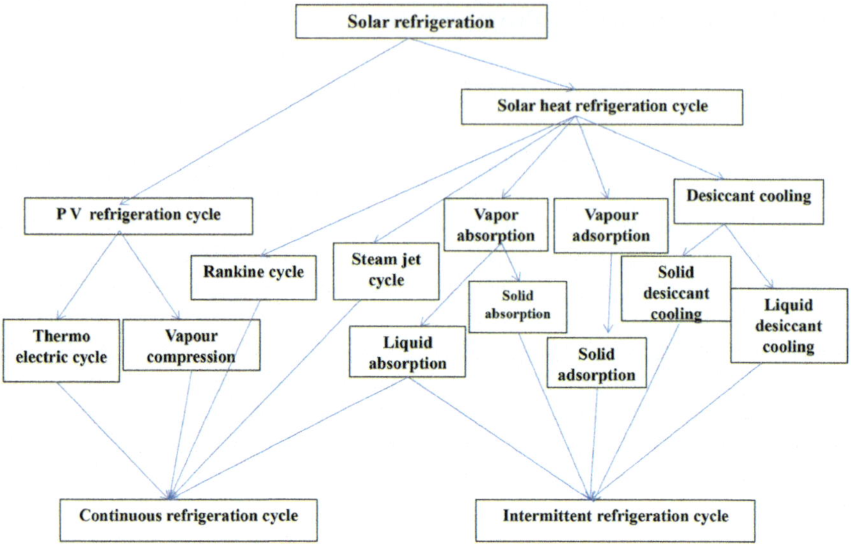

Fig. 2.6 Classification of the solar refrigeration cycle

(VCR) driven by electric power, vapour absorption (VAB) and vapour adsorption refrigeration (VAD), normally driven by a low-grade heat source such as the solar energy, waste heat etc. In this section, various solar cooling systems will be briefly described. These systems will be discussed in detail in the subsequent chapters.

Generally, there are two methods of utilisation of solar energy: (i) converting solar energy into electricity by means of a photovoltaic cell. The electricity is used to run an ordinary refrigeration system; (ii) the utilisation of solar energy to run a thermal compressor in vapour adsorption refrigeration or vapour absorption refrigeration systems. The solar-powered adsorption cooling system is proved to be an efficient method of achieving cooling with low cost, high-efficiency and simple in manufacture.

In solar vapour compression refrigeration systems, solar energy is converted into mechanical energy using a solar collector and a power cycle like Rankime cycle. The compressor of the VCR system can be driven by this mechanical energy and called as Solar Rankine cycle cooling system. The other version is based on the inverse Brayton cycle for producing cooling using solar energy, where, solar energy is converted into mechanical energy with which the air is compressed and cooled to remove the heat, the cold air is then expanded through the expander before being discharged into the air conditioning space.

Another option for cooling is steam jet ejector system where the mechanical compressor is replaced by the convergent/divergent nozzle known as jet ejector pump. The primary fluid, which receives heat as input from the solar collectors is expanded within the jet pump thereby affecting compression of the refrigerant, and suction is obtained in an evaporator where the cooling occurs. Thus, a jet ejector can be used

Table 2.2 Existing solar-driven refrigeration technologies

Refrigeration cycle	Solar technology		COP_{cycle}	Available application		
	Thermal collectors (T_{gen}/ T_{reg}), °C	PV cells powered (Power for 30 L cooling box in W)		Refrigeration	Airconditioning	Example applications
Electricity/ work driven cycles						
Vapour compression		(15–40)	3–5	✓		HR, SR, CB
Thermoelectric		A few in mW	0.5	✓		VT, CB
Stirling		(8–50)	3	✓		CB, LT
Heat driven cycles						
Absorption	(80–190)		0.6–0.8 (Single effect)	✓	✓	AC, IP
Adsorption	(80–300)		0.3–0.6	✓		CB, IP, VS
Chemical reaction	(80–300)		0.1–0.2	✓		IP, VS, FS
Duplex rankine	>20		0.3–0.5		✓	AC
Desiccant cooling	(40–80)		0.5–1.5		✓	AC

AC: air-conditioning, CB: cooling box, FS: food storage, IP: ice production, HR: household refrigerator, LT: low-temperature applications, SR: small refrigerator, VS: vaccine storage, VT: vaccine transportation

to reduce the pressure over a liquid refrigerant. This causes evaporation and hence, cooling of the liquid which is circulated through a heat exchanger to achieve the desired space cooling. Operation of the ejector is accomplished by high-temperature and pressure refrigerant vapour from a boiler through the ejector and passing to the condenser. Pumps are required to maintain the proper refrigerant volumes in the evaporiser, condenser and the evaporator.

Solar energy can also be used to produce cooling by using closed/open cycle adsorption cooling systems and/or absorption dehumidification cycles based on the solar regeneration of sorbents and liquid desiccants. In vapour absorption cooling systems, a mixture of an absorbent and the refrigerant is used to replace the mechanical compression by the thermal compression through the generator-absorber assembly and the solar-heated water is used to supply low-grade heat input in the generator. Cooling is produced in the evaporator and heat is produced from the

Table. 2.3 Various solar cooling options

Vapour compression refrigeration	Vapour absorption cooling	Vapour adsorption cooling	Non-conventional cooling
Conventional electricity operated VCR systems: (a) Solar thermal conversion based VC refrigeration (b) Photovoltaic conversion based VC refrigeration	1. Closed cycle systems: (a) Continuous absorption solar refrigeration systems (b) Intermittent absorption solar refrigeration system	1. Open cycle systems: (a) Solid adsorption (b) Liquid adsorption	1. Solar thermo-electric refrigeration
Rankine cycle solar cooling (heat engine coupled VC cycle)	2. Open cycle adsorption cooling	2. Closed cycle adsorption systems	2. Metal hydride refrigeration
Jet ejector compression solar refrigeration system	3. Multi-stage absorption systems	3. Hybrid cooling systems	
Gas compression cycle solar refrigeration systems	4. Advanced absorption cycles		

condenser and absorber. The coefficient of performance (COP) of absorption systems is rather low because the input energy is the heat energy rather than the work.

The adsorption refrigeration cooling process is similar to that of the absorption cooling except that the refrigerant is adsorbed onto the surface of the adsorbing material rather than being taken into the chemical solution. The refrigerant is then driven out by heating, where solar energy can be used, the liberated refrigerant is then subject to condensation and throttled to produce the cooling effect. The refrigerant is then re-adsorbed to the carrier material to have a closed cycle operation. Generally in vapour absorption and adsorption cooling systems, large heat energy input is required to produce a refrigerating effect than the vapour compression refrigeration cycle. However, the energy input to the former is mainly in the form of heat whereas in the VCR systems, it is the compressor work using high grade electrical energy. According to the second law of thermodynamics, heat and work are not equivalent forms of energy. Heat is a low-grade energy and only a part of it is available, whereas work is high-grade energy and hence, no fair comparison can be made between the VCR and VAB/VAD systems in terms of the performance.

The basic consideration in refrigeration and cooling systems should be a high COP or lower cooling temperature or higher cooling capacity. This depends on specific applications and localized conditions. Achieving an ideal COP is not possible because of the irreversibilities involved in different processes. Another important factor for the cooling systems is economical operation, hence, a compromise has to be made between the higher COP and operational cost. It is desirable to study the comparative performance of various cooling systems for typical climatic conditions and assess the overall power consumption and efficiency. It is also possible to improve the COP of VAB/VAD systems by multi-stage cycles and advanced absorption/adsorption cycles based on double effect generation/condensation/evaporation processes.

A non-conventional method of cooling is thermoelectric cooling. The basic difference between the thermoelectric cooling and other cooling devices is that all cooling devices make use of mechanical shafts or thermal energy as input source while in the thermoelectric cooling, the electric energy is directly utilised for cooling. For the solar operation of the device, it is desirable to use a thermoelectric generator to first generate the electricity and then coupling it with the thermoelectric refrigerator. The overall efficiency seems to be a low but simple design, lightweight and low cost are some of the advantages of thermoelectric cooling systems.

Table 2.2 illustrates an overview of some key characteristics for different types of solar cooling systems. The thermally driven cooling cycles are absorption, adsorption, chemical reaction, desiccant cooling, ejector and Rankine refrigeration cycles, whereas, *PV* driven cycles are Stirling, thermoelectric and vapour compression refrigeration. Each of these cooling systems (except the adsorption cooling system), maintains the advantages of its conventional type but suffers from certain limitations. Further as compared to the solar refrigerator, the conventional *PV* refrigerator has a high installation cost.

2.6 Solar Energy Utilisation in India

According to the Solar Energy Corporation of India (SECI) and Ministry of New and Renewable Energy (MNRE), the total installed power generation capacity in India is 336833.2 MW (as on 31/07/2018), of which 64% is contributed by thermal energy sources and 19% by renewable sources (Fig. 2.7). India is the world's 5th largest electricity market. India has one of the world's fastest rates of renewable energy growth and the total installed capacity based on renewable sources is about 156 GW as of March, 2022. India's renewable energy scenario in Fig. 2.7a and b clearly shows that wind energy is contributing a lot to India's renewable energy sector. Solar energy contribution is 31% (Paladugula et al. 2018), biomass 12%, small hydro 7% and wind energy 49%. Following are the Government of India's ongoing solar mission projects (MNRE) (Fig. 2.8):

(i) Grid Connected Solar PV power plants with VGF support under JNNSM Phase II

  (a) 750 MW (2014–15)
  (b) 2000 MW (2015–16)
  (c) 5000 MW (2015–16 to 2018–19)

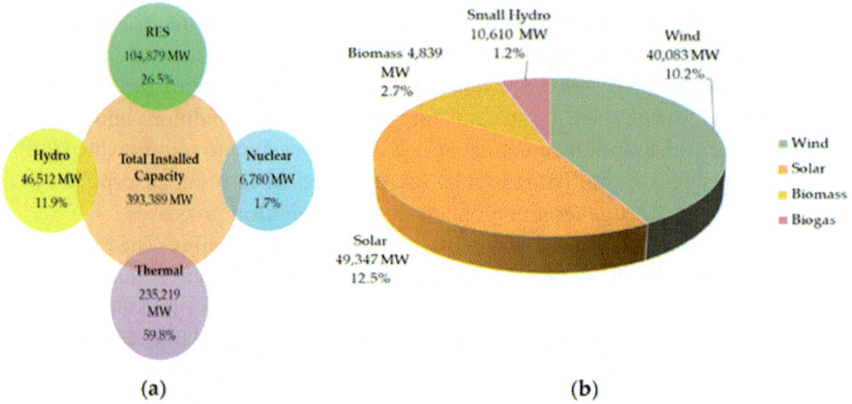

Fig. 2.7 **a** Total installed capacity as on 31 December 2021. **b** Renewable energy sources status as at 31 December (Singh et al. 2022)

Fig. 2.8 Charanka solar power plant, Gujarat, India

(d) Indo-Pak Border solarisation scheme
(e) 100 MW canal-top and canal-bank scheme
(f) 300 MW for defense/paramilitary forces
(g) 1000 MW of grid-connected projects for CPSUs

(ii) Grid-connected solar PV projects through bundling
(iii) 1700 MW grid-connected rooftop scheme, and
(iv) 20,000 MW solar parks scheme.

References

Acharya, R. H. & Sadath, A. C. (2019). Energy poverty and economic development: Household-level evidence from India. *Energy and Buildings, 183*, 785–791.

Chaney, L., Thundiyil, K., Chidambaram, S., Abbi, Y. P. & Anderson, S. (2007). *Fuel savings and emission reductions from next-generation mobile air conditioning technology in India* (No. NREL/CP-540-41154). National Renewable Energy Lab. (NREL), Golden, CO (United States).

Charters, W.W.S. & Pryor, T.L. (1980). Solar energy—An introduction to principles and applications, Mechanical Engineering Department, University of Melbourne, Australia.

Dixit, D.K. (1984). Cooling with the Sun, *Science Today*, October: 70–73.

Kaushik, S. C., Gadhi, S. M. B., Agarwal, R. S. & Kumari, R. (1988). Modeling and simulation studies on single/double-effect absorption cycle using water-multicomponent salt (MCS) mixture. *Solar energy, 40*(5), 431–441.

Mohanty, S., Patra, P. K., Sahoo, S. S. & Mohanty, A. (2017). Forecasting of solar energy with application for a growing economy like India: Survey and implication. *Renewable and Sustainable Energy Reviews, 78*, 539–553.

Mullick, S. C. (1976). Solar air conditioning analysis of systems and performance characteristics of absorption tower, PhD Thesis, IIT Madras.

Paladugula, A. L., Kholod, N., Chaturvedi, V., Ghosh, P. P., Pal, S., Clarke, L. & Wilson, S. A. (2018). A multi-model assessment of energy and emissions for India's transportation sector through 2050. *Energy Policy, 116*, 10–18.

Singh, V., Phuleria, H. C. & Chandel, M. K. (2020). Estimation of energy recovery potential of sewage sludge in India: Waste to watt approach. *Journal of Cleaner Production, 276*, 122538.

Singh, U., Rizwan, M., Malik, H. & García Márquez, F. P. (2022). Wind energy scenario, success and initiatives towards renewable energy in India—A review. *Energies, 15*(6), 2291.

Sukhatme, S. P. & Nayak, J. K. (2017). Solar Energy. McGraw Hill Education.

Tyagi, S. K. (2022). Biomass pellet-based combustion devices. *Indian Patent, 397919.*

Chapter 3
Solar Energy Collection Systems

Solar energy can be used directly or indirectly and it has been identified as one of the promising alternative sources in future. A broad classification of solar energy collection is given in Fig. 3.1. As can be seen from Fig. 3.1, there are two main roots for conversion of solar energy into useful form, direct and indirect. The direct route includes thermal and photovoltaic conversion while the indirect means include water power, wind energy, biomass (renewable organic fuels) and the ocean thermal energy conversion. Since solar energy is diffused and hence dilute, its collection is very important before it is to be converted into a useful form. In general, the collection systems can be classified into two forms: (i) natural and (ii) man-made. The natural collection occurs on the surface of the land and oceans. They give rise to the wind, weather, ocean currents and growth of plants while the man-made collection involves passive collection on a technological phenomenon, where collection and conversion occur with a collection device. For each of the solar energy application, a solar energy collection device is needed to tap the solar energy. Necessarily, the essential parts of a solar collector system are:

(i) A collector plate to collect solar energy and an absorber to absorb the collected solar radiation.
(ii) A transparent cover to permit solar radiation into the device and reduce upward heat loss.
(iii) An insulated base to reduce the bottom loss, and
(iv) A network of tubes attached beneath the collector plate through which the fluid flows.

Conventionally, the fluid takes away the heat from the hot collector plate and is stored in a separate tank. There have been many improvements in the design and choice of the collector materials to improve the collection efficiency. The solar energy has been used to perform power generation, hot water production, desalination, cooling etc. (Kalogirou 2014).

S. C. Kaushik et al., *Solar Cooling*, https://doi.org/10.1007/978-3-031-42410-6_3

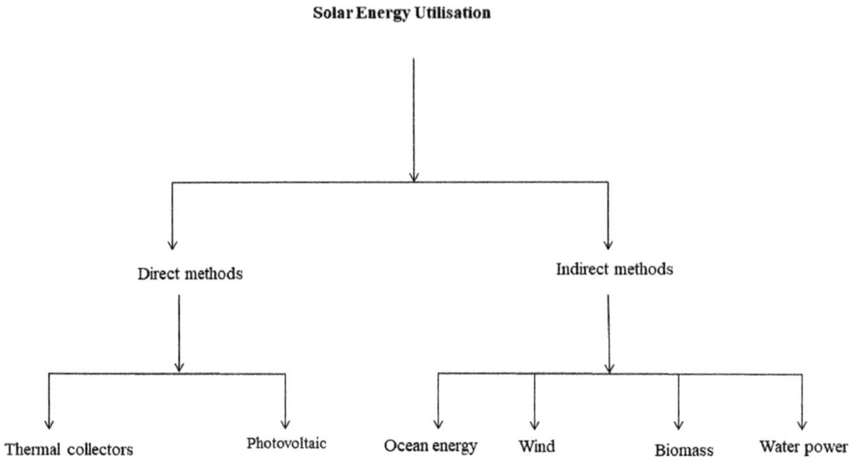

Fig. 3.1 Classification of the method of solar energy collection

3.1 Solar Energy Collection Devices

The main component of the solar system is the collection part and therefore the function of the solar collector is to collect solar radiation and converts it into the form of thermal/heat energy. This heat energy is then transferred to the fluids such as air, water or oil, which is flowing through the collector.

Solar collectors are classified into two main categories, concentrating and non-concentrating, as given below:

(a) Concentrating collectors

These collectors redirect and optically concentrate solar radiation to a point or line, and usually need a tracking mechanism. The concentrating collectors have concave reflecting surfaces and the Sun's beam radiation incident on the concave surface will be intercepted to a small receiving area of the collector (Kalogirou et al. 2003).

(b) Non-concentrating collectors

These collectors are permanently fixed in position and do not track the Sun. There are three types of the non-concentrating type of collectors, such as flat plate collectors, evacuated tube collectors and compound parabolic collectors. Table 3.1 shows the types of collectors available in the market with the concentration ratio.

Table 3.1 Types of solar collectors (Evangelisti et al. 2019; Kalogirou 2004)

Collector type	Movement type	Absorber	Temperature range (×C)	Concentration ratio
Stationary	(i) Flat plate collector	Flat	20–90	1
	(ii) Compound parabolic collector	Tubular	50–280	1–5
	(iii) Evacuated tube collector	Flat	40–210	1
Single axes tracking	(i) Parabolic trough collector	Tubular	50–300	20
	(ii) Cylindrical trough collector	Tubular	50–300	15
	(iii) Linear Fresnel reflector	Tubular	50–250	15
Two axes tracking	(i) Parabolic dish collector	Point	100–1000	100–500
	(ii) Heliostat field collector	Point	100–2500	100–1500

3.2 Flat Plate Collector (FPC)

A flat plate collector consists of a metal box with glass or plastic cover at the top of it. A dark colour absorber plate is placed on the bottom of the metal box. To minimize the heat loss from the sides and bottom of the collector, these parts are usually well insulated with suitable insulating materials.

Flat plate solar collector is a device in which solar energy is absorbed for heating a fluid of which water and air are the most common. In many parts of the country, warm water is obtained by placing a bucket of water in the Sun, the temperature of water thus heated is rather low. This is due to the large loss of heat, poor absorption of solar energy by the water, and the container acting as an absorber. The air-fluid collectors are less efficient than liquid-fluid collectors.

A typically commercially available solar collector is commonly known as a flat plate collector. This combines good absorption of solar energy with acceptable losses of heat. The collector consists of a black absorbing plate, in thermal contact with a series of tubes, which carry water, placed in a box with insulation below the plate and one or more transparent covers above it. When sunlight falls on this device a large fraction is absorbed by the black absorbing plate which gets heated. This heat is transferred to water flowing through the tubes. In a good collector, the absorption of solar energy by the plate is large and the loss of heat by the plate to the atmosphere is low. The former is achieved by suitable blackening of the surface, sprayed blackboard paint can give a surface, which absorbs more than 90% of the solar energy falling on

Fig. 3.2 Flat plate collector (Chung et al. 2018)

it. The black plate loses energy on account of conduction, convection, and radiation heat transfer. The schematic view of a flat plate collector is shown in Fig. 3.2.

The main components of a flat plate collector are:

Absorber Plate: The absorber plate can be flat, corrugated or grooved, and may be integrated into the tube. Plastic, steel and copper are the materials usually employed to make the absorber plate. Copper is used because it maximises the absorptance of solar energy and minimises the radiation emitted by the plate (Fig. 3.3).

Cover Plate: The convective and radiative heat losses from the absorber can be reduced by using the cover plate. The cover plate is usually manufactured from glass or plate.

Fluid flow lines: The fluid flow pipes or lines are integrated to allow the working fluid to pass through the collector.

Insulation: The materials, such as fibreglass, are placed at the back and sides of the collector to reduce heat losses.

Enclosure: The enclosure is a box, which covers all the components.

Normally for the maximum use, the collectors should focus towards the equator, facing south in the northern hemisphere and facing north in the southern hemisphere. The collectors are placed at an angle equal to the latitude of the location. In practice the angle varies from 10 to 15° depending upon the application.

The collectors widely differ from each other in the choice of materials of the plate, tube cover, insulation and box. The most commonly used heating fluids are water, water plus antifreeze additive or air. Their major purpose is to collect the maximum

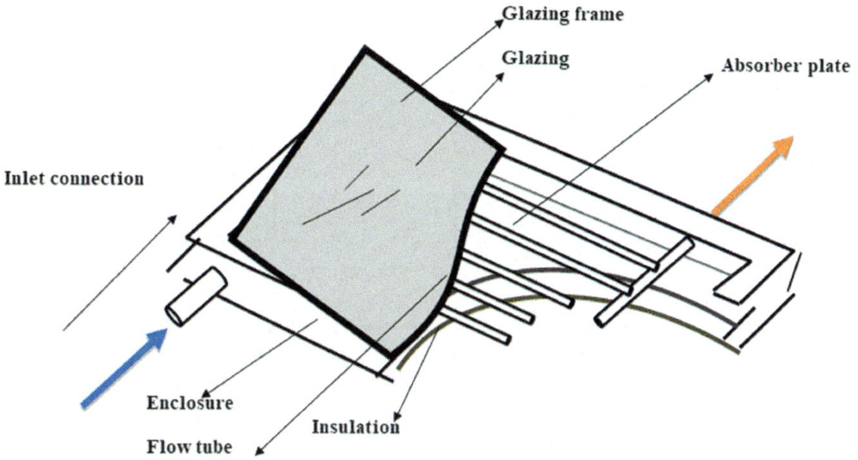

Fig. 3.3 Schematic view of a flat plate collector (Kaushik et al. 1988)

solar insolation at the lowest possible cost. The collector should have a long life. The considerations should be given to reduce the corrosion, dust deposit and the possibility of the breakage of glass should be avoided as well. Some typical choices are as follows:

Cover: Glass and plastic.

Glass is the most widely used glazing material. The transmissivity of the glass is of the order of 90%. Plastic films and sheets also possess high shortwave transmittance. Even though plastic is also a wise choice, its application is limited by the temperature.

The transmittance of the commercially available window and greenhouse glasses is in the order of 0.87 and 0.85, respectively. Research proved that the anti-reflective coatings and surface texture can improve the transmittance of glazing material considerably. An ideal glazing material should admit as much as solar radiation as possible.

Plate: Copper, aluminium, galvanized iron.

Besides the mentioned plate materials UV resistant plastic extrusions are also used as the absorber plate materials.

Insulation: Glass, wood, foams.

Box: Metal, wood, asbestos, cement.

Performance Analysis of Flat Plate Collectors

The conservation of heat regarding solar collectors shown in Fig. 3.4 can be expressed as

$$q_s = q_u + q_L \tag{3.1}$$

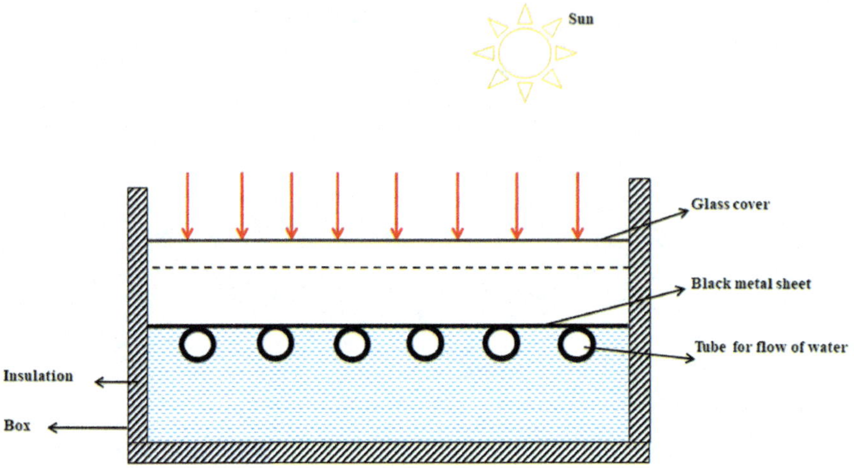

Fig. 3.4 A flat plate solar collector (Kaushik et al. 1988)

where

$q_s =$ solar energy absorbed by the plate per unit time per unit area of the collector, which is equal to the product of absorptance of absorbing plate (α), transmittance of the glass cover (τ), and incident solar energy per unit area of the collector (I_s)

$q_u =$ useful heat transferred to the working fluid per unit time per unit collector area

$q_L =$ energy losses per unit time per unit collector area from the collector to the surroundings

The overall performance of a collector is measured by the collection efficiency, thus defined as

$$\eta_{\text{collection}} = \frac{q_{\text{u}}}{I_{\text{s}}} = \text{useful heat transfer to the fluid/incident solar energy} \quad (3.2)$$

In a flat plate collector, q_{L} and q_{u} may be expressed as

$$q_L = h_i(T_{\text{p}} - T_{\text{c}}) + \lambda_{\text{e}}(T_{\text{p}} - T_{\text{a}}) \quad (3.3a)$$

$$q_u = h_{\text{f}}(T_{\text{p}} - T_{\text{a}}) \quad (3.3b)$$

and thus conservation of energy with reference to Eq. (3.1) may be expressed as

$$q_s = h_{\text{i}}(T_{\text{p}} - T_{\text{c}}) + h_{\text{f}}(T_{\text{p}} - T_{\text{f}}) + \lambda_{\text{e}}(T_{\text{p}} - T_{\text{a}}) \quad (3.3c)$$

where h_i is the total heat transfer coefficient between the plate and glass cover, T_p the steady-state absorbing plate temperature, T_c the glass cover temperature, h_f the heat transfer coefficient between the absorbing surface and the fluid, T_f the average temperature of the fluid passing through the tubes below the absorbing plate, λ_e the heat loss from the absorbing plate to the ambient air through the insulation (bottom and sides) per unit area of the plate per degree difference of temperature and T_a the ambient air temperature.

The energy balance of the glass cover assumed to have negligible thickness is given by

$$h_i(T_p - T_c) = h_o(T_c - T_a) \tag{3.4}$$

where h_o is the heat loss coefficient from the glass cover to the ambient air; this implies that the heat lost by the plate to the cover is equal to that lost by the cover to air. The mean fluid temperature may be determined from the fact that heat transferred from the absorbing plate to the fluid is equal to the heat taken out by the flowing water.

Thus,

$$Ah_f(T_p - T_f) = \dot{m}_f C_f (T_{fo} - T_{fi}) \tag{3.5}$$

where A is the area of absorbing plate/collector, $\dot{m}_f$ the mass flow rate of fluid and, T_{fo} and T_{fi} the outlet and inlet temperature of the working fluid respectively.

From Eq. (3.5), the mean fluid temperature may be expressed as

$$T_f = \frac{T_{fi} + T_{fo}}{2} = \frac{h_f T_p + \frac{2\dot{m}_f c_f \times T_{fi}}{A}}{h_f + \frac{2\dot{m}_f c_f}{A}} \tag{3.6}$$

Solving for T from Eqs. (3.3) to (3.5), the plate temperature is given by

$$T_p = \frac{[q_s + (U_t + \lambda_e)T_s + H_e T_{fi}]}{(U_t + \lambda_e + H_e)} \tag{3.7}$$

where $U_t = \frac{h_0 h_1}{h_0 + h_1}$ is the overall top loss coefficient between the absorbing plate and the ambient air and

$$H_e = \frac{\left(\frac{2\dot{m}_f C_f h_f}{A}\right)}{\left(h_f + \frac{2\dot{m}_f C_f}{A}\right)} \tag{3.8}$$

It can be seen here that when there is no fluid flow then $m_f = 0$.

$$T_p = \frac{q_s + U_L T_a}{U_L} = T_s \text{ is known as the stagnation plate temperature} \tag{3.9}$$

where $U_L = U_t + \lambda_e$ is the overall heat loss coefficient of the collector system.

Substituting for T_p and T_f as above in Eq. (3.3b), the amount of heat used in heating the fluid is given by

$$q_u = F\left[a\tau I_s = U_L(T_a - T_{fi})\right] = FU_L(T_s - T_{fi}) \quad (3.10)$$

where $F = \frac{1}{1+\frac{U_L}{H_e}}$, is known as the heat removal efficiency factor.

Thus, the collector efficiency is given by

$$\eta = \frac{q_u}{I_s} = F\left[\alpha\tau - U_L\left(\frac{T_{fi} - T_a}{I_s}\right)\right] \quad (3.11)$$

The factor F shows the performance of the fluid circulating system. $\alpha\tau$ represents the optical performance of the system.

Thus, the collector efficiency equation is of the form $y = mx + c$ and known as Hottle-Whiller-Bliss (H-W-B) equation. Knowing the variation of η with parameter $\left(\frac{T_{fi}-T_a}{I_s}\right)$ optical efficiency ($\alpha\tau$) and overall heat loss coefficient (U_L) can be evaluated.

Features of a Flat Plate Collector

(a) Flat plate collector utilises both beam and diffuse components of the solar radiation.
(b) The design of a flat plate collector is simple.
(c) Flat plate collector requires less maintenance.
(d) Because of the absence of solar concentration, the area from which the heat is lost is large. Hence, the efficiency is low.
(e) Its application is limited to the temperature of less than 110 °C.
(f) The present cost of a typical flat-plate collector is in the range of Rs. 4500–5000 per m^2.

3.3 Evacuated Tube Collectors (ETC)

The flat plate solar collectors show their ability to operate at a temperature around 80 °C only with reasonable efficiency and are generally used for domestic water and space heating. However, the temperature in the range of 110–150 °C is required for the power generation, air conditioning and refrigeration applications. The performance of flat plate collectors becomes considerably reduced during cold, cloudy and windy days. Furthermore, the condensation and moisture will cause early deterioration of internal materials resulting in reduced performance and the system failure. To meet with the requirements of high temperature and efficiency, evacuate the annular space between the absorber plate and cover glass, which can reduce the natural convection losses from the absorber plate of the collector. As the evacuation of the annular space of a flat plate collector is technically difficult, the evacuated tube collectors have been proposed and designed. Water is the most commonly used working fluid for solar thermal applications.

The most important component in the evacuated tube collectors (ETC) is the heat pipe integrated to the collector. The evacuated tube collectors have been a proven technology, with the vacuum encapsulated inside. The vacuum that surrounds outside the tube greatly reduces losses due to convection and radiation, which favours the enhancement in the efficiency of the solar thermal collectors. In 1909, Emme suggested various designs and configurations of evacuated tube thermal collector technology.

The essential features for the high performance of these evacuated tube collectors are as follows:

(a) To decrease the convection and conduction losses, pressure in the annular space is reduced, this is the basic working principle of an evacuated tube collector. The magnitude of the thermal conduction loss depends upon the gas composition, temperature of the glass envelope, the absorber element and the nature of their surfaces.
(b) Spectrally selective black absorber surfaces are used in evacuated tube collectors to reduce radiation heat loss from absorber plate/tube. The absorbance of the surface is very high and have low emittence in the infrared region at high operating temperature.
(c) To reduce reflection of the light from the glass surfaces, anti-reflection coating on the upper side of the glass envelope is used thereby increasing the transmittance through the glass.
(d) Heat mirror surfaces through transparent selective coating are used on the inner side of the glass envelope, which has a high transmittance for the incoming solar radiation and high reflectance for the thermal infrared radiation.
(e) Heat pipes are used as heat transfer system contributing high performance of evacuated collectors due to diode action and maintaining a much lower thermal capacity in the collector.

These evacuated tube solar collectors have a high strength to withstand external pressure and are known as advanced solar collectors. The conventional liquid collector shows higher efficiency at the lower values of collector variable, an evacuated tube collector shows high collection efficiency even at higher collector variable.

The evacuated tube collector consists of a heat pipe, heat pump, absorbance layers and copper heat exchangers (Fig. 3.5). The vacuum pipe of evacuated tube collector is integrated with a heat pipe. The working fluid used in the evacuated tube collector, (water) then transfers to the manifold. It is made using stainless steel or copper. The heat pump is used to transfer heat. The main function of the heat pump is to circulate the solar-heated fluid or water to the header pipe and take back heat to the storage tank. The absorbance layers are integrated into the inside of the glass tubing and are made up of a unique coating. It has small tubes attached to it, which helps to minimise the loss of heat. Heat exchangers made up of copper are integrated into the system to transfer heat to the fluid through the heat pipe. The incident solar energy is being

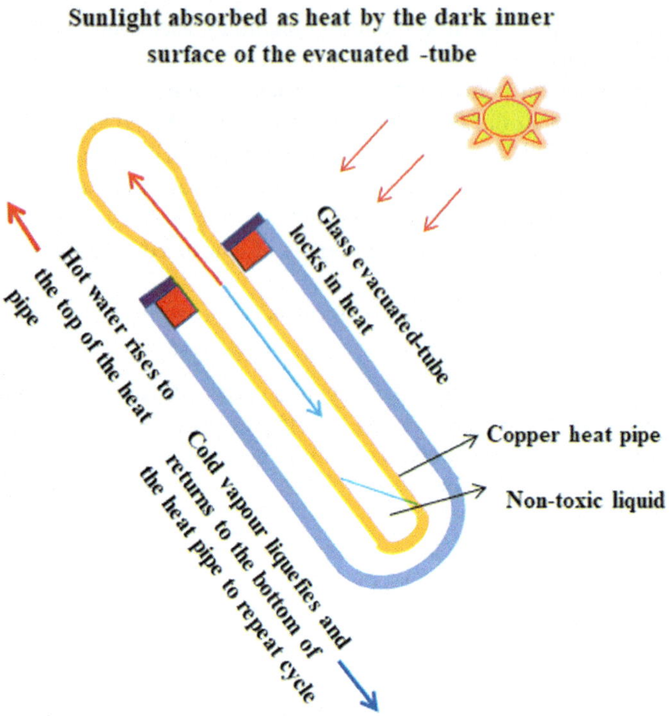

Fig. 3.5 Evacuated tube collector structure (Kalogirou 2003)

absorbed by the glass integrated at the outer layer of the heat pipe. The heat energy thus collected is then transferred to the copper heat exchanger. This solar- heated water is then transferred back to the storage tank. The schematic of an evacuated tube collector is shown in Figs. 3.5 and 3.6 represents the photograph of ETC.

3.3.1 Thermal Analysis of an Evacuated Tube Collector

The following assumptions are made for performing the thermal analysis.

(1) The heat loss coefficient from the header tube is constant.
(2) Thermal resistance through the outer glass tube wall is negligible.
(3) Convective heat transfer coefficient is assumed to be constant.
(4) Heat flux is uniform.
(5) Steady state heat transfer.

Overall Loss Coefficient

Thermal loss:

Fig. 3.6 Evacuated tube collectors installed at SMVDU, Katra

$$Q_L = S - Q_U \tag{3.12}$$

where S is the solar energy absorbed by the selective absorber coating and Q_U the net heat gain absorbed by the working fluid.

$$U_L = U_t + U_e \tag{3.13}$$

where U_e is the edge loss coefficient of the header tube and it depends on the insulation conductivity, insulation thickness and the surface area of the header tube and U_t is the loss coefficient from the absorber tube to the ambient.

$$U_e = \frac{(UA)_{\text{edge}}}{A_c} \tag{3.14}$$

$$A_c = \pi DL \tag{3.15}$$

where D is the outer diameter of the absorber tube and L the length of the absorber tube

$$(UA)_{\text{edge}} = \text{edge loss coefficient} \times \text{area}$$
$$D_r = h_1\beta_o + h_2(1 - \beta_o) + U_o$$

$$U_t = \frac{1}{\frac{1}{h_{g-a}} + \frac{1}{h_{p-g,r}+h_{p-g,c}}} \tag{3.16}$$

where $h_{g\text{-}a}$ is the convection heat transfer coefficient between the outer glass tube and the surrounding, $h_{p\text{-}g}$ the radiation heat transfer coefficient between the absorber tube and outer glass tube, $h_{p\text{-}g,c}$ the heat transfer coefficient due to the conduction by the bracket and the close up of the glass evacuated tube solar collectors.

$h_{p-g,r}$ = changes with the temperature of the absorber tube is given by

$$h_{p-g,-r} = \frac{\sigma\varepsilon_p}{1 + \frac{\varepsilon_p D_g (1-\varepsilon_p)}{\varepsilon_g}} \left(\gamma_p^2 + \gamma_g^2\right) T_p + T_g \tag{3.17}$$

$T_p =$ temperature of the outer glass tube
$T_g =$ temperature of the glass
$D_g =$ diameter of glass (inner diameter of outer glass tube)
$\varepsilon_p =$ emissivity of the selective absorbing coating
$\varepsilon_g =$ emissivity of the inner surface of the outer glass tube

The heat loss of the glass evacuated tube can be expressed as

$$U_t(C - Ta) = h_{p-g,r}\left(T_p - T_g\right) + h_{p-g,c}\left(T_p - T_g\right) \tag{3.18}$$

Collector efficiency factor

$$Q_u = \frac{\left(T_p - T\right)}{\frac{\delta_1}{\lambda_1} + \frac{\delta_2}{\lambda_2}} = \frac{\left(T_p - T\right)}{\frac{1}{C_b}} \tag{3.19}$$

$$T_p = \frac{S + U_L \times T_a + C_b T}{U_L + C_b} \tag{3.20}$$

where C_b is the synthetic conductivity, λ_1, λ_2 the conductivity of the absorber tube and air layer and δ_1, δ_2 are the thickness of the absorber tube and air layer.

The total heat gain is given by

$$\begin{aligned} q_u &= q_{fin} + q_{tube} \\ q_{fin} &= \frac{((W-d)(S - U_L(T_b - T_a)) \times F)}{\left(1 + \frac{U_L}{C_b}\right)} \end{aligned} \tag{3.21}$$

$$\begin{aligned} q_{tube} &= Q_u \times d \\ q_u &= \frac{\left(T_b - T_f\right)}{hf, i\pi d \frac{1}{+Cb}} \end{aligned} \tag{3.22}$$

$$\eta = \frac{mC_p(T_o - T_i)}{A_p \times I_o} \tag{3.23}$$

where

$$A_p = 2DL \tag{3.24}$$

3.3.2 Solar Collector with a Phase Changing Fluid

To overcome the intermittency, the solar collector systems are using phase change material. A simple heat transfer analysis of a solar collector with a phase change material is presented in this section (Aramesh and Shabani 2020 and Soin et al. 1979).

The high boiling and condensation heat transfer coefficients and easy transport of the vapour (as the heat transfer medium) and elimination of corrosion, fouling and freezing are some of the advantages offered by the phase changing fluid. The analysis of a solar collector with a phase change fluid is having great importance. Figure 3.7 shows a flat plate collector with a phase change fluid. The A and B are the inlet and outlet and β_o is the fraction of the collector length wetted by the fluid (known as liquid length fraction).

The conservation of heat for the solar collector is given by

$$(\alpha\tau)I_sA_c = Q_u + U_L \tag{3.25}$$

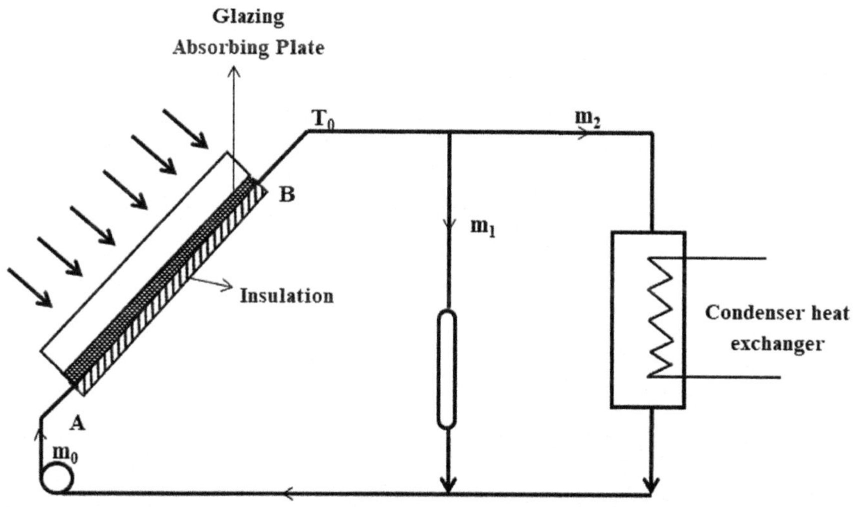

Fig. 3.7 Solar collector with a phase change fluid (Kaushik et al. 1988)

where A_c is the collector area and

$$Q_u = Q_{u1} + Q_{u2}$$

$$\text{or} \quad Q_u = h_1(T_p - T_1)\beta_o A_C + h_2(T_p - T_v)(1 - \beta_o)A_C \tag{3.26}$$

$$\text{and} \quad Q_L = U_O(T_p - T_A)A_C \tag{3.27}$$

where h_1, h_2 are the heat transfer coefficient from the absorbing plate to the liquid/vapour phases; U_O is overall heat loss coefficient of the collector and T_1 and T_v are the mean liquid/vapour phase temperatures.

Assuming

$$T_1 = (T_i + T_s)/2 \qquad T_v = (T_o + T_s)/2 \tag{3.28}$$

where T_s is the saturation temperature. The heat balance equation for liquid and vapour phases are:

$$h_1(T_p - T_1)A_c = \dot{m}_o c_1(T_s - T_i) + \dot{m}_2 \Delta H \tag{3.29}$$

$$h_2(T_p - T_v)(1 - \beta_o)A_c = \dot{m}_2 C_v(T_o - T_s) \tag{3.30}$$

$$\dot{m}_o = \dot{m}_1 + \dot{m}_2 \tag{3.31}$$

where ΔH is the latent heat of vaporization of fluid. Substituting T_p from Eqs. (3.30) to (3.29), we get

$$\dot{m}_2 = \frac{1}{\Delta H}\left[h_1\left\{\alpha\tau I_s + h_2(1 - \beta_o)\left(\frac{T_s - T_i}{2}\right)\right.\right.$$
$$\left.\left. + U_O\left(T - \frac{T_i + T_s}{2}\right)\beta_o A_c\right\} - m_o c_1(T_s - T_i)\right] \tag{3.32}$$

Solving for T_o, from Eq. (3.30), by using Eq. (3.32), we get

$$Q_u = (\alpha\tau)I_s A_c[1 - U_o/D_r] - \left(\frac{U_o}{D_r}\right)$$
$$\left[h_1\beta_o A_c\left(\frac{T_1 + T_s}{2}\right) - T_A + h_2(1 - \beta_o)A_c(T_s - T_A)\right] \tag{3.33}$$

where $D_r = h_1\beta_o + h_2(1 - \beta_o) + U_o$

Hence collector efficiency is given by

$$\eta = \frac{Q_u}{I_s A_C} = (\alpha\tau)\left(1 - \frac{U_o}{D_r}\right) + \left(\frac{U_L}{D_r}\right)\frac{h_1\beta_o\left(T_A - \frac{T_i+T_s}{2}\right) + h_2(1-\beta_o)(T_A - T_s)}{D_r I_s} \tag{3.34}$$

It can be seen that for $T_i = T_s$, the above equation can be written as (no subcooling in the condenser):

$$\eta = (\alpha\tau)_{\text{eff}} - (U_o)_{\text{eff}} - \frac{(T_1 - T_A)}{I_s} \tag{3.35}$$

which is the modified H-W-B equation for a solar collector with phase changing fluid. It is further seen that if β_o is unity it reduces to conventional Hottle-Bliss equation, as given by Eq. (3.11).

3.4 Compound Parabolic Collector (CPC)

The compound parabolic collector consists of two curved segments, which are parts of two parabolas. The CPC has a very high acceptance angle. This is an example of a non-imaging type of collector.

The aperture of the compound parabolic collector is typically tilted towards the equator at an angle equal to the latitude (Fig. 3.8). It is also clear that when the solar collector is oriented along the north–south direction the collector must track the Sun so as to face the Sun continuously. The absorber can take a variety of configurations such as the cylindrical or flat. In Fig. 3.8, CPC has two portions of the parabola AB and CD and BC is the absorber aperture area with width a_r and AD is the aperture area with width a_n of CD and AB, respectively. The axes are oriented in such a way that the point B is the focus of parabola CD and C as the focus of parabola AB. Also, the height of the collector is so chosen that tangents at A and D are parallel to the axis of the collector. The acceptance angle is obtained by joining the focus of the opposite aperture edge. The height and aperture area are designed based on the desired operating temperature. The acceptance angle is also kept as large as possible to avoid the requirement of continuous tracking. The general concentration ratio of CPC is between 3 and 10. The optical efficiency is 65% which is slightly more than the other PTC. Figure 3.9 shows a CPC located at NISE.

Thermal Analysis of CPC

The analysis of the compound parabolic collector given by Sukhatme and Nayak (2017) is presented here.

A compound parabolic collector having an aperture W, length L and acceptance angle ($2\theta_a$) is considered as shown in Fig. 3.10. The absorber surface has a width b. The heat collected at the absorber surface is transferred to a fluid flowing through N tubes, having an outer diameter D_o and an inner diameter D_i, attached to the bottom

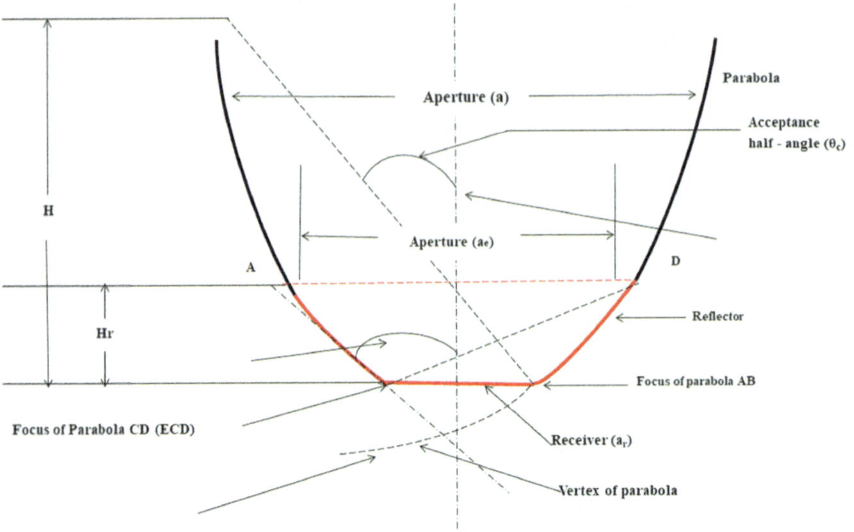

Fig. 3.8 A compound parabolic collector (Sukhatme and Nayak 2017)

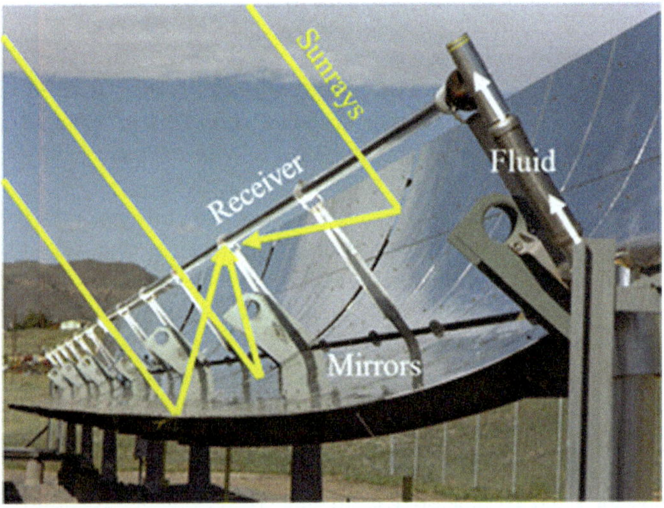

Fig. 3.9 Compound parabolic collector (Retrieved from the NREL website: https://images.nrel.gov)

side. The fluid enters at a temperature, leaves at a temperature T_{fo}, and has a mass flow $\dot{m}$. CR is the concentration ratio.

We first derive an expression for the flux S absorbed at the absorber surface. Because of its large acceptance angle, a CPC accepts both beam and diffuse radiation.

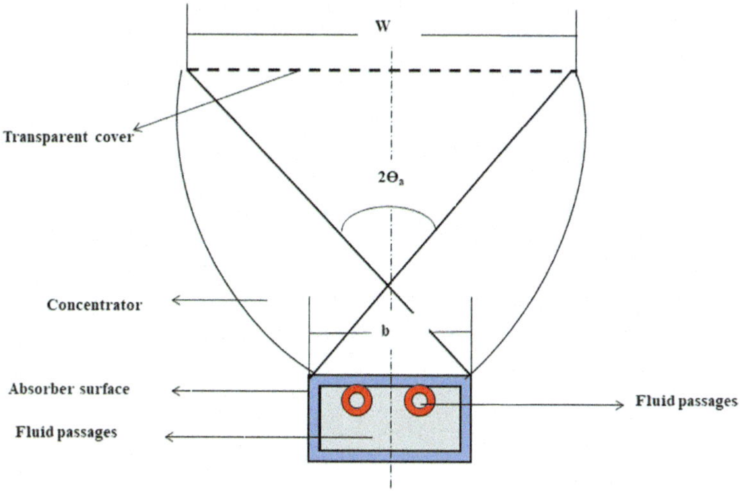

Fig. 3.10 Schematic diagram illustrating the acceptance angle (Sukhatme and Nayak 2017)

The beam radiation flux falling on the aperture plane is $I_b\ r_b$, while the diffuse radiation flux within the acceptance angle is given by

$$\left(\frac{I_d}{CR}\right)$$

Thus, the total effective flux entering the aperture plane is $\left[I_b r_b + \frac{I_d}{CR}\right]$ and

$$S = \left(I_b r_b + \frac{I_d}{C}\right)\tau \rho_e \alpha \tag{3.36}$$

where τ is the transmissivity of the cover, ρ_e the effective reflectivity of the concentrator surface for all radiation and α is the absorptivity of the absorber surface. The values of ρ_e and α are assumed to be the same for beam and diffuse radiation. It is to be noted that the flux *S* is based on the area of the aperture.

To obtain an expression for the useful heat gain rate, we take an energy balance on an elementary slice *dx* of the absorber surface at a distance *x* from the inlet. This yields the equation.

$$dq_u = \left[S - \frac{U_1}{CR}\left(T_p - T_a\right)\right] \tag{3.37}$$

Then, following the steps as in the case of a cylindrical parabolic collector, the useful heat gain is given by

$$Q_u = F_R \mathrm{WL}\left[S - \frac{U_1}{CR}(T_{fi} - T_a)\right] \quad (3.38)$$

Heat removal factor,

$$F_R = \frac{\dot{m}C_p}{bU_1L}\left[1 - \exp\left\{\frac{-F'bU_1L}{\dot{m}C_p}\right\}\right] \quad (3.39)$$

$\frac{1}{F'} = U_1\left[\frac{1}{U_1} + \frac{b}{N\pi D_i h_f}\right]$ where N is the number of tubes.

The instantaneous collector efficiency is given by the expression

$$\eta_1 = \frac{Q_u}{(I_b r_b + I_d r_d)WL} \quad (3.40)$$

where r_b and r_d are tilt factor for beam and diffused radiation.

The main difficulty in the use of equation for *qu* is the estimation of overall heat loss coefficient *Ul*. The estimation is not accurate as in the case of flat plate collector and the cylindrical parabolic collector. The estimation involves the calculation of radiation heat transfer coefficients which is more cumbersome. The advantages, disadvantages, and some of the major applications of CPC are given below.

(i) *Advantages of CPC*

- Solar tracking system not needed
- Ease of fabrication
- Simple support structure

(ii) *Disadvatages of CPC*

- Low concentration ratio
- Multiple reflections for many rays
- Height increases rapidly with the aperture

(iii) *Applications of compound parabolic collectors*

- Solar water heaters
- Solar stills
- Solar cooking

3.5 Solar Collector Reflector System

The performance of a solar collection system can also be enhanced by using reflection mirror boosters which can make more energy incident on the absorbing surface. The heat transfer rate and its output can be increased by integrating a reflector or a booster with flat plate collectors. The output mainly depends upon the shape of mirrors; the largest output can be obtained from the curved mirrors. These mirrors are normally oriented in an east–west direction. A mirror on the north side of a collector can be an inexpensive way of increasing the heat collected. If a collector is much longer in the east–west direction than in the north–south, end effects will be negligible.

Concentrated Solar Power (CSP) technology generates solar power by using mirrors or lenses to concentrate a large area of sunlight, or solar thermal energy, onto a small area. The concentrated light is converted to heat, which drives a heat engine (usually a steam turbine) connected to an electrical power generator to generate the electricity (Fig. 3.11) In this case, the solar radiation is directed to the absorber by the optical system provided in it. The presence of optical systems leads to certain losses like reflection or absorption in the lenses and geometrical imperfections. The combined effects of all these losses leads to the new term called the optical efficiency. In recent years, significant development has been made in the area of concentrating collectors and a number of them have been commercialised. Kaushik et al. (1988) presented a comprehensive review in the field. The advantages of reflector type collectors as compared with the conventional flat-plate type are as follows:

(a) The thermal efficiency of concentrating collectors is higher.
(b) The higher thermodynamic efficiencyof the concentrator system is mainly due to the ability of the working fluid to achieve higher temperatures as compared with the flat-plate system.
(c) A thermodynamic match between temperature level and the task can be achieved by the concentrator system.

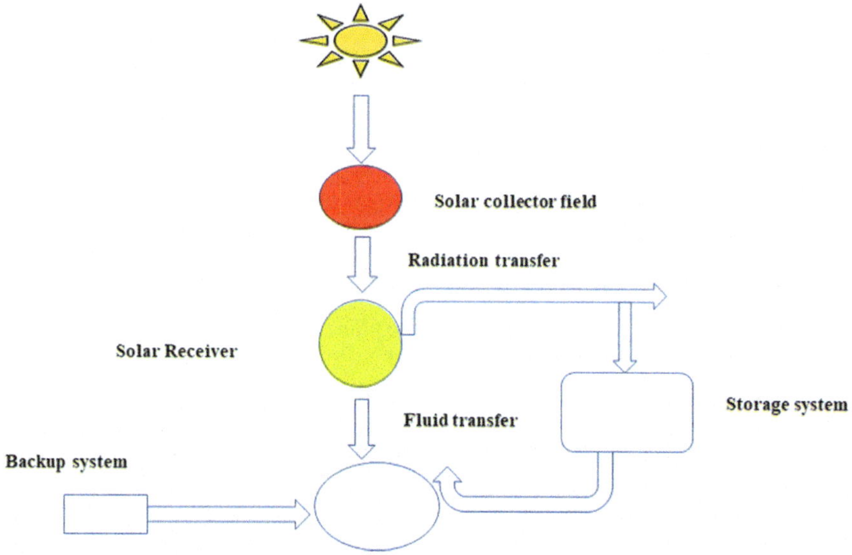

Fig. 3.11 Flow diagram of a typical concentrated solar power plant

(d) The relatively small area of the receiver and vacuum insulation made the thermal losses to the minimum value. The selective surface treatment is also provided to improve the collector efficiency.
(e) The cost per unit area of the collector is less as compared to the non-concentrating such as FPC.

Limitations

(a) The tracking system is required so as to enable the collector to follow the Sun, this leads to the high overall cost of the system.
(b) As compared to the FPC, the concentrator systems can collect very little diffuse radiation depending on the concentration ratio.
(c) A collector may require periodic cleaning and refurbishing to protect the selective reflecting surfaces.
(d) Concentrators can have concentration ratio from a low value of 1.5 or 2 to high values of the order of 10,000. Increasing ratio means increasing temperatures at which energy can be delivered. Flat plate collectors are generally fixed in position for convenience and can utilise both diffuse and beam components of the radiation. Only beam radiation can be focused in concentrating devices, but the daily beam irradiation passing the aperture with continuous adjustment about two axes maybe 20% more than the daily total irradiation passing the aperture of a flat plate collector inclined at a fixed angle. The choice between flat plate collectors must be made on performance and economic factors. At the higher temperature required by solar cooking and cooling, and industrial processes concentrating collectors will be desirable.

The concentrating collectors are grouped into the following types:

(a) Parabolic Trough Collector (PTC)
(b) Dish collector
(c) Linear Fresnel Reflector (LFR)
(d) Central receiver type

Terms used in the concentrating solar collectors:

(a) *Area of aperture*: The aperture area is the plane opening of the collector through which the incident solar flux is accepted.
(b) *Area of absorber*: The area of an absorber is the total area receiving the concentrated radiation.
(c) *Concentration ratio* (CR): The concentration ratio is defined in two ways. One, the area concentration ratio and the other, local concentration ratio or flux concentration ratio. The most commonly used concentration ratio is the area concentration ratio given by the ratio of the effective area of aperture to the absorber tube area.

    A flux concentration ratio is defined as the ratio of average energy flux on the receiver to that on the aperture, but has limited utility as in many cases there are substantial variations in energy flux over the surface of a receiver.

(d) *Intercept factor* (γ): The intercept factor is the fraction of the focused energy intercepted by the absorber of the given size. Its value depends on the size of the absorber. Its typical value is 0.9.

(e) *Optical efficiency* (η_o): The ratio of energy absorbed by the absorber to that of the incident on the collector is called as the optical efficiency of a collector, that is,

$$\eta_o = \frac{q_u}{\alpha \tau I_b}$$

where q_u is the rate of useful energy per unit aperture area, I_b the incident solar energy and α, τ are the function of angle of incidence of radiation on the aperture.

(f) *Thermal efficiency or collector efficiency* (η_c): The collector efficiency can be defined as the ratio of useful energy delivered to the absorber to the energy incident on the aperture.

$$\eta_C = \frac{q_u}{I_b}$$

The geometry of a CPC collector is shown in Fig. 3.12, where θ_C is the acceptance angle.

The concentration ratio is limited and it depends on the shape factor. The limiting value of the concentration ratio of cylindrical parabolic collectors is given by the value, CR = 1/sin θ_{max}. Let us consider for the case, when the acceptance angle is equal to θ_{max}, the concentration ratio becomes 210. Similarly, for a 3D collector as a dish, the concentration ratio can be high up to 10,000.

The value of concentration ratio CR is unity for the flat plate collectors with no reflectors. For concentrators, CR is always greater than 1. For a single-axis tracking collector, the maximum possible concentration is given by

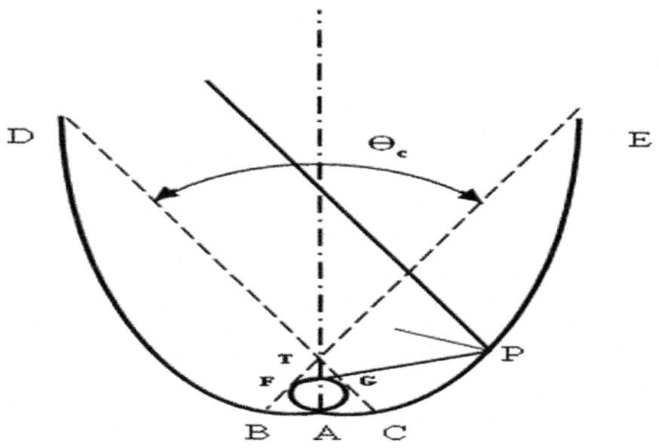

Fig. 3.12 Schematic of the geometry of a CPC collector (Kaushik et al. 1988)

$$\text{CR} = 1/\sin\theta_m \tag{3.41}$$

For the two-axes tracking collector (Kalogirou 2003):

$$C_{\max} = \frac{1}{\sin^2(\theta_m)} \tag{3.42}$$

where θ_m is the half acceptance angle.

3.6 Parabolic Trough Collector (PtC)

Parabolic trough collector is also referred to as cylindrical parabolic collector or the linear parabolic collector. It is also desirable for many applications to deliver energy at a temperature higher than those possible with ordinary flat plate collector. If we decrease the area from which heat losses occur, then energy delivery temperature can be increased. This is done by imposing an optical device between the sources of radiation and the energy-absorbing surface to raise the level of incident radiation on the relatively small absorber. The collector consists of a receiver and a concentrator. The receiver of the collector absorbs solar radiation and converts into useful form of thermal energy. The concentrator is the optical part of the collector that directs solar insolation onto the receiver. PTC can effectively produce heat at temperature between 50 and 400 °C (Fig. 3.13).

Parabolic trough collectors are made by bending a sheet of reflective material into a parabolic shape as shown in Fig. 3.14.

Working Principle of PTC

A solar PTC is a linear-focus concentrating collector that basically comprises a highly reflective parabolic trough concentrator. The direct solar radiation falling on the reflective surface concentrates onto a receiver tube located at the parabolic focal axis. A silver or aluminium film is often coated over the parabolic surface to make a specular surface. The concentrator requires single-axis tracking to concentrate direct solar radiation on the linear receiver.

The steel receiver has a selective coating that does have a high solar absorptivity (>90%) and a low emissivity in the infrared wavelength range (<30%), decreasing heat losses due to radiation. A concentric glass envelope commonly encircles the receiver tube to reduce convective heat losses from the hot receiver. The annular gap between the receiver tube and the glass envelope is evacuated to improve PTC performance, particularly at higher temperatures. The receiver tube absorbs the concentrated solar radiation and heats the heat transfer fluid (HTF) that flows through it, which is often a high boiling point synthetic fluid. Hence, solar radiation transforms into thermal energy in the form of the sensible heat of the working fluid. The working fluid can be heated effectively up to 400 °C. The schematic of a PTC is shown in Fig. 3.14.

Fig. 3.13 The photograph of PTC located at Energy research lab TKMCE, Kollam

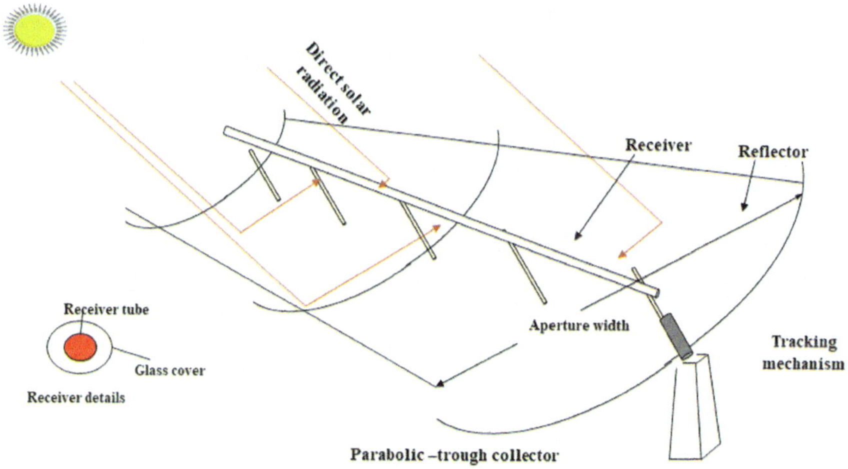

Fig. 3.14 Schematic of a PTC (Sharma 2011)

These collectors are ideally suited for large-scale power generation in locations with intense direct solar radiation, such as the desert areas in the north Indian states. The solar field consists of hundreds of parabolic collector component assemblies in a parabolic trough-based power plant. Because these collector systems have a considerable installation space, they are often ground-mounted. They are also highly vulnerable to structural stress transmission of wind loads, and ground-mounted supports the structural requirements. Roof-mounted small-scale standalone systems, on the other hand, are rarely used for domestic applications at medium temperatures.

The orientation of PTC to be selected depends on the application. The amount of energy collected by a parabolic collector mainly depends on the orientation and weather conditions. Studies reveal that in summer, north–south field gives lots of energy and during winter PTC gives lesser energy. The east–west field collects more energy in the winter than a north–south field and less in the summer.

3.7 Thermal Performance of PTC

The thermal performance parameters such as useful heat gain, outlet temperature of the working fluid, instantaeous efficiency, absorbed flux and overall heat loss coefficient are derived in this section (Fig. 3.15).

3.7.1 Useful Heat Gain in PTC

The instantaneous heat energy absorbed by the thermal energy fluid during its flow is termed as the useful heat gain and is given by

$$q_u = \dot{m}C_p(t_{fo} - t_{fi}) \tag{3.43}$$

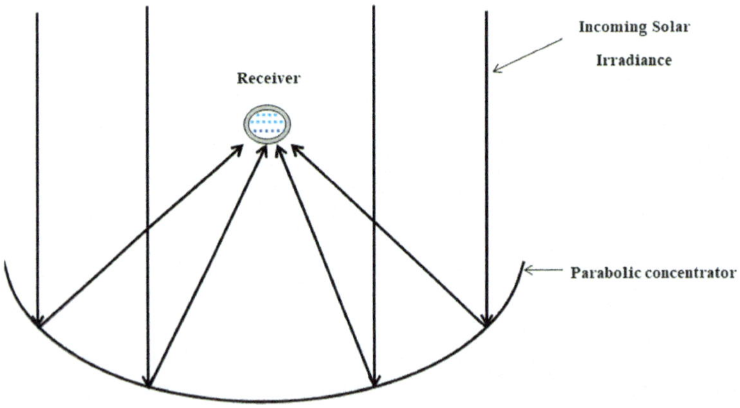

Fig. 3.15 Schematic of the geometry of a parabolic trough collector (Kaushik et al. 1988)

3.7.2 Outlet Temperature

The outlet temperature of the heat transfer fluid gives us the heat gained by the thermal fluid while passing through the absorber tube of the parabolic trough collector. It is given by

$$t_{fo} = t_{fi} + \frac{q_u}{\dot{m}C_p} \tag{3.44}$$

3.7.3 Instantaneous Beam Efficiency

The efficiency of a parabolic trough collector is normally expressed in terms of the instantaneous beam efficiency and is given by (Sukhatme and Nayak 2017)

$$\eta_{ib} = \frac{q_u}{I_b \times r_b \times L \times W} \tag{3.45}$$

Declination

$$\delta = 23.45 \sin\left(\frac{360}{345}(284 + n)\right) \tag{3.46}$$

Here, the arrangement of the parabolic collector is E–W and horizontal.

So,

$$\begin{aligned}
\beta &= \delta - \phi \text{ for } \gamma = 180^\circ \\
\gamma &= 0^\circ \text{ or } 180^\circ (\text{surface is horizontal}) \\
\cos w &= \sin \phi (\sin \delta \cos \beta + \cos \delta.\cos \gamma.\cos w.\sin \beta) \\
&\quad + \cos \phi (\cos \delta.\cos w.\cos \beta - \sin \delta.\cos \gamma.\sin \beta) \\
&\quad + \cos \delta.\sin \beta.\sin \gamma
\end{aligned}$$

3.7.4 Absorbed Flux

This represents the solar beam radiation per unit effective aperture area absorbed in the absorber

$$S = I_b \times r_b \times \rho \times (\tau\alpha)\beta \times I_f + I_b \times r_b \times \rho$$

$$\times (\tau\alpha)\beta \times \left(\frac{D_o}{W - D_o}\right) \tag{3.47}$$

3.7.5 Overall Heat Loss Coefficient

The overall heat loss coefficient is determined based on the convection and re-radiation losses. The heat transfer rate can be given by

$$\frac{q_1}{l} = h_{pc}\left(T_{pm} - T_c\right)\pi D_o + \frac{\sigma\pi D_o\left(T^4{}_{pm} - T^4{}_c\right)}{\frac{1}{\varepsilon_p} + \frac{D_o}{D_{ci}}\left[\frac{1}{\varepsilon_c} - 1\right]} \tag{3.48}$$

$$\frac{q_1}{l} = h_w(T_c - T_a)\pi D_{co} + \sigma\pi D_{co} t_c\left(T^4{}_c - T^4{}_{sky}\right) \tag{3.49}$$

where T_c represents the cover temperature, h_w the wind heat transfer coefficient (W/m^2 K), ε_c represents the emissivity of the cover for long-wavelength radiation, D_o the outer diameter of the absorber tube, D_i the inner diameter of the absorber tube, D_{co} the outer diameter of the glass tube, D_{ci} the inner diameter of the glass tube and σ the Stephan-Boltzmann coefficient.

3.7.6 Convective Heat Transfer Coefficient

(i) Between the absorber tube and cover plate

The convective heat transfer coefficient between a horizontal absorber tube and the concentric cover (in enclosed annular space) is given by

$$R_a = \frac{l^3 g\beta\Delta T}{\rho^2} \times \Pr \tag{3.50}$$

$$h_{pc} = \frac{2k}{D_o \ln\left(\frac{D_o}{D_{co}}\right)} {}^*Ra^{1/4} = \frac{\ln\left(\frac{D_{ci}}{D_o}\right)}{b^{3/4}\left(\frac{1}{D_o^{3/5}} + \frac{1}{D_{ci}^{3/5}}\right)^{5/4}} Ra^{1/4} \tag{3.51}$$

$$k_{\text{eff}} = 0.317\text{k}\left(\text{Ra}^*\right)^{1/4} \tag{3.52}$$

(ii) On the inner surface of the absorber tube

For higher Reynolds number (greater than 2000) the flow is turbulent and the heat transfer coefficient can be calculated by:

$$h_w = \frac{(Nu \times k)}{D_{\text{co}}} \tag{3.53}$$

The useful energy produced from the tube can be given by

$$Q_u = mc_p \Delta T \tag{3.54}$$

$$q_u = mc_p\left(T_{fo} - T_{fi}\right)$$
$$q_u = mc_p\left[\frac{CS}{U_1} + T_a - T_{fi}\right]\left[1 - \exp\left(\frac{-F' \prod D_o U_1 L}{mc_p}\right)\right] \tag{3.55}$$

where q_u represents the useful energy delivered from the concentrator (W), m is the mass flow rate (kg/s), S the incident flux absorbed in the absorber plate (W/m^2), U_1 the overall loss coefficient (W/m^2 °C), T_a the ambient temperature, T_{fi} the fluid inlet temperature, D_o the outer diameter of absorber tube, L the length of the concentrator in (m) and F' the collector efficiency factor.

The instantaneous calculated efficiency is given as

$$\eta_i = \frac{q_u}{I_b \times r_b \times W \times L} \tag{3.56}$$

3.8 Second Law Analysis of Concentrating Collectors

Kalogirou (2003) presented the second law based analysis of the concentrating collectors.

Consider a concentrating collector of area A_a.

Let Q' be the solar radiation incident on the surface:

q' be the solar energy flux on the surface

$$Q' = q' A_a \tag{3.57}$$

For the concentrating collectors, optical errors should be considered for the analysis. By considering the optical efficiency, the radiation falling on the receiver is given by

$$q_0' = \eta_o q' = \frac{\eta_o Q'}{A_a} \tag{3.58}$$

Of the total incident solar radiation, partly it is delivered to the power cycle as heat transfer Q at the receiver temperature T_r. The remaining fraction Q_o represents the collector–ambient heat loss.

$$Q_o = Q' - Q \tag{3.59}$$

Q_o is given by

$$Q_o = U_r A_r (T_r - T_o) \tag{3.60}$$

where U_r is the overall heat transfer coefficient based on the A_r, T_r the receiver temperature and T_o the ambient temperature.

Then combining Eqs. (3.59) and (3.60), it is apparent that the maximum receiver temperature occurs when $Q = 0$, that is, when the entire solar heat transfer Q' is lost to the ambient.

The dimensionless form of the maximum collector temperature can be expressed as

$$\begin{aligned}\psi_{\max} &= \frac{T_{\mathrm{rmax}}}{T_o} \\ &= 1 + \frac{Q'}{U_r A_r T_o}\end{aligned} \tag{3.61}$$

where T_{rmax} is also called as the stagnation temperature of the collector. This is defined as the temperature that can be obtained at no-flow condition. Combining Eqs. (3.59), (3.60) and (3.61)

$$\psi_{\max} = 1 + \frac{q'_o A_{\mathrm{a}}}{\eta_o U_r A_r T_o} \tag{3.62}$$

It can be rewritten in the following form:

$$\psi_{\max} = 1 + \frac{q'_o C}{\eta_o U_r T_o} \tag{3.63}$$

where $C = \frac{A_{\mathrm{a}}}{A_{\mathrm{r}}}$ is the concentration ratio.

The collector temperature $\psi = \frac{T_r}{T_o}$ (in dimensionless form) will vary between 1 and 0 and the term $\psi_{\max}$ depending on the heat delivery rate Q.

From Eq. (3.63), it is clear that $\psi_{\max}$ is directly proportional to the geometrical concentration ratio C of the collector. In the cases where there is no flow through the collector $\psi_{\max}$ is the parameter that describes the performance of the collector. Here, it is assumed that the receiver collects all the solar energy incident on it and is utilised for increasing the temperature of the working fluid to the stagnation temperature.

Thus, the collector efficiency can be written in the following form:

$$\eta_{\mathrm{o}} = \frac{Q}{Q'} = 1 - \frac{\psi - 1}{\psi_{\max} - 1} \tag{3.64}$$

From Eq. (3.64), it is clear that efficiency is a linear function of the collector temperature.

3.9 Dish Collector

A dish collector is similar to a large satellite dish. A dish collector consists of mirror-like reflectors and an absorber at the focal point. It uses dual-axis suntracker. The optimum temperature for different solar collectors is given in Table 3.2.

The dish system has a dish, which focuses the solar radiation on a thermal cavity receiver. The receiver heats water or the thermal fluid circulated through the cavity. The heated fluid then transfers its heat to end industrial applications.

A dish collector shown in Fig. 3.16 is capable of delivering very high temperature above 1500 °C. The dish collectors are similar to the parabolic trough collectors, where receivers are distributed throughout the collector field and often called distributed-receiver systems. The dish can have a reflector area of up to 100 m^2. Such a system can produce more than 60 kW of thermal output at 1000 W/m^2 solar insolation. In most of the cases, a solar tracking system is enabled to track optimum sunlight. The average solar to thermal efficiencies of a dish system is about 65%.

Components of a Dish Collector

The basic components include a collector, receiver, a hot water pump, an expansion tank and thermal storage tank.

The collector is used to collect sunlight. The receiver placed at the focus point is used to absorb and transmit the concentrated heat energy continuously to a fluid flowing through it. To maintain the water flow rate, a hot water pump is also integrated into the unit. An expansion tank is also integrated with the dish collector system to maintain the pressure of circulating fluid in a closed system. The function of a thermal storage tank integrated is to make the system in continuous operation mode. The thermal storage tank stores excess heat energy collected and is used during the non-solar hours.

The advantages of the dish collectors are:

(a) High efficiency as compared to the other collectors.
(b) They are integrated with the modular collector and receiver units. Hence, they can operate either independently or as a part of the larger systems.

Table 3.2 Optimum temperature for different solar collectors (Kalogirou 2003)

Collector type	C (Concentration ratio)	Stagnation temperature (°C)	Optimal temperature (°C)
PTC	50	575	229
Dish collector	500	1290	409
Central receiver	1500	1780	503

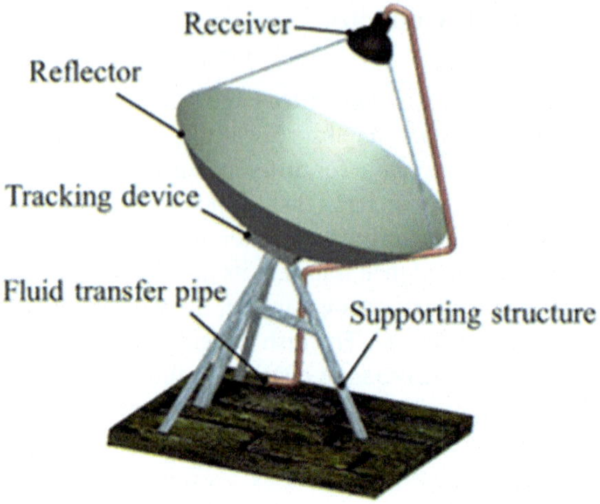

Fig. 3.16 Dish collector (Suman et al. 2015)

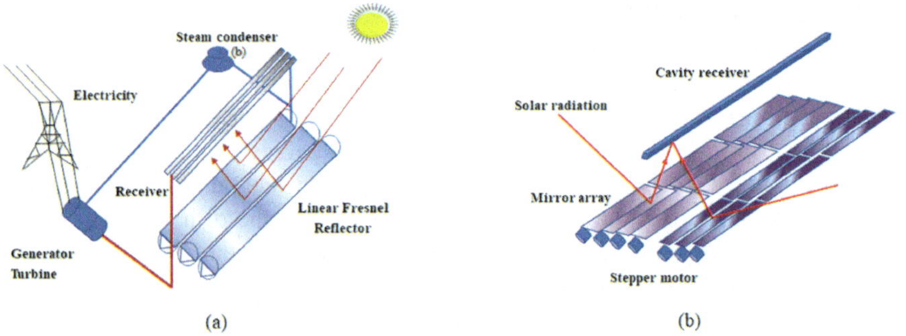

Fig. 3.17 **a** Schematic of the Linear Fresnel collector. **b** Linear Fresnel power plant (Behar et al. 2013)

(c) The concentration ratio of dish collectors is very high (in the order of 500–1500).

3.10 Linear Fresnel Reflector (LFR)

The linear fresnel mirror technology is the perspective method of concentrating solar power. LFR relies on an array of mirror strips mounted on it. It basically uses a cheaper curved typed reflector as compared to the parabolic glass reflectors. Generally, an LFR system consists of three components, viz. a mirror field, tracking and the receiver system as shown in Fig. 3.17.

The linear Fresnel receiver (LFR) is a low-priced solar system with adequate thermal performance, and so it is regarded as an attractive and beneficial option for harvesting solar energy. This mirror-based device works on the same theory as a Fresnel lens, which employs flat plane mirrors to follow the sun and reflect radiation onto a receiver. The concentration ratio ranges from 10 to 50, and the working temperature ranges from 50 to 300 °C. Despite PTC's high efficiency, the LFR benefits from a lower investment cost and fewer mechanical issues during operation. Linear Fresnel reflectors are made from straight mirrors or reflectors. As illustrated in Fig. 3.17a, these mirrors concentrate the incident solation radiation onto a metallic receiver tube located above the mirrors. The mirrors are moved such that they always reflect the solar radiation onto the absorber tube. The mirrors are grouped into multiple sets to enhance optical and ground-use efficiency, each with its receiver. This enables a larger mirror packing density, improving optical efficiency while reducing land consumption. The schematic of LFR power plant is shown in Fig. 3.17b.

3.11 Central Receiver System or Heliostat Field Collectors

Central receivers (or power towers) use thousands of individually sun- tracking mirrors called heliostats to reflect solar energy onto a receiver located on top of a tall tower. A fluid flowing through the receiver absorbs the concentrated radiation. The molten salt is the heat transfer fluid used in the central receiver collector. The receiver collects the Sun's thermal energy and transfers it to the molten salt. The energy of molten salt is used to make steam to generate electricity. The molten salt can store energy for hours or even days before being used to generate electricity. The steam generated or the hot molten fluid can be used to operate the thermodynamic cycles like Rankine cycle or the Brayton cycle. The idea of the central receiver system is suggested by Baum et al. (1957).

Figure 3.18 shows the schematic of a typical central receiver system using molten salt as a heat transfer fluid. The molten salt commonly used is a mixture of 60% sodium nitrate and 40% potassium nitrate.

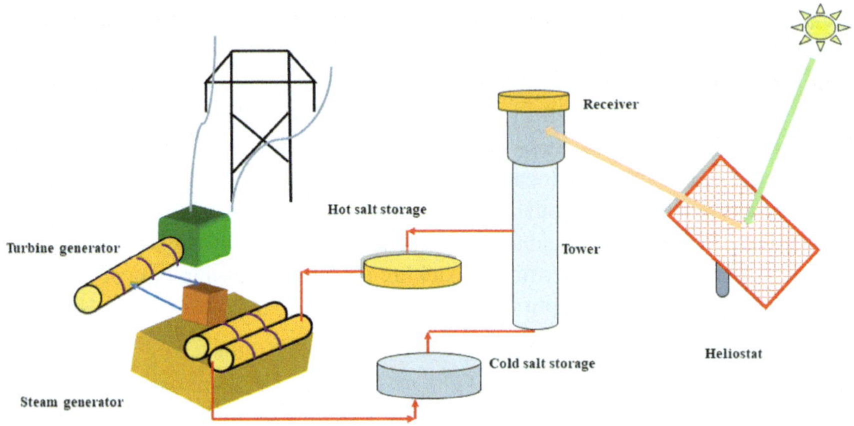

Fig. 3.18 A central receiver collector system (Kalogirou 2003)

Central receivers have the following advantages:

1. They can effectively collect solar energy and then convert into thermal energy with minimum loss.
2. Effective thermal energy storage.
3. The concentration ratio is very large (ranging from 400 to 2000), and are highly efficient.
4. Suitable for high capacity thermal power plants, usually greater than 10 MW capacity hence benefit from economics of sale.

Heliostats are the central part of the CRS. The heliostats form an array of circular arcs around the central tower. They intercept, reflect and concentrate the solar radiation onto the receiver. Nowadays most heliostats are used for the daylighting or the production of concentrated solar power.

The most important and difficult task in the design of such systems is the proper selection of heat transfer fluid, thermal storage medium and power conversion cycle that defines the central receiver plant. The thermal storage method can be thermochemical methods such as the use of oil mixed with the molten nitrate salt, crushed rock or liquid sodium. Nowadays, most of the systems are based on Rankine cycle. Research in the direction of the advanced cycles that would use air as a heat transfer fluid, ceramic bricks for the thermal storage and open cycle Brayton cycle for the power conversion, is in progress (Figs. 3.19 and 3.20).

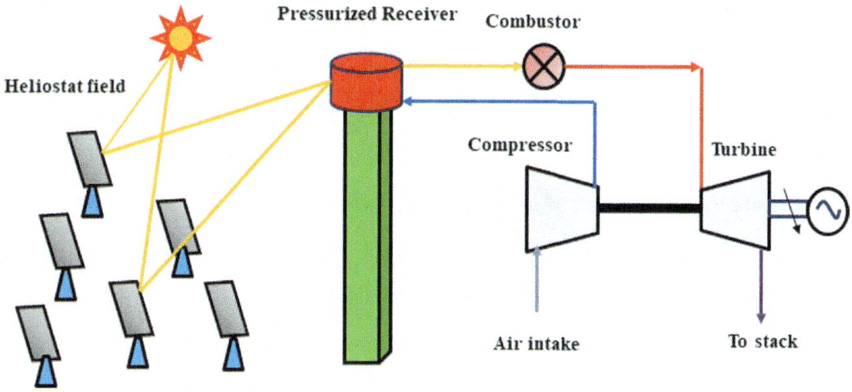

Fig. 3.19 Solar central receiver Brayton system (Kalogirou 2003)

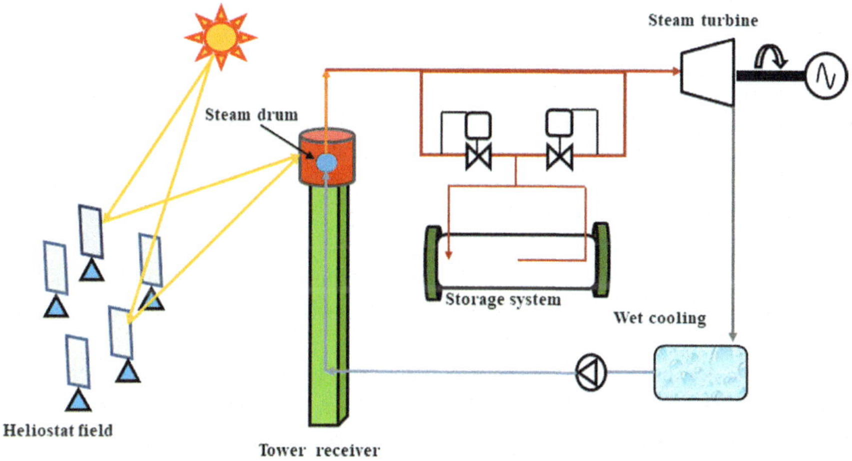

Fig. 3.20 The central receiver power system with storage (Kalogirou 2003)

3.12 Material Considerations in Solar Collectors

The cost of energy produced by solar thermal collectors depends on the cost of various materials which make up the system, maintenance costs and the amount of solar energy collected. Serious efforts are being done towards making solar collector technology economically more competitive. One way is to provide materials having high specific strength that leads to the cost reduction of the solar collectors, e.g. the use of composite materials or polymers.

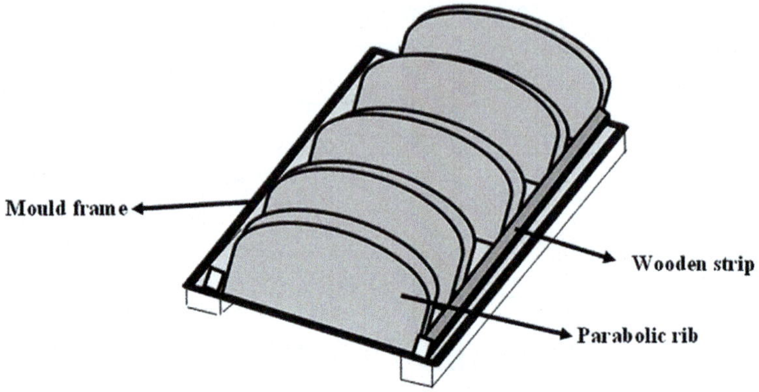

Fig. 3.21 View of the mould

3.12.1 Fibre Reinforced Parabolic Trough Collector

A composite material is made from two or more of the constituents with significantly different physical or chemical properties. The material produced thus will have characteristics different from the individual constituents. The material is less expensive, stiffer, high specific strength (strength-to-weight ratio) etc.

Thus, composite material is of proven rigidity under severe weather conditions. Parabolic trough solar collectors are now made up of composite material because they offer several outstanding properties as compared to conventional materials. A fibre reinforced parabolic trough collector made by Baiju and John (2016) is shown in Fig. 3.21.

The most commonly used composite material is the E-glass fibreglass-reinforced plastic (FRP), which has been selected as a material for a collector. The wooden mould is constructed with plywood ISI BWR IS 303. The mould fabricated by hand lay-up method is shown in Fig. 3.22.

3.12.2 Polymers for Solar Collectors

Polymers are lightweight, widely available and low-cost material (Asmussen and Vallo 2018). The main advantages are being corrosion resistant, reliability and long durability.

From the various literatures, it can be concluded that the most suggested materials for various components of thermal collectors are:

(a) For the transparent cover low-iron glass, which is relatively inexpensive
(b) Insulation: vacuum insulation—a no-cost material
(c) The selective coating material, which is relatively cheap, as compared to the high output of energy of solar collector.

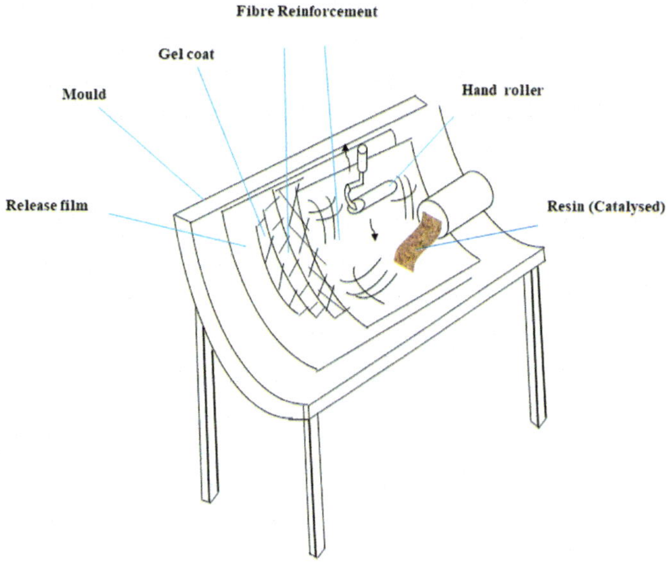

Fig. 3.22 Hand lay-up method

3.13 Current Status and Applications of Concentrated Solar Collectors

The selection of collectors is always a difficult task. The final selection is to be done only after conducting detailed thermodynamic analyses of the complete system. A preliminary screening can be done based on the performance parameters to identify those which best matches with the load.

The expected value of the term $\frac{\Delta T}{G}$ to be considered for the load and climate, where G is the global radiation and ΔT the $T_i - T_a$.

The variation of efficiency as a function of the heat loss parameter is shown in Fig. 3.23.

Various applications of CSP technology are illustrated in Fig. 3.24. The application includes food processing, solar cooking, textile industries, solar heating, electricity production etc.

The parabolic trough collectors are selected as the best option for low-temperature range. The important applications include detoxification, liquid waste recycling and heating water.

All three systems are suitable for the mid-temperature range applications. The central tower is the most suitable system for very high-temperature systems because the temperature of more than 1000 °C can be easily sustained.

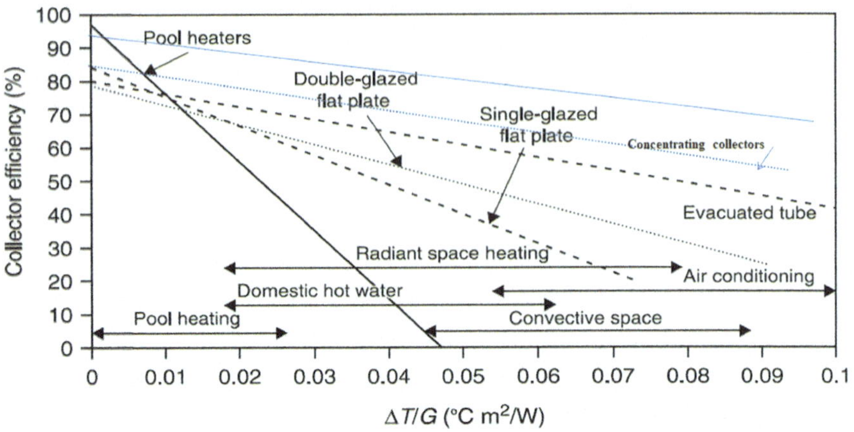

Fig. 3.23 Collector efficiencies for various liquid collectors (Kalogirou 2003)

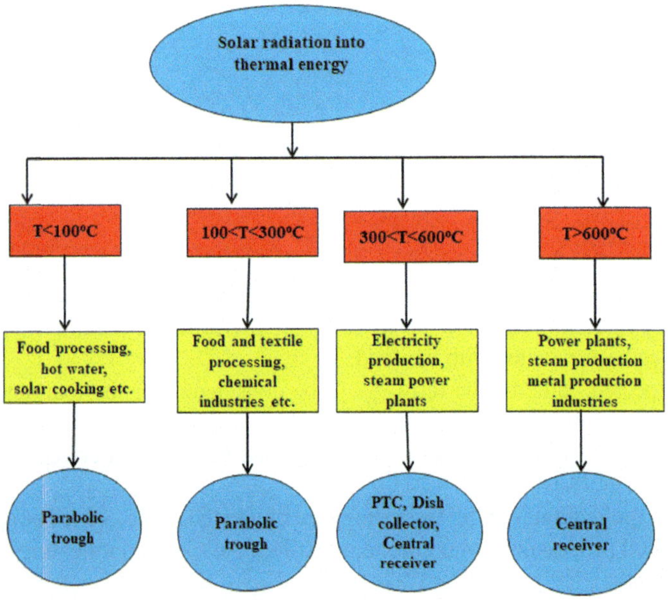

Fig. 3.24 Applications of concentrated solar power technology

Table 3.3 shows the world's largest concentrating solar thermal power stations. Although the applications mentioned here are concerned with the generation of electricity, the essential product of each system is heat, which could potentially be used to provide process heat for industrial applications.

Table 3.3 World's highest capacity solar collectors (Behar et al. 2013)

Capacity (MW)	Technology type	Name	Country
354	Parabolic trough	Solar energy generating systems	USA
75	Parabolic trough	Martin next-generation solar energy centre	USA
64	Parabolic trough	Nevada solar one	USA
50	Parabolic trough	Andasol 1	Spain
20	Solar power tower	PS20 solar power tower	Spain
11	Solar power tower	PS10 solar power tower	Spain

References

Aramesh, M. & Shabani, B. (2020). On the integration of phase change materials with evacuated tube solar thermal collectors. *Renewable and Sustainable Energy Reviews, 132,* 110135.

Asmussen, S. V. & Vallo, C. I. (2018). Absorber materials based on polymer nanocomposites containing silver nanoparticles for solar thermal collectors. *Solar Energy, 174,* 640–647.

Baiju, V. and John, L. (2016). Design, fabrication and performance studies of fibre reinforced glass structured parabolic trough collectors, M.Tech. Thesis, University of Kerala.

Baum, V. A., Aparasi, R. R., & Garf, B. A. (1957). High power solar installation. *Solar Energy, 1*(1), 6–12.

Behar, O., Khellaf, A. & Mohammedi, K. (2013). A review of studies on central receiver solar thermal power plants. *Renewable and sustainable energy reviews, 23,* 12–39.

Chung, K. M., Chen, C. C. & Chang, K. C., 2018. Effect of diffuse solar radiation on the thermal performance of solar collectors. *Case studies in thermal engineering, 12,* 759–764.

Evangelisti, L., Vollaro, R. D. L. & Asdrubali, F. (2019). Latest advances on solar thermal collectors: A comprehensive review. *Renewable and Sustainable Energy Reviews, 114,* 109318.

Kalogirou, S. (2003). The potential of solar industrial process heat applications. *Applied Energy, 76*(4), 337–361.

Kalogirou, S. A. (2004). Solar thermal collectors and applications. *Progress in Energy and Combustion Science, 30*(3), 231–295. https://doi.org/10.1016/j.pecs.2004.02.001.

Kalogirou, S. A. (2014). Solar Energy Engineering: Processes and Systems: Second Edition. *Solar Energy Engineering: Processes and Systems: Second Edition,* 1–819.

Kaushik, S. C., Gadhi, S. M. B., Agarwal, R. S. & Kumari, R. (1988). Modeling and simulation studies on single/double-effect absorption cycle using water-multicomponent salt (MCS) mixture. *Solar energy, 40*(5), 431–441.

Sharma, A. (2011). A comprehensive study of solar power in India and World. *Renewable and Sustainable Energy Reviews, 15*(4), 1767–1776.

Soin, R. S., Rao, K. S., Rao, D. P. & Rao, K. S. (1979). Performance of flat plate solar collector with fluid undergoing phase chance. *Solar Energy, 23*(1), 69–73.

Sukhatme, S. P. & Nayak, J. K. (2017). Solar Energy. McGraw Hill Education.

Suman, S., Khan, M. K., & Pathak, M. (2015). Performance enhancement of solar collectors—A review. *Renewable and Sustainable Energy Reviews, 49,* 192–210.

Chapter 4
Solar Energy Storage Systems

Solar energy is intermittent, variable and unpredictable source of energy and hence, after the collection through suitable collectors, it needs to be stored using proper storage for further usage. The energy storage system may store excess solar energy when the availability is more than the requirement, and discharges for later use. The energy storage devices can be classified into several categories such as mechanical, chemical, biological, magnetic and thermal energy storage, as shown in Fig. 4.1.

As far as solar energy is concerned, the suitable thermal energy storage (TES) system is required for almost (low cost and high energy density) all energy applications.

The following parameters are to be kept in mind while designing an energy storage system:

(a) The thermal capacity of the system
(b) Operation strategy
(c) Maximum load
(d) Nominal temperature
(e) Specific enthalpy drops in load and integration into the utility plant.

However, prior study about the technical properties, cost of the storage system and environmental properties of the storing materials etc. in detail are required for the design of a highly efficient thermal energy storage system. There are some important thermal properties that are required for the ideal thermal storage system:

(a) *High energy density in the storage materials*: A high thermal storage capacity is required to reduce the system volume and increase the system efficiency.
(b) *High heat transfer rate*: The rate of heat transfer is an important parameter in thermal energy storage. It should be as high as possible to ensure quick release and absorption of energy, so that charging and discharging time can be as low as possible.
(c) *Chemical stability*: The storage material should be chemically stable. The mechanical and chemical degradation should be avoided.

S. C. Kaushik et al., *Solar Cooling*, https://doi.org/10.1007/978-3-031-42410-6_4

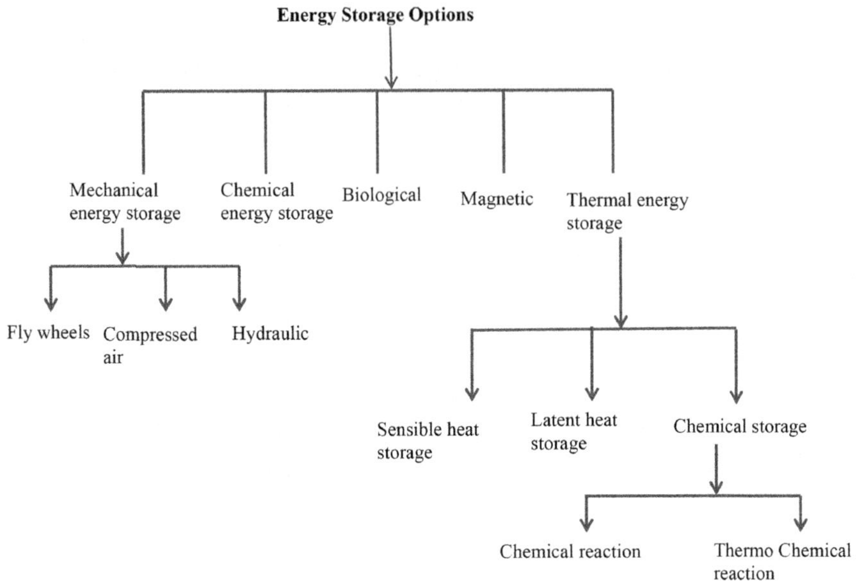

Fig. 4.1 Classification of different energy storage options

(d) *Energy storage*: The thermal energy storage systems should be insulated properly to avoid the heat loss.
(e) *Capital and operational costs*: The capital and operational cost of the system should be as low as possible The cost of a solar thermal energy storage system mainly consists of three parts namely storage material, heat exchanger and land cost (Gil et al. 2010).

In this chapter, various thermal energy storage methods applicable to solar systems are discussed.

4.1 Methods of Thermal Energy Storage

According to physics of materials, different storage mechanisms are used in the thermal energy storage systems and are classified into three main groups as given below and shown in Fig. 4.2:

(a) Sensible heat storage
(b) Latent heat storage
(c) Chemical storage methods.

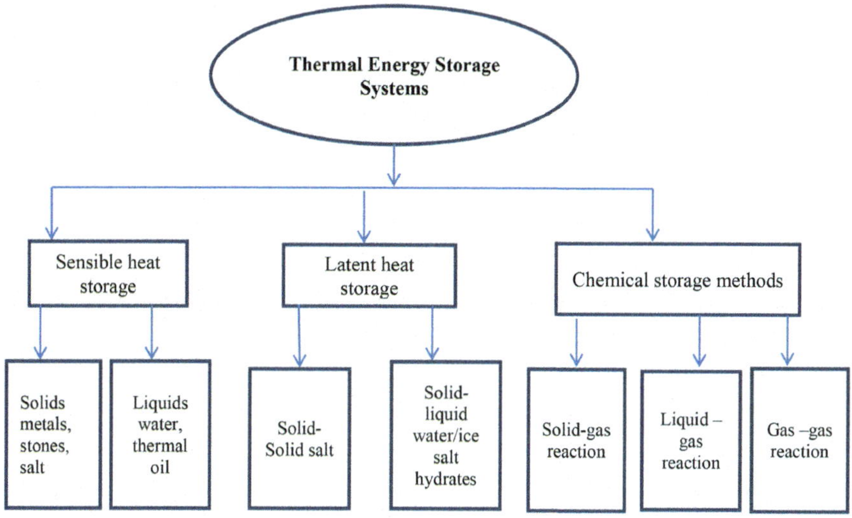

Fig. 4.2 Classification of different methods of the thermal storage system

4.1.1 Sensible Heat Storage

In the sensible heat storage systems, solar energy is collected and stored or extracted by heating or cooling of a liquid or solid material without phase change. The sensible heat storage has been used to store and release thermal energy passively. However, the problem with sensible heat storage systems is their large size. Thus, a larger volume of material is required to store the same amount of energy in comparison to the other methods of thermal energy storage. The storage media can either be a liquid or a solid (Dheep and Sreekumar 2014). Few commonly used solid storage materials are shown in Tables 4.1 and 4.2.

Liquid Sensible Heat Storage System

Water is the most commonly used medium in the liquid storage system particularly, for the solar water heating and space heating applications use water as storage media in the energy storage systems. Water is cheaply available and having higher specific heat than other materials and chemically stable. For low-temperature applications, water is one of the best storage media, however it requires costly insulation along with a pressure containment system for high temperature applications.

Steel and concrete are the most commonly used materials for making the water tank. To avoid the thermal losses, the tanks are well insulated with the glass wool, mineral wool, puff etc. (Fig. 4.3).

The proposed substitutes for water are petroleum-based oils and molten salts as they have very low vapour pressure and are able to operate at higher temperature than water. The oils are limited to less than 400 °C due to stability and safety reasons and can be quite expensive. Therefore, water has been substituted effectively with

Table 4.1 Commonly used liquid storage materials (Gil et al. 2010)

Storage materials	Working temperature (°C)	Density (kg/m^3)	Thermal conductivity (W/(m K))	Specific heat (kJ/(kg °C))	Specific heat (kW h/(m^3 °C))	Cost per kg (US$/kg)
Hitec solar salt	220–600	1899	–	1.5	0.79	0.93
Hitec XL solar salt	120–500	1992	0.52	1.4	0.77	1.19
Mineral oil	200–300	770	0.12	2.6	0.56	0.3
Synthetic oil	250–350	900	0.11	2.3	0.58	3
Silicone oil	300–400	900	0.1	2.1	0.53	5
Nitrite salts	250–450	1825	0.57	1.5	0.76	1
Liquid sodium	270–530	850	71	1.3	0.31	2
Nitrate salts	265–565	1870	0.52	1.6	0.83	0.5
Carbonate salts	450–850	2100	2	1.8	1.05	2.4

Table 4.2 Solid-state sensible heat storage materials (Tian and Zhao 2013)

Storage materials	Working temperature (°C)	Density (kg/m^3)	Thermal conductivity (W/(m K))	Specific heat (kJ/(kg °C))	Specific heat (kW h/(m^3 °C))	Cost per kg (US$/kg)
Sand-rock minerals	200–300	1700	1	1.3	0.61	0.15
Reinforced concrete	200–400	2200	1.5	0.85	0.52	0.05
Cast iron	200–400	7200	37	0.56	1.12	1
NaCl	200–500	2160	7	0.85	0.51	0.15
Cast steel	200–700	7800	40	0.6	1.3	5
Silica fire bricks	200–700	1820	1.5	1	0.51	1
Magnesia fire bricks	200–1200	3000	5	1.15	0.96	2

therminol and Caloria-HT and for high temperature applications (>400 °C), molten mixtures of inorganic salts are considered. One of the most commonly used molten mixture is sodium hydroxide, which has a melting point of 320 °C and could be used for temperatures up to 800 °C (Srikanth and Balaji 2017).

The other possibility is the use of liquid metals as most of the liquid metals are similar to that of the water and when mixed with the petroleum-based oils, their specific heat capacity is considerably reduced and they have a high potential for the reactivity with the container. Besides, the liquid storage system has ability to

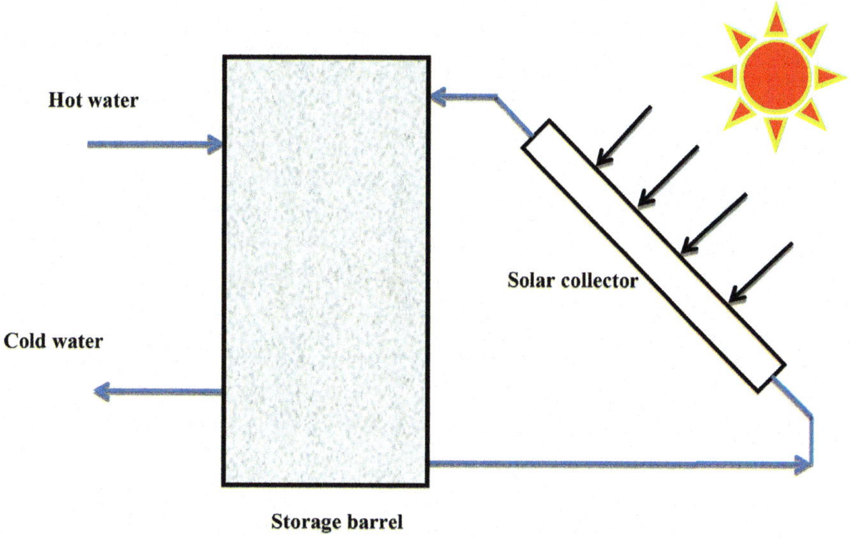

Fig. 4.3 A typical solar water heater

maintain natural thermal stratification within the storage system, due to the density difference which exists between the hot and cold fluid. The hot fluid is circulated to the upper part of the storage media during charging, and the cold fluid is extracted from the bottom part during discharging. The thermal stratification devices such as the floating entry, a mantle heat exchanger are used to avoid the mixing of hot and cold fluids (Gil et al. 2010).

Solid Sensible Storage

The solid sensible heat storage system can be used for low and high-temperature applications. It is generally employed along with a solar air heater (Fig. 4.4). The most commonly used solid materials are rocks, pebbles, sand, brick, concrete etc. The commonly used solid storage materials are listed in Table 4.2. The rock or pebble bed is preferred as the solid materials for the storage since they have low cost. Figure 4.5 represents the schematic of a rock bed sensible heat storage system where the pebble bed or loosely packed rock material is stored inside the container. The heat transfer fluid is allowed to flow through the bed, the thermal energy is stored in the bed by forcing the heated air in the bed. This stored thermal energy is then retrieved back by recirculating the ambient air. The other possibility includes integrating the solid sensible heat storage system with buildings. The suitable storage materials used in this type of systems are magnesite, ceramic brick, microtherm and metals including aluminium, magnesium and zinc.

The sensible heat storage systems are simple in design and construction, easy to control and maintain and integrate with solar collector system. However, the temperature of the working fluid always keep on changing and therefore, bigger size storage

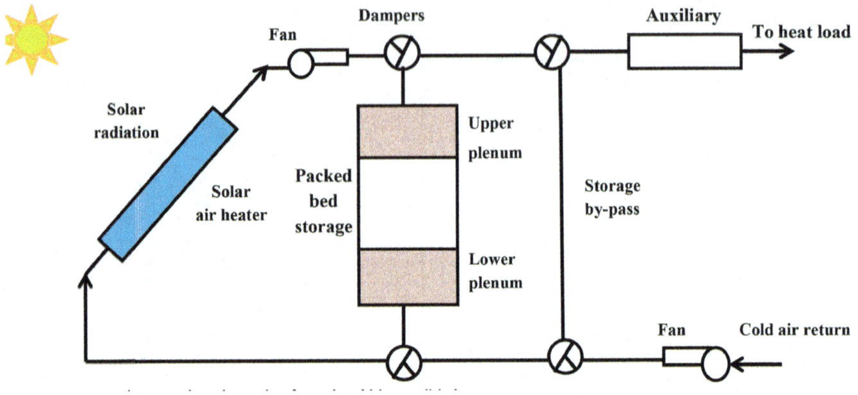

Fig. 4.4 The schematic of a solid (rock pebble) sensible heat storage system

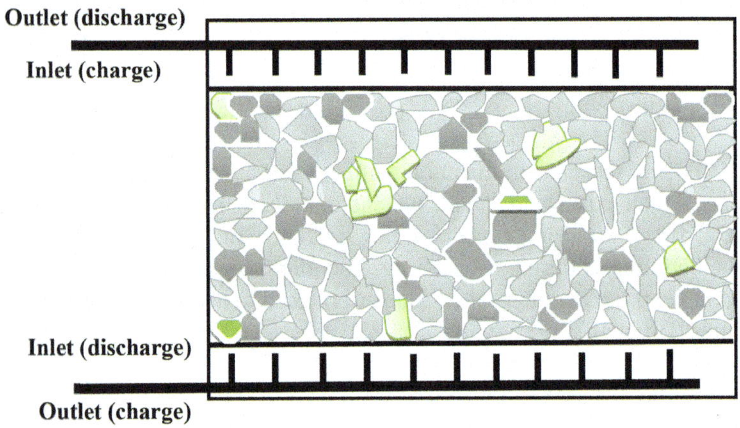

Fig. 4.5 Schematic of the rock bed sensible heat storage system

system is required. Further, the heat loss to the surroundings is high due to high temperature operation.

4.1.2 *Latent Heat Storage*

In the latent heat storage systems, heat is stored in a material when it melts and is extracted from the material when it freezes. This is a more attractive technique since it can provide high energy storage density. In this type of storage, the thermal energy can be stored nearly isothermally in most of the substances as the latent heat. In the solid–liquid transition the energy is stored as the heat of fusion, while in certain

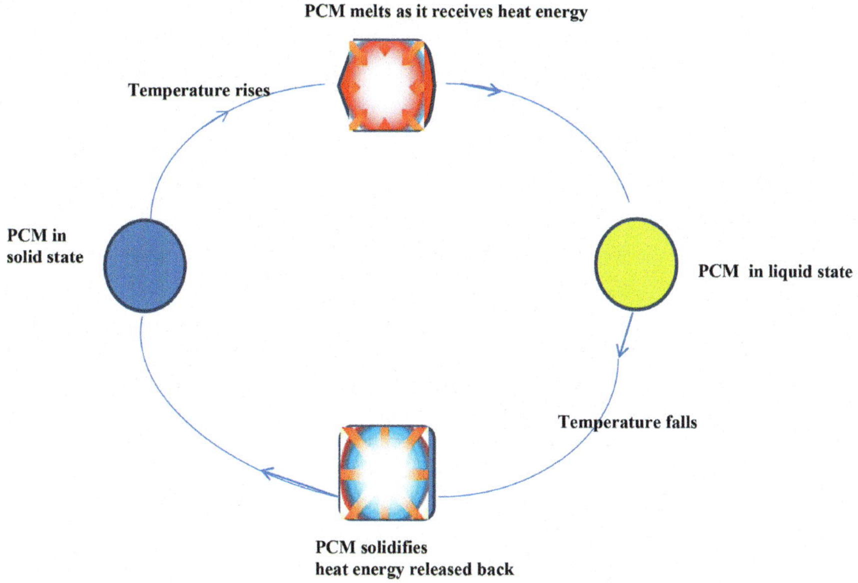

Fig. 4.6 Different processes in a PCM material storage

conditions, it is stored as the heat of vapourisation (liquid–vapour transition). The phase change material (PCM) uses mainly the solid–liquid transition (Fig. 4.6).

Phase Change Materials

The latent heat storage system is considered to be one of the most effective means of thermal energy storage. The phase change materials (PCMs) used in it absorbs or releases heat during their phase transition. These systems can provide heat energy at nearly a constant rate, as the temperature remains constant during the phase change process. The storage capacity of PCMs is far better than the sensible heat storage material because of their high latent heat value which is nearly 5–10 times higher than the sensible heat storage materials. Commonly, solid to liquid transition type PCMs are widely used for medium applications, such as, waste heat recovery, building material, pharmaceutical, etc. However, liquid to vapour type PCMs are also used in high-temperature applications, the total heat energy that can be stored in such materials, (Q) is given by:

$$Q = m\left[\int_{t_i}^{t_m} C_{p,s} dT + H_f + \int_{t_m}^{t_f} C_{p,l} + H_e\right] \tag{4.1}$$

where m is the mass of the energy storage material used $C_{p,s}$, $C_{p,l}$ are the specific heat of material in its solid and liquid phases, respectively. H_f the latent heat of fusion, H_e the latent heat of evaporation and t_m the melting point of the material. Favourable characteristics of PCMs are shown in Fig. 4.7.

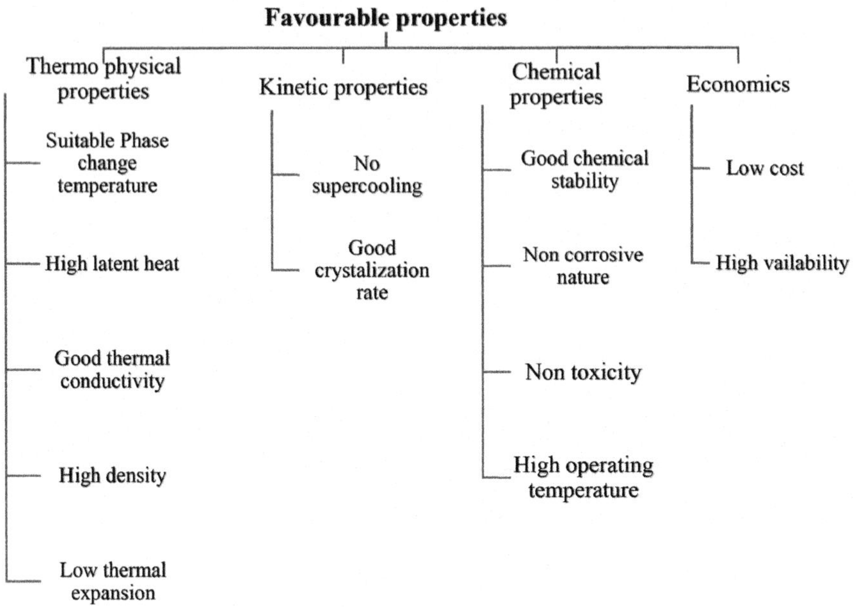

Fig. 4.7 The properties of PCM

There are number of applications, where PCMs are being used including but not limited to refrigeration and air conditioning, waste heat recovery, active and passive cooling of electronic devices, preservation of food and pharmaceuticals, insulating materials for human thermal comfort, cold chain for transportation of different food items (fruits and vegetables), water heating and space conditioning applications. The comparison of sensible heat storage and phase change materials is given in Table 4.3.

The main advantages of PCMs are high storage density, comparably less quantity of material is needed, almost constant temperature heat storage and heat release process, and low operating cost. However, there are certain limitations such as low

Table 4.3 Comparison of phase change materials with sensible heat storage material

Phase change materials	Sensible heat storage material
• High storage capacity	• Comparably low storage capacity
• Store/release energy with less temperature swings	• Large changes in temperature occur during heat storage/release
• Less quantity of material is needed	• Large quantity of material, as well as containers, is needed
• Heat source temperature must be greater than the melting point of PCM	• High-temperature heat source is needed to store the same quantity of heat energy

thermal conductivity, high investment cost, subcooling and supercooling, corrisiveness and phase separation. Different types of PCMS are shown in Table 4.4 (Tyagi et al. 2022).

The PCMs should exhibit certain thermal, physical and chemical properties that make them suitable material for thermal energy storage systems. The desired thermal properties includes, phase transition temperature, high latent heat of transition, high thermal conductivity and better heat transfer. On the other hand, the desired physical and chemical properties are, constant volume during phase change, better phase equilibrium, high density, low vapour pressure, no supercooling, high crystallization rate, non-toxic, chemically stable, non-corrosive, and non-flammable.

Classification and Characteristics of PCMs

PCM materials are generally classified into three categories viz. organic, in-organic and eutectic as shown in Fig. 4.8.

(i) Organic PCMs

The organic type of PCM comprises paraffins and non-paraffins compounds, the most commonly used organic PCM is the paraffin wax. They are readily available in the temperature range of 35–64 °C, which makes them suitable for space

Table 4.4 Different types of PCMs used (Tyagi et al. 2022)

Storage materials	Density (kg/m^3)	Thermal conductivity (W/mK)	Specific heat (kJ/kg K)	Latent heat (kJ/g)	Phase change temperature (°C)
RT100 (paraffin)	880	0.2	–	124	100
RT110 (paraffin)	–	–	–	213	112
KNO_3–$NaNO_2$–$NaNO_3$	–	–		275	141
E117 (inorganic)	1450	0.7	2.61	169	117
A164 (organic)	1500	–	–	306	164
$NaNO_3$	2260	0.5	–	172	307
KNO_3	2110	0.5	–	226	333
KOH	2044	0.5	–	149.7	380
$AlSi_{12}$	2700	1.6	1.04	560	576
$MgCl_2$	2140	–	–	452	714
NaCl	2160	5	–	492	800
LiF	–	–	–	–	850
Na_2CO_3	2533	2		275.7	854
K_2CO_3	2290	2		235.8	897
48%$CaCO_3$–45%KNO_3–7% $NaNO_3$		–		–	130
$LiNO_3$–$NaNO_3$	–	–		252	195
$MgCl_2$–KCl–NaCl	2044	0.5		149.7	380

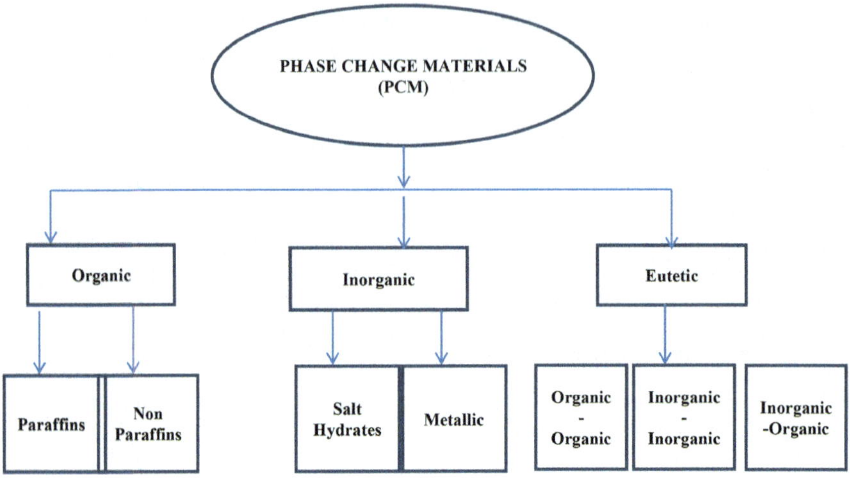

Fig. 4.8 Classification of different PCMs used

heating and cooling applications. The organic type of PCM materials are subjected to congruent melting and self-nucleation and hence, the organic types of PCMs are almost free from the problem of phase segregation and supercooling. Characteristic properties of paraffins are (Tyagi et al. 2022) safety, cost effectiveness, reliablitity, non-corrosiveness, chemical stability and good nucleating properties. Undesirable characteristics include low thermal conductivity, non-compatibility with plastic containers and its moderate flammability. The most commonly used non-paraffin materials are esters, fatty acids, alcohols, eutectics etc. Among these, fatty acids have better properties. The characteristics of non-paraffin organic PCMs are varying level of toxicity, inflammability and heat of fusion. Besides, they have very low thermal conductivity and instability as compared to paraffin type PCMs and low flash point. Also, the cost of non-paraffin organic PCM is approximately three times higher than the commercial paraffin PCMs. Properties of some of the non-paraffin type PCM are shown in Table 4.5.

(ii) Inorganic PCMs

Inorganic type of PCMs includes salt hydrates, salts, metals and alloys. The salt hydrates are more attractive PCMs because of their relatively high thermal conductivity, and high volumetric storage capacity. In addition to these, the inorganic PCMs do not supercool appreciably and the chemical degradation is also lesser as compared to other PCMs. Most commonly used inorganic type of PCM is shown in Table 4.6 (Tyagi et al. 2022).

The general formula of a salt hydrate is X_nH_2O. X can be any metal or an alloy, n represents the number of water molecules in it. The high thermal conductivity, high density and compactivity with the other plastics are major advantages of the salt hydrates.

Table 4.5 Properties of non-paraffin type phase change materials (Tyagi et al. 2022)

Material	Melting point (°C)	Latent heat (kJ/kg)
Acetic acid	16.7	2184
Xylitol pentapalmitate	18.75	170.05
Erythritol tetrastearate	30.35	208.84
Glycerol trimyristate	31.96	154.3
Capric acid	36	152
Polyethylene glycol 600	20–25	146
Eladic acid	47	218
Lauric acid	49	178
Pentadecanoic acid	52.5	178
Palmatic acid	55	163
Tristearin	56	191
Myristic acid	58	199
Glycerol tripalmitate	58.50	185.9
Heptaudecanoic acid	60.6	189
Bee wax	61.8	177
Glyolic acid	63.0	109
Stearic acid	69.4	199
Xylitol	93.20	–
Erythritol	118.4	379.57
Galactitol	187.41	401.76

The other type of inorganic PCMs include metallics. The metallic means low melting metals. For example, gallium–gallium, antimony eutectic, Bi–Cd–In eutectic, cerrobend eutectic Bi–Pb–In eutectic etc. The advantages of metallics include very high thermal conductivity, very high latent heat of fusion per unit volume. But their latent heat of fusion per unit weight and specific heat are very less. The disadvantage of such materials is its very high weight.

(iii) Eutectics

The metallurgy definition of the eutectics is a combination of two or more substances in desired proportions that freezes and melts at a single temperature that is lower than the melting point of the different substances or of any other combination of them without segregation. Few of eutectics that can be used as PCM materials are given in Table 4.7.

Supercooling and Phase Segregation

Supercooling is the process of solidification of liquids below its normal freezing point and involves three phases, viz. (a) induction, (b) crystal growth and (c) crystal regrowth phase. The formation of nuclei and growth to sufficient size is called the

Table 4.6 Inorganic type PCM (Tyagi et al. 2022)

Material	Melting point (°C)	Latent heat (kJ/kg)
$LiBO_2 \cdot 8H_2O$	25.7	289
$Mn(NO_3)_2 \cdot 6H_2O$	25.8	125.9
$FeBr_3 \cdot 6H_2O$	27	105
$CaCl_2 \cdot 6H_2O$	29	190.8
$LiNO_3 \cdot 2H_2O$	30.0	296
$KFe(SO_4)_2 \cdot 12H_2O$	33	173
$CaBr_2 \cdot 6H_2O$	34	138
$Na2CO_3 \cdot 10H_2O$	34.2	146.9
$Zn(NO3)_2 \cdot 6H_2O$	36.2	246.5
$Mn(NO_3)_2 \cdot 4H_2O$	37.1	115
$Na2HPO_4 \cdot 12H_2O$	40.0	279
$Ca(NO_3)_2 \cdot 3H_2O$	51	104
$Na(NO3)_2 \cdot 6H_2O$	53	158
$FeCl_3 \cdot 2H_2O$	56	90
$CO(NO_3)_2 \cdot 6H_2O$	57	115
$MnCl_2 \cdot 4H_2O$	58	151
$A12(SO_4)_3 \cdot 18H_2O$	88	218
$Sr(OH)_2 \cdot 8H_2O$	89	370
$(NH_4)Al(SO_4) \cdot 6H_2O$	95	269
$LiCl \cdot H_2O$	99	212
$Mg(NO_3)_2 \cdot 6H_2O$	89–90	162–167

induction phase. The crystal growth involves diffusion of PCM towards the nuclei to be adsorbed onto the surface and hence grow the nucleus. These small crystals continue to grow and eventually, it becomes large and numerous enough to sustain a rapid rate of the crystal growth. The supercooling is significant in salt hydrates. The organic PCMs like paraffins and non-paraffins have less effect on supercooling, which are reduced furthur by controlling the size of the PCMs, and adding nucleating agent or metal additives before encapsulation and/or by optimising the composition and structure of the capsule shells. The phase segregation can be prevented by using thickening agents, rotating storage devices and the direct contact heat transfer etc. The other methods used for avoiding the phase segregation includes the use of excessive water so that the melted crystals do not produce the supersaturated solution and modification of the chemical composition of the system.

In recent decades, the researchers identified various active heat transfer enhancement techniques in PCMs. These includes use of agitators, vibration techniques, use of composite materials, nano powders etc. However, these active techniques greatly increases the complexity and the overall cost of the system. The most promising technique to overcome these drawback is the use of *encapsulated PCM* as a storage material and a packed bed as a heat exchanger. The value of thermal conductivity of

Table 4.7 Eutectic PCM materials (Sharma et al. 2009)

Material	Melting point (°C)	Latent heat (kJ/kg)	Type
Tetradecane + docosane	41.5–5.7	234.33	Organic eutectics
Tetradecane + geneicosane	3.54–5.57	200.28	Organic eutectics
Pentadecane + heneicosane	6.23–7.21	128.25	Organic eutectics
Ga + (Ga–NH_4)	29.8	–	Inorganic eutectics
47% $Ca(NO_3)_2{\cdot}4H_2O$ + 53%	30	136	Inorganic eutectics
$Mg(NO_3)_2{\cdot}6H_2O$	31.5	226	Inorganic eutectics
60% $Na(CH_3COO){\cdot}3H_2O$ + 40% CO(NH2)2	40.5	255	Organic eutectics
50% $Na(CH_3COO){\cdot}3H_2O$ + 50% $HCONH_2$	58–59	132	Inorganic eutectics
50% $Mg(NO_3)2{\cdot}6H_2O$+ 50% MgC12·$6H_2O$	61	25	Inorganic eutectics
Bi–Cd–In	65	218	Organic eutectics
50% CH_3CONH_2 + 50% C17H35 COOH	67	123	Organic eutectics
67.1% napthalene + 32.9% benzoic acid	70	29	Inorganic eutectics
Bi–Pb–In	72	180	Inorganic eutectics
14% $LiNO_3$ + 86% $Mg(NO_3)_2{\cdot}6H_2O$	77.2–77.9	150.7	Inorganic eutectics
$Mg(NO_3)_2{\cdot}6H_2O$ + 7% $MgCl_2{\cdot}6H_2O$	96		Inorganic eutectics
Bi–Pb–tin			

PCMs lies in the range of 0–2 W/mK, which are considered to be their main disadvantage because low thermal conductivity results in a low heat transfer rate which decelerate the storage process.

To solve this, PCMs are also coated with good thermal conductivity material and these coatings are known as encapsulations which increases the heat transfer area, controls volume changes happening during phase transition and protects the material from its working environment. There are two ways of encapsulations: micro encapsulations and macro encapsulations. In macroencapsulation, a certain amount of PCM is sealed within a shell or tube or sphere-shaped container, while macro encapsulations are comparably easy to manufacture and widely used in heat storage applications. However, poor heat transfer characteristics, leakage issues and thermal stratifications are other problems of macro encapsulation technology. In microencapsulation, the PCM is filled inside a capsule with size in the order of few micrometres,

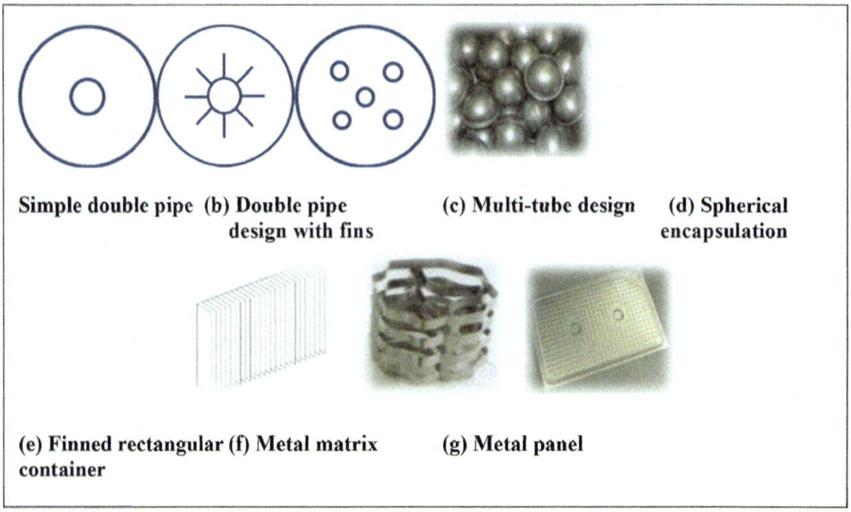

Fig. 4.9 Macroencapsulation methods

and these capsules are generally made up of polymers and the PCMs used in it are of either organic or inorganic type.

(a) Macro Encapsulation Designs

The macro encapsulation is one of the encapsulation technique used for PCMs which generally requires a metal container, in the shape of a shell, tube, plate, or sphere and is insulated externally. The entire container is made as a heat exchanger, with internal passage ways for circulating heat transfer fluids. Most of the designs involve fins to improve the heat transfer process. The addition to nanoparticles, metal matrix and metal forms are added to the PCM to enhance the heat transfer rate. Some of the examples are shown in Fig. 4.9.

(b) Microencapsulation Design

The microencapsulation is the process of covering the droplets of PCMs with a thin layer of a different material having good thermal conductivity. Generally, they are spherical shaped, with a thick shell covering (polymer) and the core (PCM). The shell material controls the volume changes in core material during the phase change process. Different types of microencapsulations are presented in Fig. 4.10 which include single-core, multi-core, multi-walled and continuous matrix type encapsulations. Production of microencapsulated PCMs can be done by either physical methods or chemical methods or by the physical–chemical methods.

Physical methods include spray-drying, electrostatic encapsulation, and one-step method.

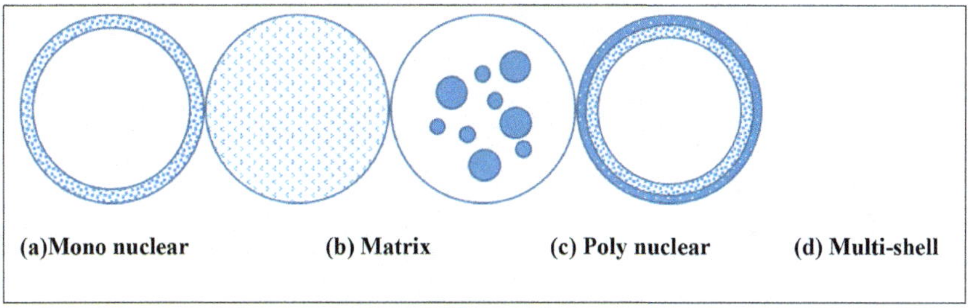

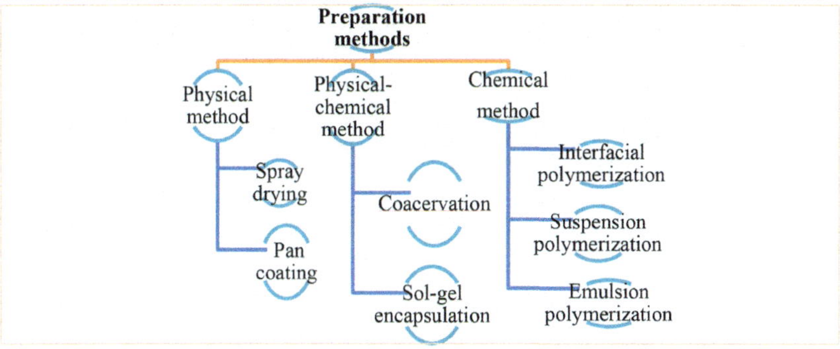

Fig. 4.10 Microencapsulation methods

Spray Drying: The phase change material is dispersed in a polymer solution (shell material) and then this emulsion is sprayed as small droplets in a hot chamber. The hot, dry gas in the chamber evaporates the solvent, thereby forming a film of encapsulated PCM droplets at the bottom of the chamber. Spray drying is the most common and least expensive method of encapsulating PCM. Different types of microencapsules of PCMs are shown in Fig. 4.12 (Tyagi et al. 2011).

Pan Coating: In this method, the solid-state core particle is mixed with liquid-state shell materials inside a pan. Warm air is then used to dry the coated core material. It is one of the oldest encapsulating techniques, and it is used to encapsulate larger solid materials.

Tyagi et al. (2011) investigated various aspects in microencapsulation of PCMs in which individual particles or droplets of solid or liquid material (core) are surrounded or coated with a continuous film of polymeric material (shell) to produce capsules in the micrometre to millimetre range, known as microcapsules (Fig. 4.11). The microcapsules may be spherically shaped, with a continuous wall surrounding the core, while others are asymmetrically and variably shaped, with a quantity of smaller droplets of core material embedded throughout the microcapsule.

All three states of matter are shown in Fig. 4.12.

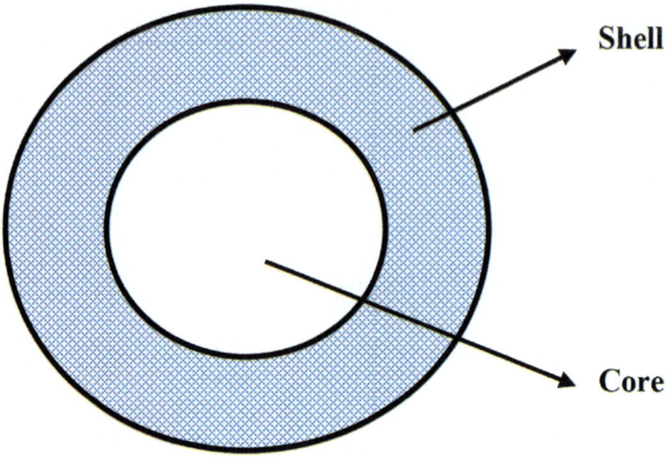

Fig. 4.11 Description of micro-capsule

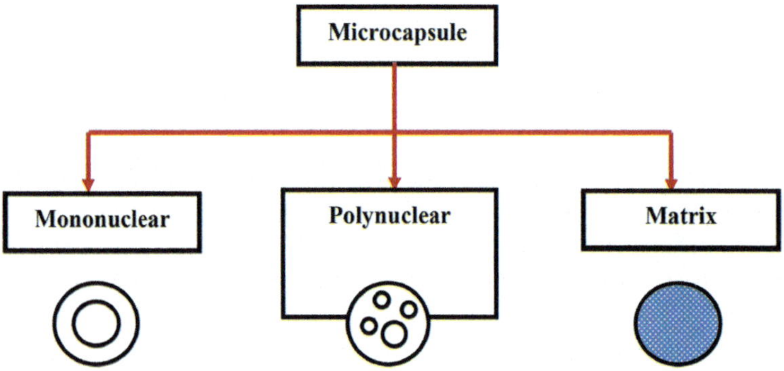

Fig. 4.12 Types of micro-capsule

Different forms viz. solids, liquids, and gases may be microencapsulated and processed as aqueous dispersion or powder that improves the formulation of phase change materials as many building materials go through a powder state during processing (Pinel et al. 2011).There are two types of microencapsulation methods available, that is, physical and chemical methods. The physical methods include pan coating, air-suspension coating, centrifugal extrusion, vibrational extrusion and spray drying. Various chemical methods include interfacial polymerisation, in situ polymerization, and matrix polymerization as shown in Fig. 4.13.

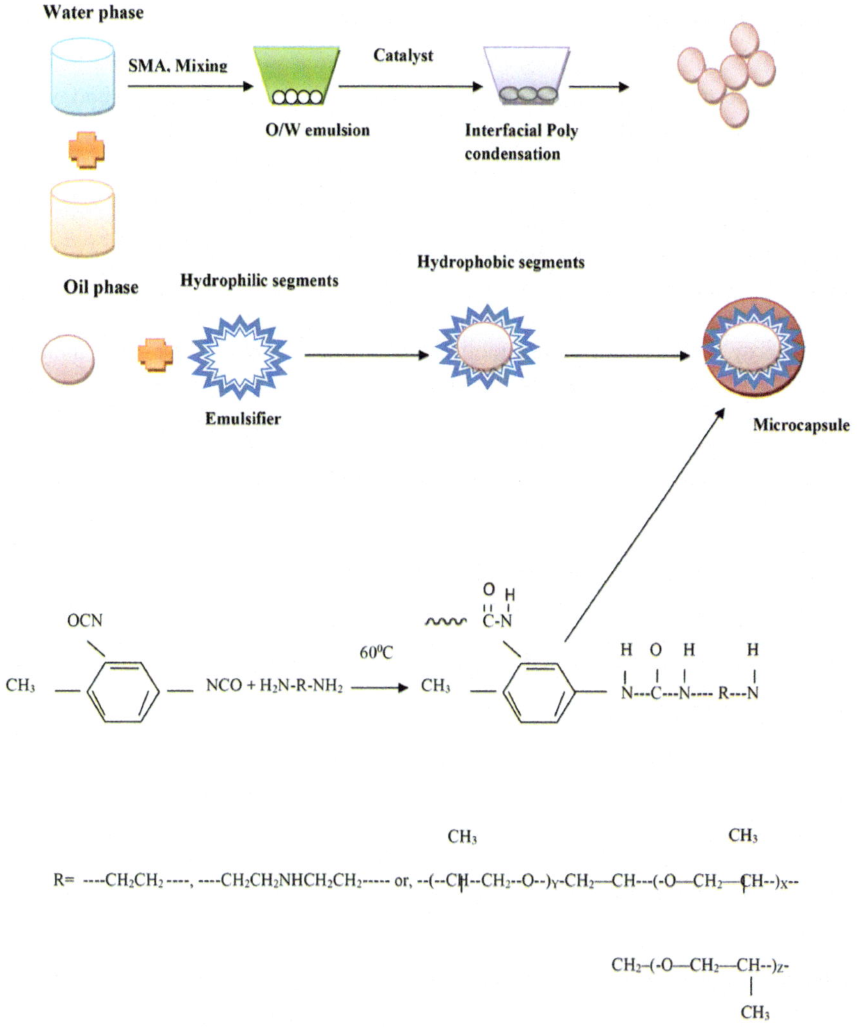

Fig. 4.13 Schematic of the formation of the microencapsulated *n*-octadecane with the polyurea shells containing different soft segments through interfacial polycondensation (Tyagi et al. 2011)

4.1.3 Chemical Storage Methods

The third thermal energy storage mechanism is the chemical storage method, which has classified the chemical storage system into two categories (Bales et al. 2008) namely Chemical reaction method and Thermochemical (sorption process) method.

(a) Chemical Reaction Method

This system involves the use of chemical reactions to store and release energy. In this type, the chemical reactions involved are completely reversible. Certain chemicals can absorb or release a huge quantity of heat while forming or breaking the bond. The solar energy is used to excite an endothermic chemical reaction. The thermal energy can be recovered during the reverse process sometimes by adding a catalyst. The factors to be considered in the design of a chemical storage system are its chemical stability, high chemical enthalpy and simple reaction conditions.

Table 4.8 represents the most commonly used chemical reaction materials in chemical storage systems.

From Table 4.8, it can be seen that silicon dioxide is more attractive due to its very high storage density. But the problem with silicon dioxide is its complexity of the endothermic part of the reaction. Similarly, chemical reaction involves the separation and recombining of the $MgSO_4{\cdot}7H_2O$ (Fig. 4.14). The adsorption–desorption process involves theoretically about 2.8 GJ/m3 of energy (Pinel et al. 2011) and the chemical reaction is given by

$$MgSO_4 \cdot 7H_2O(s) + heat \Leftrightarrow MgSO_4(s) + 7H_2O(g)$$

The advantages of the chemical storage system include high energy storage density, high pumping capacity and indefinitely long storage duration at nearby ambient temperature.

Table 4.8 Potential candidates for chemical storage (Pinel et al. 2011)

Material	Dissociation reaction	Storage density (GJ/m^3)	Turnover temperature (°C)
Magnesium sulphate	$MgSO_4{\cdot}7H_2O \Rightarrow MgSO_4(s) + 7H_2O$	2.8	122
Silicon oxide	$SiO_2 \Rightarrow Si + O_2$	37.9	4065 + HF: 150
Iron carbonate	$FeCO_3 \Rightarrow FeO + CO_2$	2.6	180
Iron hydroxide	$Fe(OH)_2 \Rightarrow FeO + H_2O$	2.2	150
Calcium sulphate	$CaSO_4{\cdot}2H_2O \Rightarrow CaSO_4 + 2H_2O$	1.4	89

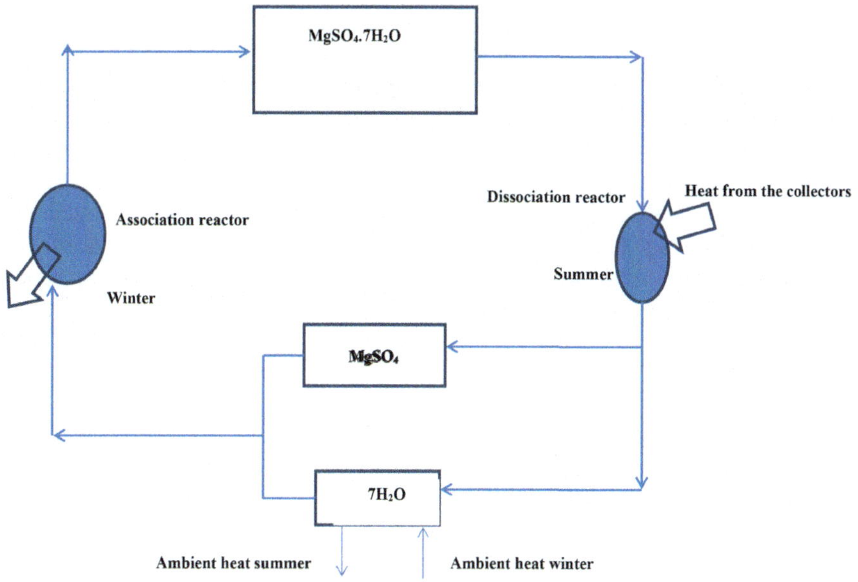

Fig. 4.14 Thermochemical storage system

(b) Thermochemical (Sorption Process) Methods

The thermochemical storage methods are based on the sorption process, which involves desorption and adsorption processes. The desorption process involves breaking the bonding of water and storing energy with a relevant substance. The heat is recovered during the reverse process, by re-evaporating the condensed product and re-bonding it (sorption) with the other substance and the reverse reaction, sorption, is exothermic. Table 4.9 gives the potential thermochemical storage materials that can be used for solar heating applications.

General chemical formula of the materials used in the thermochemical storage is $A_mX_pO_{2p}{\cdot}nH_2O$ where A represents the chemical elements Ca, Na, K, Ba or Sr; X the chemical elements, Al or Si, m, n and p are the number of atoms of each element in a molecule. The thermochemical storage systems are generally operated in two modes, viz. closed-loop and open-loop. In the closed-loop operation heat is exchanged with the ambient. In this process, the sorbate (water) condenses in the charging phase

Table 4.9 Thermochemical storage materials (Onder et al. 2008)

Absorption	Adsorption	Solid–Gas reaction
NH_3–H_2O 0.40 GJ/kg	H_2O–Zeolite 4A 0.08 GJ/kg	H_2O–Na_2S 1.27 GJ/kg
H_2O–NaOH 1.00 GJ/kg	H_2O–Silica gel 0.14 GJ/kg	H_2O–$MgCl_2$ 0.84 GJ/kg
		H_2O–$CaCl_2$ 0.98 GJ/kg
		H_2O–LiCl 0.71 GJ/kg

and evaporated during the discharging mode (Fig. 4.15). The evaporation process in the discharging phase requires heat input. This can be a low-grade heat source from the building thus cooling it or maybe solar energy (Lin and Al Kayiem 2016). The sorption process is exothermic, and liberates a large amount of the heat so that it can be further used for heating the buildings. Thus, thermal storage can be used for providing either heating or cooling.

On the other hand, in the open-loop cycle mode, the system is directly integrated with the building as in the case of a desiccant cooling system (Fig. 4.16). The sorbent is made as a honey-comp structure in which the ventilation air is circulated from the building.

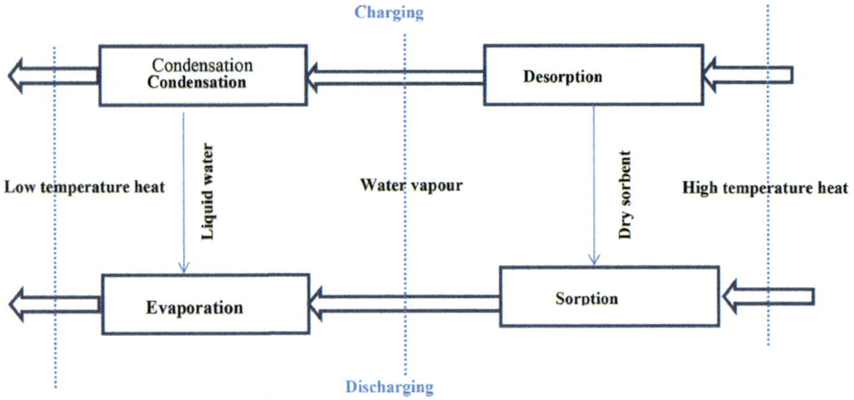

Fig. 4.15 Closed-loop system

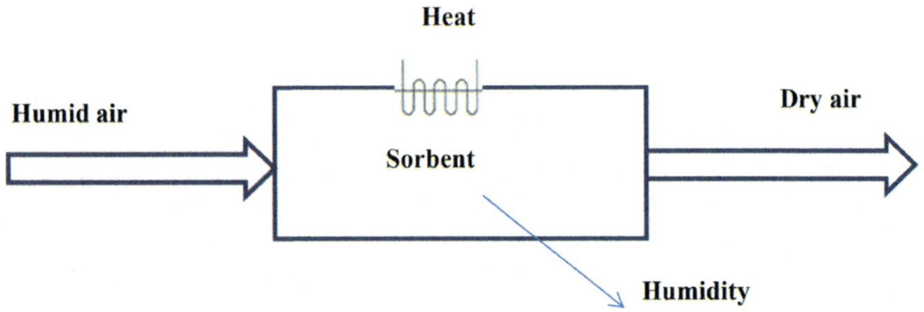

Fig. 4.16 Open-loop cycle system

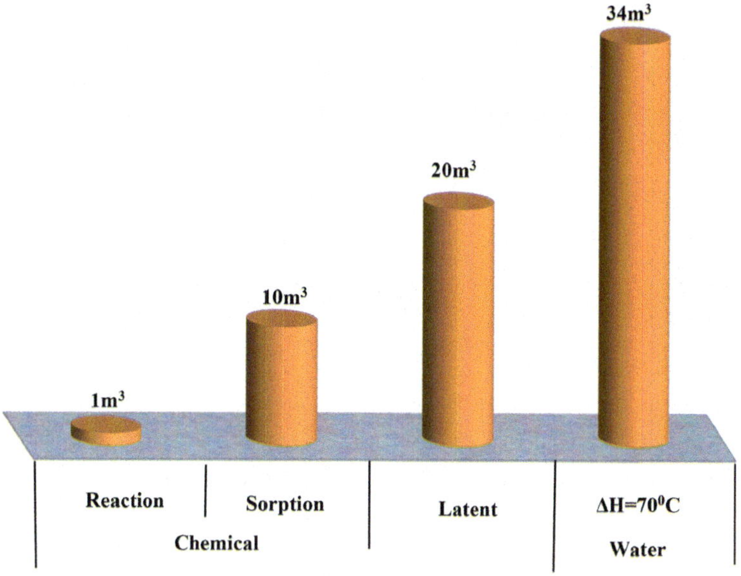

Fig. 4.17 Comparison between different storage systems

4.2 Comparison Among Various Storage Systems

A comparison between the theoretical storage volumes of different storage systems presented by Hadorn (2008) is shown in Fig. 4.17. The comparison is for the theoretical storage volume required to store 6.7 MJ of heat assuming 25% of heat loss using the different mechanisms. From Fig. 4.17, it is evident that the storage volume required is 1–10 m^3 for chemical methods whereas for water it becomes 34 m^3. Hence, the size of the chemical storage and latent heat storage systems is much lesser as compared to the water storage systems.

Sharma et al. (2009) compared the performance of PCM water storage tank and the rock-based system as shown in Fig. 4.18.

4.3 Application of PCMs

Nowadays phase change materials are used in so many thermal systems such as conditioning of buildings, cooling of heat and electrical engines, medical applications including hot–cold therapy, transportation of blood etc., human body cooling under bulky clothing or costumes. The phase change materials are also used in waste heat recovery, heat pump, solar power plants, and the thermal systems in spacecraft and thermal comfort in vehicles.

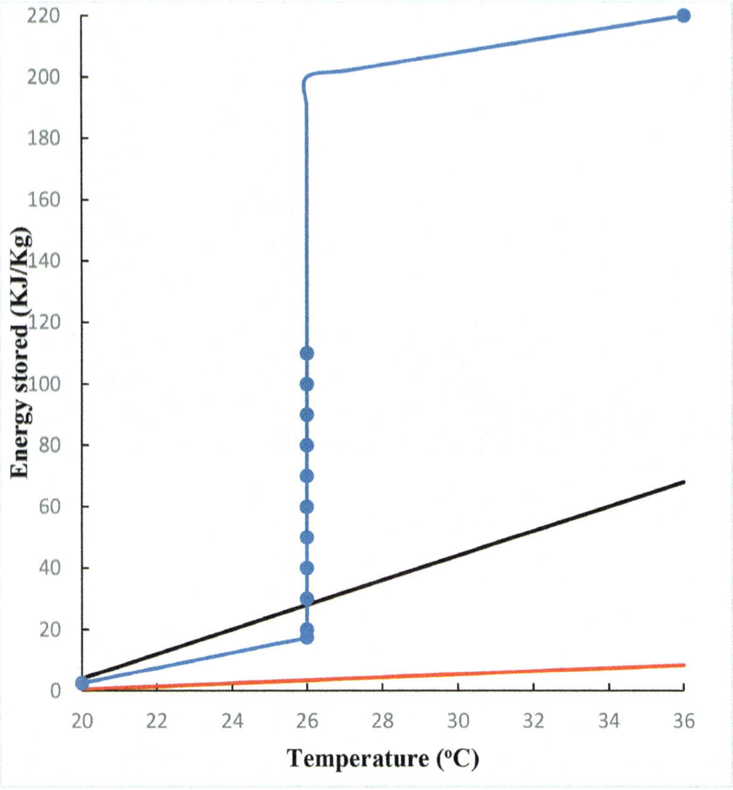

Fig. 4.18 Performance comparison of PCM, water and rockstorage system (Sharma et al. 2009)

References

Bales, C., Gantenbein, P., Jaenig, D., Kerskes, H. & van Essen, M. (2008). Final report of Subtask B "Chemical and Sorption Storage" The overview. Report of IEA Solar Heating and Cooling Programme–Task, 32.

Dheep, G. R. & Sreekumar, A. J. E. C. (2014). Influence of nanomaterials on properties of latent heat solar thermal energy storage materials–A review. *Energy conversion and management, 83*, 133–148.

Gil, A., Medrano, M., Martorell, I., Lázaro, A., Dolado, P., Zalba, B. & Cabeza, L. F. (2010). State of the art on high temperature thermal energy storage for power generation. Part 1—Concepts, materials and modellization. *Renewable and sustainable energy reviews, 14*(1), 31–55.

Hadorn, J. C. (2008, October). Advanced storage concepts for active solar energy—IEA SHC Task 32 2003–2007. *In Eurosun 1st International Conference on Solar Heating, Cooling and Buildings, Lisbon* (pp. 1–8).

Lin, S. C. & Al-Kayiem, H. H. (2016). Evaluation of copper nanoparticles–Paraffin wax compositions for solar thermal energy storage. *Solar Energy, 132*, 267–278.

Onder, E., Sarier, N. & Cimen, E. (2008). Encapsulation of phase change materials by complex coacervation to improve thermal performances of woven fabrics. *Thermochimica Acta, 467*(1–2), 63–72.

Pinel, P., Cruickshank, C.A., Beausoleil-Morrison, I. & Wills, A. (2011). A review of available methods for seasonal storage of solar thermal energy in residential applications. *Renewable and Sustainable Energy Reviews, 15*(7), 3341–3359.

Sharma, A., Tyagi, V. V., Chen, C. R. & Buddhi, D. (2009). Review on thermal energy storage with phase change materials and applications. *Renewable and Sustainable energy reviews, 13*(2), 318–345.

Srikanth, R. & Balaji, C. (2017). Experimental investigation on the heat transfer performance of a PCM based pin fin heat sink with discrete heating. *International Journal of Thermal Sciences, 111*, 188–203.

Tian, Y. & Zhao, C. Y. (2013). A review of solar collectors and thermal energy storage in solar thermal applications. *Applied Energy, 104*, 538–553.

Tyagi, V. V., Kaushik, S. C., Tyagi, S. K. & Akiyama, T. (2011). Development of phase change materials based microencapsulated technology for buildings: A review. *Renewable and sustainable energy reviews, 15*(2), 1373–1391.

Tyagi, V. V., Chopra, K., Sharma, R. K., Pandey, A. K., Tyagi, S. K., Ahmad, M. S., ... & Kothari, R. (2022). A comprehensive review on phase change materials for heat storage applications: Development, characterization, thermal and chemical stability. *Solar Energy Materials and Solar Cells, 234*, 111392.

Chapter 5
Vapour Compression Refrigeration System and Its Solar Cooling Options

The most commonly used method of cooling is the vapour compression refrigeration (VCR) cycle because they are fairly easy to construct, cheaper and employ a method which is being used for decades. The vapour compression refrigeration systems are like a heat engine where the heat energy is delivered from a cold reservoir to a hot reservoir. According to Clausius statement of second law of thermodynamics, heat transfer takes place from higher temperature level to the lower temperature level and the reverse is not possible without external work. Therefore, this device has to be powered externally to transfer heat in the opposite direction and hence, the VCR systems are work consuming devices. The VCR systems employ a working fluid called the refrigerant which is pushed through the system and undergoes a phase change from liquid to gas and vice-versa. During evaporation, it absorbs heat from the space to be cooled and converts into saturated vapour, which is then compressed through the compressor and converts into superheated vapour. The super-heated vapour is then condensed in the condenser by rejecting the heat to the circulating water/air. This machine is also known as latent heat pump as it absorbs latent heat in the evaporator and gives out in the condenser. The vapour compression refrigeration is approximately a reversed Rankine cycle as shown in the schematic, T–s and P–h diagrams in Figs. 5.1 and 5.2, respectively.

5.1 Working of an Ideal VCR System

A simple VCR system consists of a compressor, a condenser, an expansion device and an evaporator (Figs. 5.1 and 5.2). There are two pressure levels in the cycle- high pressure prevailing in the condenser and low pressure in the evaporator. It consists of four different thermodynamic processes, listed as below.

1. Isentropic compression in a compressor (Process 1-2)
2. Constant pressure heat rejection in a condenser (Process 2-3)

S. C. Kaushik et al., *Solar Cooling*, https://doi.org/10.1007/978-3-031-42410-6_5

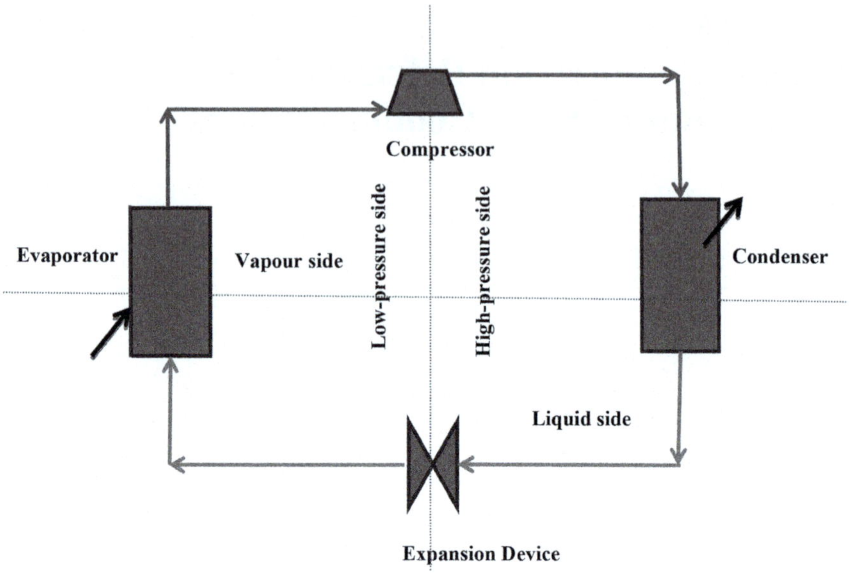

Fig. 5.1 Schematic diagram of an ideal vapour compression cooling system

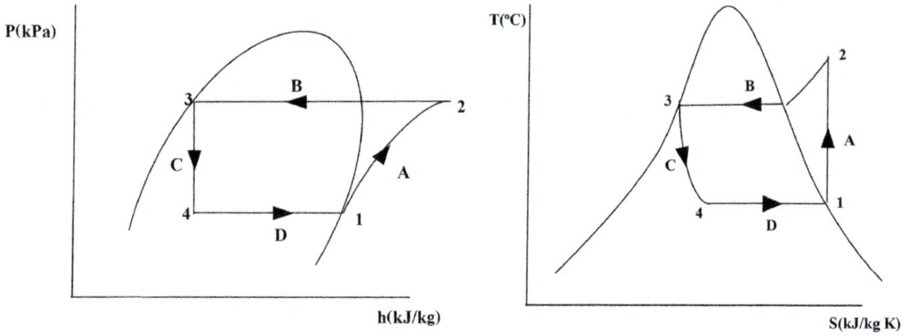

Fig. 5.2 Representation of vapour compression cooling cycle in *P–h* and *T–s* diagrams

3. Throttling in an expansion device (Process 3-4)
4. Constant pressure heat absorption in an evaporator (Process 4-1)

The low-pressure and low-temperature refrigerant (i.e., saturated) vapour coming from the evaporator is compressed isentropically in the compressor, and converts into high pressure superheated vapour. The superheated vapours are then subject to condensation in the condenser. The refrigerant leaves the condenser as high pressure liquid (at state point 3) and the pressure is reduced through an expansion device (process 3-4). As a result, the refrigerant converts into low pressure and low temperature liquid vapour mixture, because some of the liquid evaporates passing through the expansion valve.

The low-pressure and low-temperature refrigerant mixture enters the evaporator, which is in contact with the space to be refrigerated and hence boiled off at a low temperature, converting into vapour completely. Finally the low-temperature and low-pressure refrigerant vapour coming from the evaporator is then compressed by the compressor, thereby completing the cycle. Based on a reversed Carnot cycle operation, the isothermal processes of heat rejection/absorption are achieved by the condensation and evaporation of a pure substance at constant temperature and constant pressure. The compression process is isentropic whereas the expansion of refrigerant in the expansion device is ideally isenthalpic. The four processes are represented in both pressure–enthalpy (P–h) and temperature–entropy (T–s) diagrams of Fig. 5.2.

5.2 Cycle Analysis

The vapour compression cooling system consists of two pressure levels viz. high-pressure level consisting of the compressor, delivery head, discharge line, condenser and liquid line, while the low pressure level consists of expansion line, evaporator suction line and compressor suction head as shown in Fig. 5.3. The thermodynamic analysis of the vapour compression refrigeration system is presented here.

The assumptions made in the analysis of a simple vapour compression refrigeration system are:

- The flow is assumed to be at steady-state conditions.
- There is no heat transfer in the pipelines.
- Changes in kinetic energy and potential energy are neglected.
- Refrigerant vapour behaves as an ideal gas.
- Molecular weight is constant.
- Specific heat is assumed to be constant.
- Thermodynamic parameters vary at the steady-state conditions.

Ideally, the heat rejection (2-3) and heat addition processes (4-1) are isobaric and the compression process (1-2) is isentropic while the expansion is isenthalpic process (3-4) as shown in Fig. 5.3b.

As the process 1-2 is assumed to be isentropic and the required work for compression is given by

$$W_c = h_2 - h_1 \tag{5.1}$$

This work is known as specific work of compression, which is equal to the work input to the compressor per kilogram of refrigerant, where h_1 and h_2 are the specific enthalpies of the refrigerant at the inlet and the exit of the compressor, respectively. The heat rejected in the condenser during the process 2-3 is given by:

$$Q_c = h_2 - h_3 \tag{5.2}$$

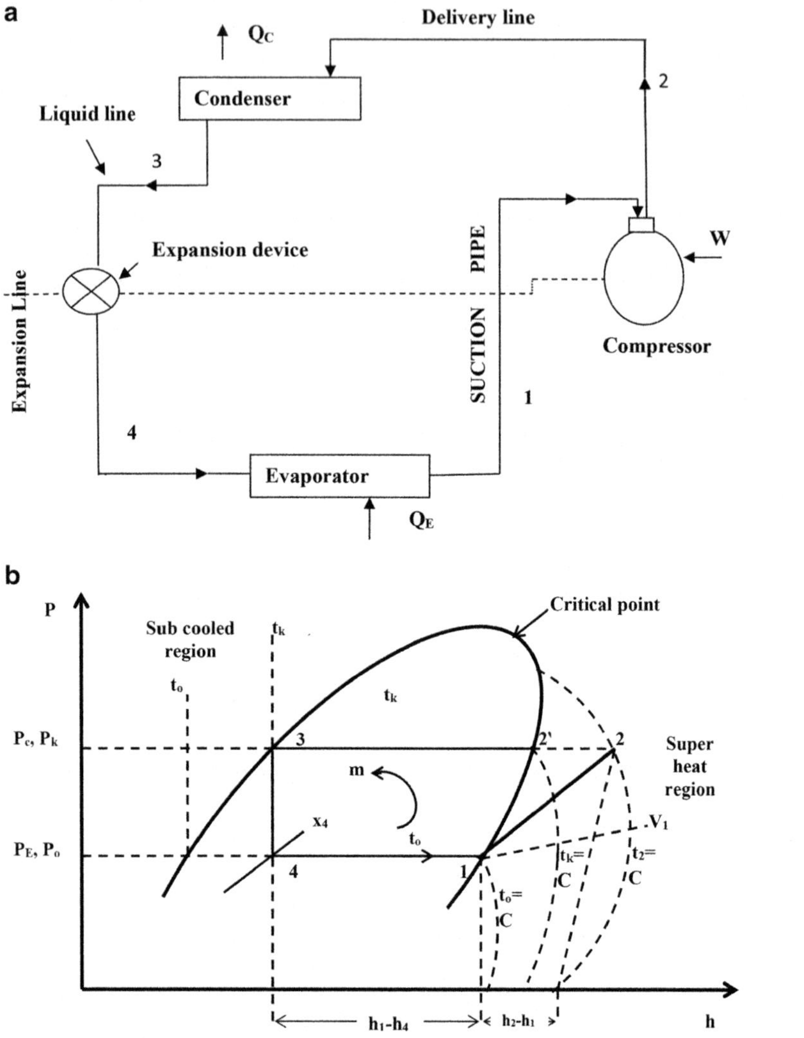

Fig. 5.3 **a** Schematic of vapour compression cooling system. **b** *P–h* diagram of VCR cooling system

where h_3 and h_2 are the specific enthalpies (kJ/kg at the exit and inlet to the condenser, respectively. The condenser pressure is the saturation pressure corresponding to its saturation temperature T_c, $P_c = f\ (T_c)$. Since, the process 3-4 is isenthalpic,

$$h_3 = h_4 \tag{5.3}$$

Since the refrigerant at state point 4 is the mixture of liquid and vapour as can be seen from Fig. 5.3b, the dryness fraction (x) of the refrigerant can be defined as:

$$h_3 = h_4 = h_{f4} + x\left(h_1 - h_{f4}\right) \tag{5.4}$$

where $x = \frac{h_3 - h_{f4}}{h_1 - h_{f4}}$, is the dryness fraction, and h_f is the enthalpy of the fluid at the state point 4.

The process 4-1 is the evaporation, where the refrigerant absorbs heat from the space to be cooled and get evaporated. The heat absorbed or cooling produced during the process 4-1 is also known as refrigeration effect (Arora 2000), and is given by

$$Q_r = (h_1 - h_4) \tag{5.5}$$

where h_1 and h_4 represent the specific enthalpies at the exit and inlet to the evaporator, respectively. Q_r is known as *specific refrigeration effect* or simply *refrigeration effect*, which is equal to the heat transferred at the evaporator per kilogram of refrigerant. The evaporator pressure is the saturation pressure corresponding to the evaporator temperature $P_e = f(T_e)$.

It must be noted that the cycle described above is a simple saturated cycle implying that both the states of liquid after condensation and vapour after evaporator are saturated and lie on the saturated liquid and saturated vapour curves, respectively. The condensation temperature T_c and evaporator temperature T_e corresponding to the saturation pressures P_c and P_e are called as the saturation discharge temperature and the saturation suction temperature, respectively. The performance of VCR systems are defined in terms of Coefficient of Performance (COP) and is given by:

$$\text{COP} = \text{Refrigerating effect/Input compression work} \tag{5.6}$$

$$\text{COP} = \frac{h_1 - h_4}{h_2 - h_1} \tag{5.7}$$

The mass flow rate of refrigerant can be written in terms of volumetric flow rate and specific volume, i.e.,

$$m = \frac{v}{\upsilon} \tag{5.8}$$

Applying this equation to the inlet condition of the compressor, $m = V_1/v_1$, where V_1 is the volumetric flow rate at the compressor and v_1 the specific volume at the inlet of the compressor. At a constant compressor speed, V_1 is an indication of the size of the compressor. The refrigeration capacity in terms of volumetric flow rate is given by (Arora 2000)

$$Q_e = m_r(h_1 - h_4) = V_1\left[\frac{h_1 - h_4}{v_1}\right] \tag{5.9}$$

This term is called the volumetric refrigeration effect (kJ/m^3) of the refrigerant.

The refrigerant delivered from the compressor is in the superheated region. Two independent properties are required to fix the state of the refrigerant at this point. One of these is the already known condenser pressure and the other property needs to be fixed. The type of refrigerant, capacity, evaporator and condenser temperature are known. The evaporator pressure and condenser pressure can be determined from the known values of its corresponding temperatures. The exit enthalpy of condenser and evaporator can be estimated from the known data. The compression process is isentropic, and hence the entropy at exit of the compressor is equal to the entropy at the inlet.

Thus, from the known pressure and entropy the exit state of the compressor could be fixed

$$h_2 = h(P_c, S_2) = h(P_c, S_1)$$
$$S_1 = S_2$$

The quality (x_4) could be obtained from the known values of h_3, h_{fe} and h_{ge}. The required properties such as mass flow rate, the volumetric flow rate COP and cycle efficiency can be easily determined from the known values of enthalpies at different state points.

5.3 Actual Vapour Compression Cycle

The practical vapour compression cooling cycle (Fig. 5.4a and b) differs from the theoretical vapour compression in many ways, some of which are due to the thermodynamic constraints.

The main deviations from the actual cycle are as follows (Arora 2000):

- The superheating of vapour refrigerant in the evaporator 1s-1r.
- Superheating in the suction line 1r-1q.
- Pressure drop in the suction line 1q-1p.
- Pressure drop in the suction valve 1p-1.
- The compression process in an ideal VCR system is assumed to be isentropic. But in real cases, the compression is polytropic with friction and heat transfer to the surroundings 1-2.
- Pressure loss at the discharge valve of compressor 2-2p.
- Pressure loss in the delivery line 2p-2q.
- 2p-2q represents the heat loss and desuperheating of the vapour in the delivery line
- Pressure loss in the condenser 2q-3.
- Liquid refrigerant subcooling 3-3p.
- Heat addition in the liquid line 3p-3q.
- Pressure loss in the evaporator 4-1s.

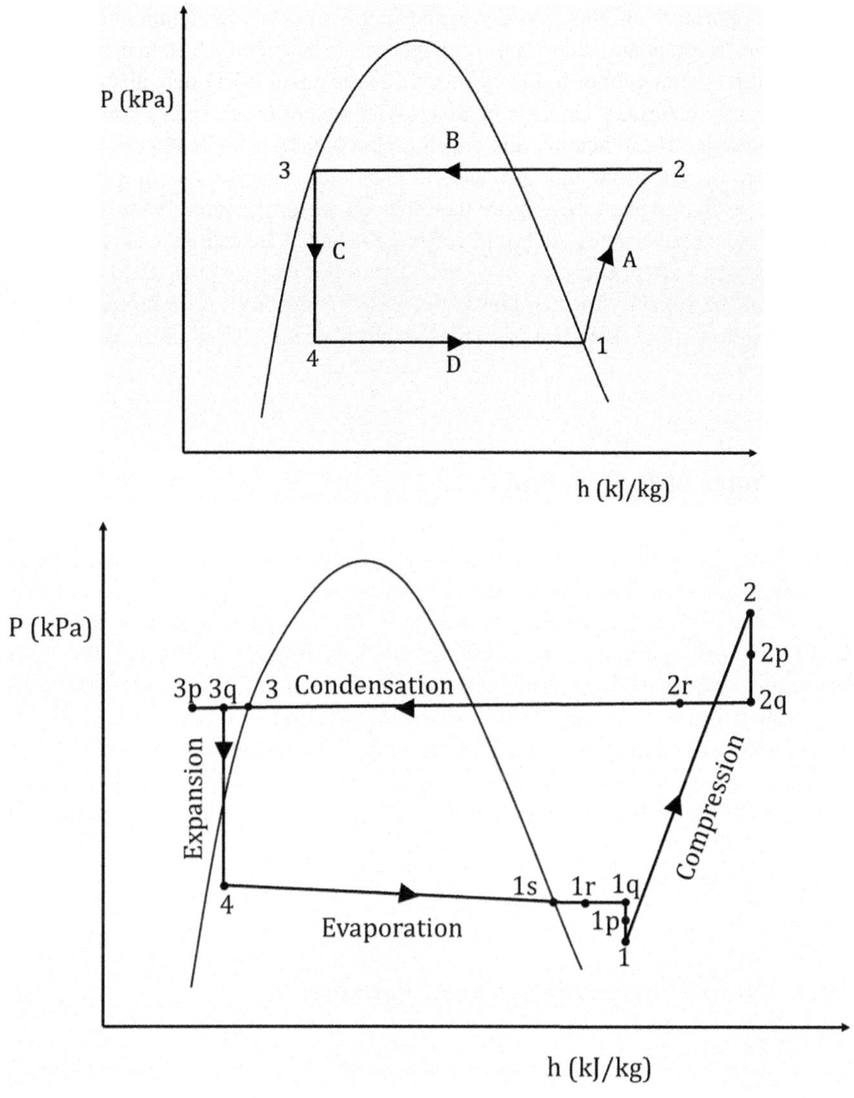

Fig. 5.4 **a** Ideal VCR and **b** actual VCR cycle

The superheating of refrigerant vapour in the evaporator (1s-1r) is due to the automatic control of expansion valve so that the refrigerant leaves the evaporator as a superheated state, picking up a large amount of heat from the evaporator.

The superheating of refrigerant vapour increases the compressor work as well as the refrigerating effect. However, the COP, depending on the nature of refrigerant, may be increased, decreased or remain unchanged. As the compressor cylinder is at a higher temperature than that of the surroundings, some heat may also be lost through its wall and hence, the heat rejected to the condenser may reduce.

The temperature of the cylinder walls is in-between the temperature of the discharge refrigerant and cold vapour refrigerant at the suction. The heat transferred from the refrigerant vapour to the cylinder walls is equal to the heat absorbed from the cylinder wall surface. Like the heating effect at suction, there is a cooling effect at the discharge. These heating and cooling effects take place at the constant pressure. Due to the frictional and resistance in flow, there is a pressure drop, therefore, the actual discharge pressure is more than the condenser pressure. Also, the process 3-3p represents subcooling of liquid refrigerant and it is desirable as it increases the refrigerating effect compressor work. Due to the various pressure drops of the refrigerant, the capacity of the plant is decreased and power consumption per unit of refrigeration effect (kW/TR) is increased and hence, the COP of the actual cycle gets reduced.

5.4 Choice of Refrigerants

A refrigerant is a substance or mixture, usually a fluid. It undergoes phase transitions from liquid to gas and back again. Any fluid having the desired thermodynamic properties can be used as a refrigerant. Nowadays the selection of a suitable refrigerant has become an important aspect in the design of the refrigeration system due to the increasing concern about ozone depletion potentials and global warming. The design factors such as the safety, reliability, initial and operating cost of the system are very much dependent on the type of the refrigerant for a given application. The air is also considered as one of the refrigerants and there are certain applications where it is the only option till now like aircraft cooling. In 1834, ether, ammonia, sulphur dioxide, methyl chloride and CO_2 came into use as the refrigerants in vapour compression refrigeration systems.

5.4.1 Desired Properties of Ideal Refrigerant

The suitability of a certain refrigerant for a certain application is determined by its physical, thermodynamic, chemical properties and by various practical factors. A refrigerant with greater advantages and lesser disadvantages is to be selected for a particular application considering the following points:

- The refrigerant having a low boiling point and low freezing point at atmospheric pressure is desirable.
- Low specific heat and high latent heat. The refrigerating effect per kg of the refrigerant is higher for the refrigerants with high latent heat of the vapourisation. High specific heat decreases the refrigerating effect per kg of the refrigerant.
- The operating pressure required to be maintained in the evaporator and condenser should be low enough to reduce the material cost and must be positive to avoid

leakage of air into the system. Selection of the compressor also depends upon the choice of the refrigerants. For example, reciprocating compressors are suitable for the refrigerants having low specific volumes and high-pressure differentials while the centrifugal compressors are preferred for the refrigerants operating under the low evaporator and condenser pressures and low-pressure differentials. The high-pressure difference results in the high compression ratio and power consumption and gives low volumetric efficiency.

- To avoid larger power requirement, high critical pressure and temperature is recommended for the refrigerants.
- The critical temperature of the refrigerant should be higher than the condensing temperature.
- Low specific volume of the refrigerant is desirable to reduce the size of the compressor except for the centrifugal compressor.
- Thermal conductivity should be higher.
- Low viscosity of the refrigerant in both the liquid/vapour phases is desirable for low pumping requirements.
- An ideal refrigerant should be chemically stable at temperature encountered in the system and so the refrigerant must be inert with all materials used in the refrigeration system.
- An ideal refrigerant should be inexpensive and easily available.

Some other practically suitable features of the refrigerants are:

(1) Nonflammable and non-toxic.
(2) The refrigerants should not produce any bad effects on the food when it leaks in the system.
(3) Miscibility with the lubricating oil should be high.
(4) From an economic point of view, the refrigerant should give high COP in the working temperature range. This is necessary to reduce the running cost of the system.
(5) The ideal refrigerant should be cheaply available.

Thus, we see that there is no substance which could be called an ideal refrigerant. The most important property of a refrigerant is its normal boiling point and other thermodynamic properties seem to be dependent on it. It is, therefore, desirable to know the significance of normal boiling point through pressure–temperature relation. For a pure substance, we have the Clausius-Clapeyron equation given by (Arora 2000)

$$\frac{dp}{dT} = \frac{h_{fg}}{T\left(v_g - v_f\right)} \tag{5.10}$$

The specific volume of liquid is very small as compared to that of the vapour, i.e., $v_f \ll v_g$.

And the specific volume of the vapour is given by the ideal gas equation of state as $v_g = RT/P$, where R is the ideal gas constant.

Hence, we have

$$\frac{dP}{dT} = -\frac{Ph_{fg}}{R}\frac{d}{dT}\left(\frac{1}{T}\right) \tag{5.11}$$

$$\frac{d(\ln P)}{d(1/T)} = -\frac{h_{fg}}{R} = +b \tag{5.12}$$

where

$$b = -\frac{h_{fg}}{R} \tag{5.13}$$

Integrating, we get

$$\ln P = a + \frac{b}{T} \tag{5.14}$$

where b represents the slope of ln P – 1/T line and maybe assumed constant for a small range of operating temperatures where the latent heat of vapourisation is constant. Also, for pure substances belonging to the same family, the dimensionless Trouton's number is constant, i.e.,

$$\frac{M(h_{fg})}{T_s} = k, \text{ a constant} \tag{5.15}$$

where M is the molecular weight and T_s the normal boiling point. Hence,

$$b = -\frac{h_{fg}}{R} = -\frac{M(h_{fg})}{R_o} = f_1(T_s)$$
$$b = b/T_s = f(T_s) \tag{5.16}$$

Thus, both the constants a and b are the functions of the normal boiling points of the refrigerant (Arora 2000).

5.4.2 *Classification of Refrigerants*

Refrigerants are generally classified into two types: (i) primary and (ii) secondary refrigerants. The ordinary refrigerants used in vapour compression and vapour absorption refrigeration systems provide cooling by undergoing phase change at the evaporator. As the name implies, secondary refrigerants are those liquids, which are used for transporting thermal energy from one location to other.

The refrigerants are broadly classified into the following four groups:

(1) Halocarbon refrigerants
(2) Hydrocarbon refrigerants

(3) Inorganic refrigerants
(4) Mixed refrigerants

(1) *Halocarbon Refrigerants*

According to ASHRAE, 42 halocarbons can be used as the refrigerants. The commonly used halocarbons as the refrigerants are given in Table 5.1. The halocarbon refrigerants are derivatives of methane or ethane. Most of the halocarbon refrigerants are now available under other trade names such as Genetron, Isotron etc. The first of the halocarbon R12 was developed in 1934 by Thomas Midgley.

(2) *Hydrocarbon Refrigerants*

The hydrocarbon refrigerants are shown in Table 5.2. Most of them are tested in industry and commercial applications. These refrigerants are highly flammable and explosive.

(3) *Inorganic Refrigerants*

The inorganic refrigerants are still in use due to their inherent physical and thermodynamic properties. The most commonly used inorganic refrigerants are as shown in Table 5.3.

Among these, ammonia is the oldest and most widely used of all these refrigerants. Its application is largely found in commercial and industrial applications with the reciprocating compressors, where safety is secondary. Its boiling point at the atmospheric pressure is −33 °C and its melting point is −78 °C. Dry air is used as the refrigerant in some aircraft airconditioning systems.

Table 5.1 Halocarbon refrigerants

Refrigerant number	Chemical name	Chemical formula
R11	Trichloromonofluoromethane	CCl_3F
R12	Dichlorodifluoromethane	CCl_2F_2
R13	Monochlorotrifluoromethane	$CClF_3$
R14	Carbontetrafluride	CF_4
R21	Dichloromonofluoromethane	$CHCl_2F$
R22	Monochlorodifluoromethane	$CHClF_2$
R30	Methylene chloride	CH_3Cl_2
R40	Methyl chloride	CH_3Cl
R100	Ethyl chloride	C_2H_5Cl
R113	Trichlorotrifluoroethane	CCl_2FCClF_2
R114	Dichlorotetrafluoroethane	$CClF_2CClF_2$

Table 5.2 Hydrocarbons

Refrigerant number	Chemical name	Chemical formula
R170	Ethane Propane	C_2H_6
R290	Butane Isobutane	C_3H_3
R600	Trichloroethylene	C_4H_{10}
R600a	Dichloroethylene	C_4H_{10}
R1120	Ethylene	$C_2H_4Cl_3$
R1130	Propylene	$C_2H_4Cl_2$
R1150		C_2H_4
R1270		C_3H_6

Table 5.3 Inorganic refrigerants

Designation of the refrigerant	Name of the refrigerant	Formula
R717	Ammonia	NH_3
R729	Air	–
R744	Carbon dioxide	CO_2
R764	Sulphur dioxide	SO_2
R718	Water	H_2O

The CO_2 refrigerant is non-toxic, non-irritating and non-flammable. It is used in some industrial applications and aboard ships.

The boiling point of sulphur dioxide is −10 °C at the atmospheric pressure. Water is the most commonly used refrigerant in vapour absorption systems.

Properties of pure refrigerants are shown in Table 5.4.

(4) *Mixed Refrigerants*

Mixing of two or more refrigerants in proper proportion will provide the properties, which are most desirable for refrigeration and airconditioning applications. The three types of mixtures used in refrigeration and airconditioning applications are (Mohanraj et al. 2009): (i) azeotropes, (ii) near azeotropes and (iii) zeotropes. Azeotrope means a stable mixture of refrigerants. Azeotrope mixture retains their vapour and liquid phases identical over a wide range of temperature. They cannot be separated into individual constituents by distillation.

ASHRAE designations for near azeotropes are in 400 series. They have properties nearly the same as that of the azeotropic mixtures. The objective of the near azeotropic mixtures is to extend the range of refrigerant alternatives beyond single compounds. The zeotropic mixture means non-azeotropic mixtures, which are blends of two or more refrigerants that deviate from azeotropic properties. The phase change process of the azeotropic mixture is isothermal whereas in non-azeotropic it is non-isothermal. ASHRAE has classified all zeotropic mixtures into a 400 series of the refrigerant number (Mohanraj et al. 2011).

Table 5.4 Properties of pure refrigerants (Calm and Hourahan 2001)

Refrigerant	Molecular weight	Temperature (°C)	Pressure (MPa)	Boiling point (°C)	ASHRAE Code	ODP (R11 = 1)	GWP 100 year
R11	137.37	198.0	4.41	23.7	A1	1	4600
R12	120.9	112.0	4.14	–29.8	A1	0.82	10,600
R22	86.47	96.2	4.99	–41.4	A1	0.034	1700
R23	70.01	25.9	4.84	–82.1	A1	0	12,000
R32	52.02	78.2	5.8	–51.7	A1	0	550
R41	34.03	44.1	5.90	–78.1	A1	0	97
R123	152.93	82.0	3.66	27.8	B1	0.012	120
R124	136.48	122.3	3.62	–12	A1	0.026	620
R125	120.02	66.2	3.63	–54.6	A1	0	3400
R134a	102.03	101.1	4.06	–26.1	A1	0	1300
R142b	100.49	137.2	4.12	–9	A1	0.043	2400
R143a	84.04	72.9	3.78	–47.2	A2	0	4300
R152a	66.05	113.3	4.52	–24	A2	0	120
R161	48.06	102.2	4.70	–34.8	A1	0	12
R170	30.07	90	4.87	–88.9	A3	0	20
RE170	46.07	128.8	5.32	–24.8	A3	0	1
R218	188.02	71.9	2.68	–36.6	A1	0	8600
R227ea	170.03	102.8	2.98	–15.6	A1	0	3500
R245fa	134.05	154.1	4.43	15.1	B1	0	950
R290	44.1	96.7	4.25	–42.2	A3	0	20
R600	58.12	152.0	3.80	–0.5	A3	0	20
R600a	58.12	134.7	3.64	–11.7	A3	0	20
R717	17.03	132.3	11.34	–33.3	B2	0	<1
R744	44.01	31.1	7.38	–78.4	A1	0	1

The alternatives to R22 reported are HFC-based mixtures such as R404A, R407C and R410A for refrigeration, airconditioning and heat pump applications. The problems with HFCs mixtures is their immiscibility with mineral oil, and lubricants, therefore HFC mixtures require synthetic lubricant like polyester. Hence, a major modification is required for HFC mixtures to retrofit in HCFC systems. Properties of commonly used HFC mixtures are compared in Table 5.5. As discussed earlier (Sect. 5.4.1), all refrigerants are not suitable for all type of applications. Apart from thermodynamic and chemical properties, environmental factors such as ODP, GWP and TEWI are to be considered before making choice, as below:

(a) **Ozone Depletion Potential (ODP):** The ozone depletion potential of an ideal refrigerant should be zero (according to the Montreal Protocol). The refrigerants such as R11, R12 having non-zero ODP have been already phased out. The

Table 5.5 Properties of HC mixtures as alternatives (Mohanraj et al. 2011)

Refrigerant designation	Composition	By mass	Replaces	Molecular weight	Critical temperature	Critical pressure (MPa)	Boiling point	ODP	GWP
R432A	R1270/RE170	80:20	R22	42.87	99.68	4.8	–43.12	0	20
R433A	R290/R1270	30:70	R22	42.68	93.69	4.54	–46.05	0	20
NRM	R290/R170	94:06	R22	43.25	96.29	4.28	–45	0	20
NRM	R290/R600a	45.2:54.8	R134a	51.78	117.52	3.91	–25.48	0	20
NRM	R290/R600	60:40	R134a	49.7	118.7	4.07	–25.2	0	20
NRM	R290/R600/R600a	60:20:20	R134a	49.7	115.36	4.038	–27.76	0	20

presence of chlorine or bromine in the molecules increases its ODP value, hence the refrigerants having either chlorine or bromine cannot be used under new regulations.

(b) **Global Warming Potential (GWP):** The global warming potential of the refrigerant should be preferably zero. The refrigerants such as R134a having zero ODP but with high GWP is likely to be regulated in future.

(c) **Total Equivalent Warming Index (TEWI):** The direct and indirect contributions of a refrigerant to global warming are expressed in terms of the TEWI. The direct contribution indicates the release into the atmosphere.

5.5 Effect of Operating Conditions on COP of VCR

It is of interest to study the effect of operating conditions like evaporator pressure, condenser pressure, subcooling of the liquid and superheating of the suction vapour in power consumption per unit tonne of refrigeration and cooling COP. It can be seen from the *P–h* diagram that the decrease in evaporator temperature (or evaporator pressure) results in decreasing the refrigerating effect and increase in the compression work and hence the decrease in COP of the system. Further, the effect of subcooling of the liquid refrigerant is to reduce the flashing of liquid during expansion and increasing the refrigerating effect as well as COP. On the other hand, superheating of the suction vapour is desirable in practice in real systems to ensure the complete vapourisation of the liquid refrigerant in the evaporator before it enters the compressor.

The effect of superheating of the suction vapour is to increase the refrigerating effect and increase in specific work. The net effect of superheating depends upon the refrigerant used and the operating conditions. In general, however, the effect of slight superheat on the volumetric efficiency of the compressor and COP is desirable on account of vapourisation of liquid refrigerant droplets in the suction vapour. A liquid vapour heat exchanger can also be used as shown in Fig. 5.5, to achieve both subcooling of the liquid refrigerant and superheating of the suction vapour.

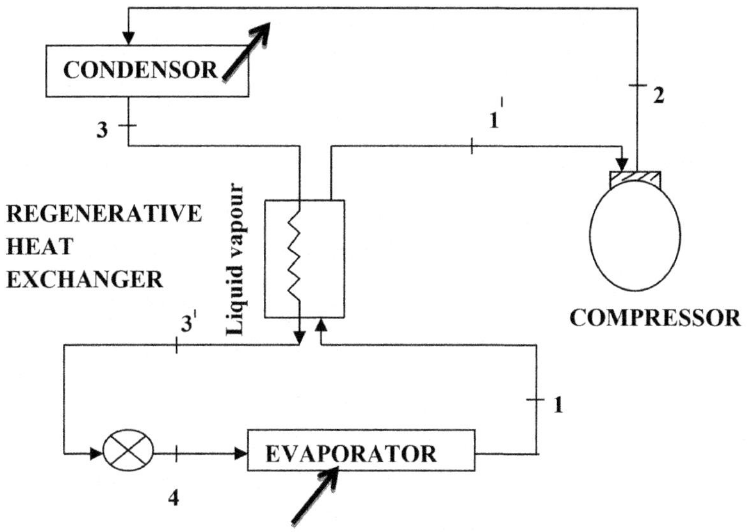

Fig. 5.5 Vapour compression cooling system with a regenerative heat exchanger

The purpose of using regenerative heat exchanger is to improve the performance of vapour compression cooling system (Table 5.6).

Table 5.6 The refrigerants and their replacements

Refrigeration applications	Descriptions	Refrigerants used
Domestic refrigeration	Preservation of foods	R134a, R600, R22
Commercial	Preservation of frozen and fresh foods	R134a, R404A, R507
Transport refrigeration	For preservation of foods during transport as in railways, air and sea	R134a, R407C, R410A
Cold storage	To preserve, store and process of food items from the source to the distribution point	R134a, R123, R407C, R410A, R507
Electronic cooling and thermal management	For cooling of large computers and processers	R134a, R404A, R507
Medical applications	For preservation of medicines and equipment	R134a, R404A, R507
Industrial refrigeration	For food storage, chemical industries, large capacity systems	R134a, R717, R507, R123
Cryogenic refrigeration		Helium, ethylene, propane, nitrogen etc.

Table 5.7 COP and HP/TR values for some refrigerants (Arora 2000), ($T_c = 30$ °C and $T_e = -15$ °C)

Refrigerant	NBP (°C)	COP	HP/TR
Carnot value	–	5.73	0.82
R11	23.7	5.09	0.93
R113	45.9	4.92	0.96
R717(NH_3)	–33.4	4.76	0.99
R12	–29.8	4.70	1.00
R22	–40.8	4.66	1.01
R114	3.6	4.49	1.05
R718(H_2O)	100.0	4.10	1.13
R744(CO_2)	–78.52	2.57	1.86

Table 5.7 shows the theoretical COP and horsepower per tonne for some of the refrigerants arranged in the order of decreasing COP. The Carnot values of COP for the given $T_c = 30$ °C and $T_e = -15$ °C is also shown in Table 5.7 (Carnot value = 5.73). It is seen from Table 5.7 that R11 has a COP value closer to the Carnot value and the other refrigerants having lower values and the CO_2 has the lowest value of 2.57 due to its low critical temperature. In terms of HP per TR, all refrigerants have almost the same range and thus there is little choice among the R12, R22 and NH_3. However, Freon has high density coupled with the low boiling point and high molecular weight. For these refrigerants, the latent heat is not so large as compared to the NH_3. On the other hand, the ratio of specific heats is lower for Freons as compared to NH_3 and hence, the isentropic compression work is also smaller thereby leading to higher COP, similar to Carnot one (Table 5.7).

It must be mentioned that a VCR system is by far the most efficient but there are some problems like failure of the compressors, wear and tear and high input work requirements.

Further, improvements in the performance of VCR systems are desirable to meet the different requirements with reduced plant size and can be achieved either by decreasing the compression work and/or by increasing the refrigeration effect or both. The compression work can be reduced by operating close to the vapour saturation line and higher refrigerating effect can be achieved by expanding the liquid refrigerant close to the saturated liquid line. The staging of compression can also be used to reduce the compression work and multi- evaporator systems can be used to provide different cooling temperatures.

The cascading of VCR system can also be used to produce very low refrigeration temperature wherein, the higher temperature cascade produces the refrigeration at a certain low temperature at which the heat is removed from the lower temperature cascade and cooling is produced even at a still further lower temperature.

5.6 Cascaded VCR System

A cascaded system was first introduced by Pietet in 1877 for the liquefaction of oxygen using SO_2 and carbon dioxide as intermediate refrigerants. A sketch of the cascaded VCR system is shown in Fig. 5.6, the coupling temperature is given by the fixed conditions, as below:

$$T_{\text{coupling}} = \sqrt{T_{k1} \times T_{o2}} = \sqrt{T_{c1} \times T_{E2}} \quad (5.17)$$

where T_{k1} or T_{c1} and T_{o2} and T_{E2} are the sink and the refrigeration temperatures for the first and second stages, respectively. The cascaded refrigeration cycle consists of two simple VCR (top and bottom) cycles in series, in a way that the condenser of the bottom cycle is the evaporator of the top. The refrigerants used are progressively having different lower boiling points.

A two-stage cascaded system using two refrigerants is shown in Fig. 5.6 where the lower cycle cools the refrigerated space and the upper cycle cools the lower cycle using a heat exchanger, as can be seen in Fig. 5.6. The evaporator of the upper part of a cycle will be the condenser of the lower cycle. This heat exchanger acts as an evaporator for the upper cycle and as a condenser for the lower cycle. The energy released by the refrigerant in the lower cycle as it condenses in the heat exchanger is absorbed by the refrigerant in the upper cycle, causing it to evaporate. The cascade cycle has a special advantage, that is, it takes the advantage of its compatibility

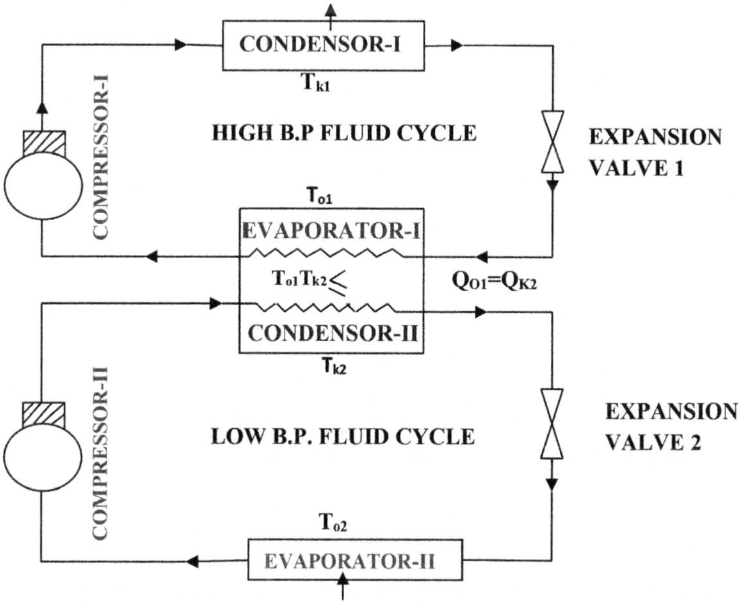

Fig. 5.6 Cascaded cooling system

with different refrigerants in two cycles. This advantage helps us to ensure that the pressure in the evaporator is above the atmospheric pressure. The high-temperature cascaded system utilises refrigerant with the high boiling temperature such as R12 or R22. The low temperature cascaded system utilises a refrigerant with the low boiling temperature such as R13 or R13B1.

The cascade refrigeration system has better perfromance when it comes to production of much lower temperature. For a given cooling/refrigeration load, Q_r, if the mass flow rate of refrigerant in the lower (bottom) cycle is m_1 and m_2 is the mass flow rate of the refrigerant in the upper (top) cycle, then the mass of the refrigerant flowing through the low temperature cascaded system can be calculated as below:

$$m_1 = \frac{Q_r}{h_2 - h_1}\text{kg/min} \tag{5.18}$$

where m_1 is the mass flow rate (kg/s) of the refrigerant in the bottom cycle and h_1 and h_2 are the enthalpies at the exit and entry of the evaporator, respectively. The mass of refrigerant m_2 required in the high-temperature cascaded system in order to liquefy the refrigerant to the low temperature of low temperature cascaded system in the cascaded condenser can be obtained from heat balance of both systems as below:

$$\begin{aligned} m_2(h_5 - h_6) &= m_1(h_3 - h_4) \\ m_2 &= \frac{m_1(h_3 - h_4)}{h_5 - h_6} = \frac{m_1(h_3 - h_1)}{h_5 - h_6} \\ m_4 &= m_1 \end{aligned} \tag{5.19}$$

The total compression work is given by

$$W = m_1(h_3 - h_2) + m_2(h_7 - h_6) \tag{5.20}$$

$$RE = Q_r \tag{5.21}$$

$$\text{COP} = \frac{Q_r}{m_1(h_3 - h_2) + m_2(h_7 - h_6)} \tag{5.22}$$

5.7 Solar Operation of VCR Systems

The VCR systems are commonly used in a number of commercial and industrial applications due to their high cooling capacity and coefficient of performance at low temperature. But to run these systems, high-grade energy such as electrical or mechanical energy is required. However, in solar operation of a VCR cooling system electricity produced either using solar photovoltaic (SPVs) or using a solar heat engine is used. In VCR systems where a refrigerant is circulated to achieve cooling, a substantial amount of high grade energy is required to run the compressor.

Further, the traditional VCR machines are leading electricity consumers, accounting for approximately 16% of the total electricity consumption by building sector 2020. Though it has some promising features such as high COP, wide operating temperature range, and reasonable control over the operation, engineers and scientists are obligated to minimize its use. On the other hand, the requirements for high grade energy are met by fossil fuels such as coal, and is responsible for number of problems including global warming, emission of hazardous gases and hence, not favourable in today's scenario. Therefore, the solar operated systems gains a considerable attention now a days.

There are two types of solar VCR systems, namely, indirect solar VCR and direct solar VCR cooling systems. In an indirect VCR cooling systems, called as thermo-mechanical systems, solar energy is used to energize the boiler of the Rankine cycle to generate mechanical energy, which can then be used to power the compressor of the VCR system. In direct solar VCR systems, termed as PV-assisted VCR systems, where photovoltaic cells convert solar energy into electricity by means of photoelectric effect is used to drive the compressor of VCR cooling systems. PV collectors can only convert less than one-third of solar radiation into electricity, but solar thermal collectors used in thermo-mechanical VCR systems can absorb majority of the solar energy spectrum. Since it needs an additional Rankine/Stirling cycle to tap mechanical energy from solar technologies, thermo-mechanical VCRs need more space and are expensive than the former. Here, we describe both the conventional solar operation options for VCR systems based on solar thermoelectric and photoelectric conversion processes.

5.7.1 Indirect Solar Operated VCR Systems

Solar thermal electric conversion (STEC) system provides the shaft power which can be used to run the compressor of a VCR cooling system and/or to operate a generator producing electricity to run a VCR system Thus, solar thermal electric conversion (STEC) is the process of production of shaft work and electricity from solar (thermal) energy. A typical STEC system consists of the following subsystems like Solar energy collection subsystem, energy transport subsystem, energy conversion subsystem (heat engine) and energy storage subsystem (thermal or electrical). In each subsystem, there are several options available such as flat plate solar collector, focusing collectors and solar ponds etc. for the solar collector subsystems, (ii) thermal, optical or electrical transport in the energy transport subsystem and (iii) Rankine, Stirling or Brayton cycle, etc. for heat to mechanical energy conversion subsystems.

Thus, solar-thermal conversion efficiency can be expressed as follows:

$$\eta_{system} = \eta_{collection} \times \eta_{transport} \times \eta_{engine} \times \eta_{storate} \tag{5.23}$$

where both the collector and engine efficiencies are dependent upon the operating temperatures and availability of solar radiation.

The collection efficiency of all solar collection systems drops from its maximum efficiency (attainable at the ambient temperature called as the optical efficiency) to lower value as the collection temperature increases. The Carnot principle limits the maximum possible efficiency of the heat engine according to the relation:

$$\eta_{\text{Carnot}} = \frac{T_{\text{L}}}{T_{\text{H}}} \tag{5.24}$$

where T_H is the maximum (source) temperature and T_L the minimum (sink) temperature. Hence, there is an optimum point for each collection subsystem and power conversion subsystem combination as shown in Fig. 5.7.

In the actual case, the power conversion efficiency is a fraction of the Carnot efficiency. The collector efficiency drops most rapidly with increasing temperature for flat plate collectors followed by the moderate concentrating system and evacuated tube collectors, and for the high concentrating collectors. The basic heat engine cycle shown in Fig. 5.8 consists of three subsystems, viz. solar thermal collector, heat

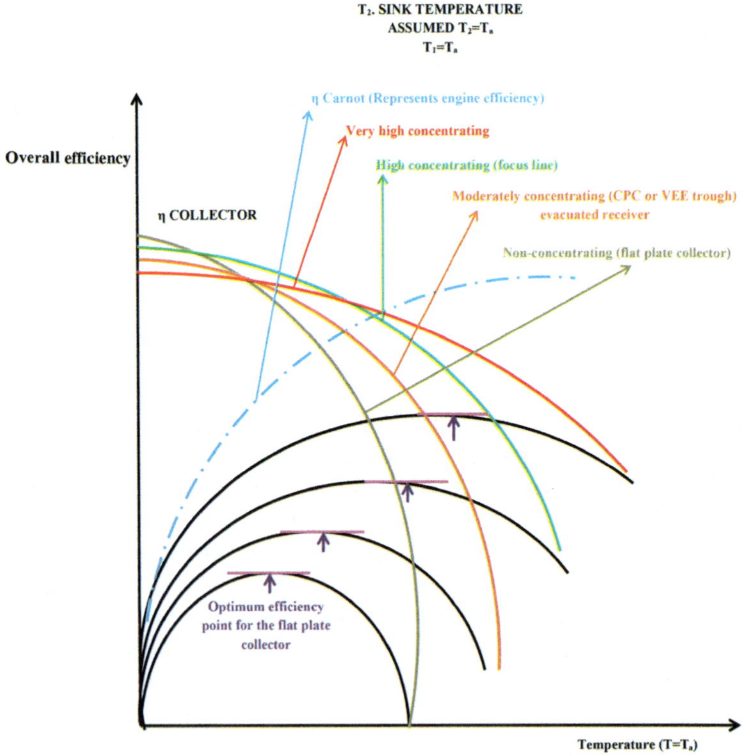

Fig. 5.7 Trends of Efficiency curves for collectors and power conversion subsystems

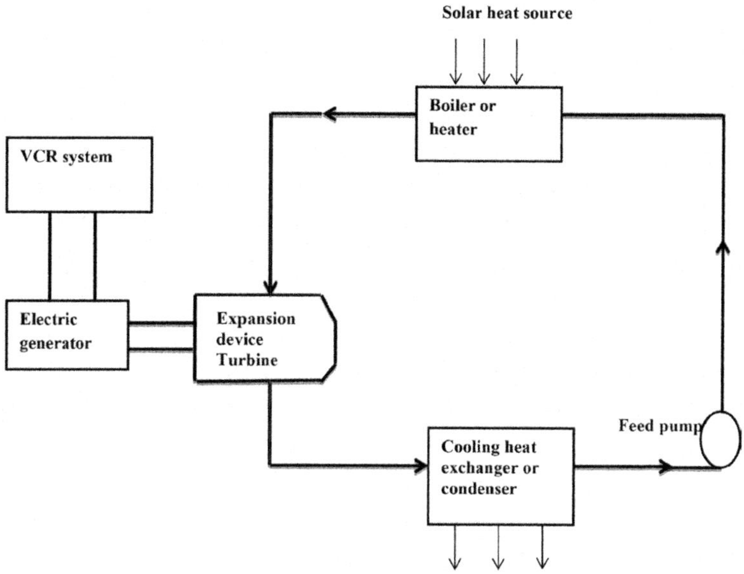

Fig. 5.8 Schematic of solar heat engine operated VCR system

engine and the VCR cooling cycle. The solar collector converts solar energy into heat and transfers to the boiler cum heat exchanger, where the working fluid (water) is being heated and converted to steam to run the turbine, and the exhaust is then cooled down to saturated in a condenser of the heat engine. The turbine produces the shaft work which can be used (directly or indirectly) to drive the compressor of the VCR cooling system in a similar way as discussed in Sect. 5.3, thereby completing the entire operation of the system. The engine efficiency is given by

$$\eta_{engine} = \eta_{cycle} \times \eta_{mechanical}$$

where η_{cycle} is the ideal work of the engine and $\eta_{mechanical}$ is the net shaft work of the actual turbine or the reciprocating engine divided by the maximum work obtainable by means of an ideal engine. The first term depends upon the type of the cycle itself (such as Rankine, Stirling or Brayton), operating temperature and regeneration processes. The second term relates to the mechanical and fluid friction losses and leakages and heat losses from the engine itself. Normally, steam is used as the working fluid for solar thermal power generation and the cycle followed is known as Rankine cycle.

In the Rankine cycle, superheated vapour or steam obtained from solar collector-cum-boiler is expanded into the turbine and the exhaust low pressure steam is condensed as liquid in the cooling heat exchanger known as condenser because of the phase change from steam to water phase change. In the gas turbine engine, a fuel–air mixture is used as a working fluid in open cycle systems where air from the environment is compressed, heated, expanded and expelled to the atmosphere thereby eliminating the need of a cooling tower. Since the working fluid is gaseous throughout the cycle, it is termed as gas power cycle. However, in solar-powered systems, Rankine engine is commonly used due to its higher efficiency than a Brayton or Stirling cycle for low-temperature differences. Another option is the Stirling cycle heat engine as proposed for the solar cooling options requiring high-temperature. The primary difference in various solar engines is the type of expander used.

However, for common temperature below 300 °C only the Rankine cycle is used for solar power generation. The engine transport efficiency is mainly dependent on the properties of heat transport fluid, flow rate, piping grid configuration, pump efficiency and the thickness as well as the thermal conductivity of the insulating material. It is usually found to be better than 90% while the electrical transport efficiency is usually higher than 95%. On the other hand, the energy storage efficiency consists of three terms as:

$$\eta_{\text{storage}} = \eta_{\text{charging}} \times \eta_{\text{discharge}} \times \eta_{\text{heat loss}}$$

where η_{charging} and $\eta_{\text{discharge}}$ are the efficiency of storage charging and discharge processes and $\eta_{\text{heat loss}}$ is due to the thermal losses and is a function of the storage period. Generally, η_{storage} is of the order of 80%. The electrical storage subsystems have the limitations and thus only some fraction of the storage capabilities of thermal and electrical storage subsystems could be utilised.

Theoretical Basis and Optimum Conditions of STEC

For any solar heat engine, the maximum efficiency is the Carnot cycle

$$\eta_C = 1 - \frac{T_L}{T_c} \tag{5.25}$$

where T_L and T_c are respectively the ambient and the collector thermodynamic temperature.

The maximum power output (W) can be obtained as

$$W = \eta_C \times Q_U \tag{5.26}$$

$$Q_u = I - \text{Thermal losses} \tag{5.27}$$

$$I, \text{Irradiance} = I_s - U_L(T_c - T_L) \tag{5.28}$$

U_{L}, is the overall heat loss. The power output per unit area is given by

$$W = \left(1 - \frac{T_L}{T_c}\right)(I_s - U_L(T_c - T_L)) \tag{5.29}$$

i.e., there is some optimum T_c at which W is maximum,

So,

$$\frac{\partial W}{\partial T_c} = 0$$

$$T_{\text{copt}} = \sqrt{T_L(I_s + U_L T_L)/U_L} \tag{5.30}$$

or

$$\frac{T_{\text{opt}}}{T_L} = \sqrt{\frac{I_s}{U_L T_L} + 1} \tag{5.31}$$

If we define the dimensionless variable $M = \frac{U_L T_L}{I_s}$

$$\frac{T_{\text{opt}}}{T_L} = \sqrt{\frac{(1+M)}{M}} \tag{5.32}$$

and the maximum power output (W_{max}) is given by

$$\frac{W_{\text{max}}}{I_s} = \left[1 - \frac{T_L}{T_{\text{opt}}}\right]\left[1 - \left(\frac{U_L}{I_s}\right)(T_{\text{opt}} - T_L)\right]$$

$$= \left(1 - \sqrt{\frac{(1+M)}{M}}\right)[1 + M - \sqrt{M(M+1)}] \tag{5.33}$$

It can be seen that as M tends to 0, $\frac{W_{\text{max}}}{I_s}$ tends to 1. At $M = 0$, $\frac{W_{\text{max}}}{I_s} = 1$, i.e., the maximum power approaches the incident power, which is an unattainable condition because $M = 0$ implies that $U_L = 0$ and the temperature of the collector is infinity. Whereas it cannot be the case, because the temperature of the Sun is 6000 K. Thus, the analysis and result are not valid at very high collector temperatures, where the losses would be due to the radiation and cannot be expressed as a linear term. However, in the vicinity of $M = 0$, the trend is that as M increases $\frac{W_{\text{max}}}{I_s}$ decreases (which may be due to either increase of U or decrease of I), which is evident. Similarly, decreasing M either by using good insulating material to decrease U_L or by concentrating the solar energy to increase I, results in an increase in $\frac{W_{\text{max}}}{I_s}$, the maximum usable fraction of the incident solar energy. So, for a flat plate collector, it is not possible to imagine condition such as M is less than unity. At $M = 1$, the maximum usable fraction is only 17% (Fig. 5.9).

Rankine Cycle Solar Cooling System

A conventional Rankine cycle cooling system consists of two subsystems as shown in Fig. 5.10. The first is the Rankine cycle heat engine, which produces mechanical energy. The working fluid used to drive a turbine gains its heat of vapourisation from input solar energy. The mechanical energy is produced by expanding the fluid in the

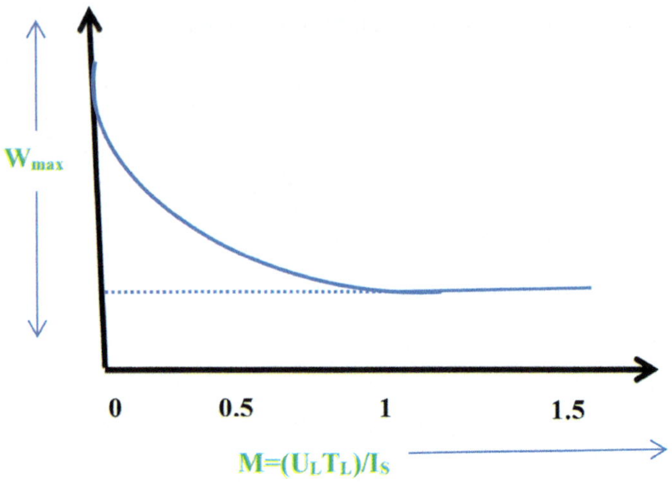

Fig. 5.9 Variation of maximum power with *M*

turbine. The working fluid is condensed and pumped back to the boiler to complete one cycle. This mechanical energy output is used to drive a conventional VCR system (the second subsystem). The subsystem consists of a compressor, condenser, expansion valve and an evaporator. The compressed hot vapour being condensed is throttled to produce its cooling effect in the evaporator at low-pressure. The evaporated refrigerant is recompressed and the cycle continues. Thus, in a Rankine heat engine cycle, working fluid is heated by the addition of heat in the boiler to evaporate it at its higher pressure. The refrigerant vapour expanded in the turbine to produce shaft power to drive the compressor of VCR loops. The unused vapour at lower pressure is then condensed in the condenser and the liquid is again pumped to the boiler by the feed pump (Fig. 5.10).

The thermodynamic efficiency of the Rankine engine is defined as the ratio of net output work to the heat input from the source

$$\eta_{\mathrm{TC}} = \frac{W_T - W_p}{Q_S} \tag{5.34}$$

where W_T is the turbine work output, W_p the pump work and Q_S the heat supplied.

The cycle efficiency is given by

$$\eta_{\mathrm{engine}} = \text{Useful shaft output power/Heat input} \tag{5.35}$$

It must be mentioned here that the fraction of Carnot value achieved by a practical system is governed by the selection of a working fluid. Ideally, all Rankine cycle heat addition should take place at the constant temperature preferably the highest temperature reached by the solar collector. In practice, this is not possible because of the thermodynamic properties of real fluids and the required temperature drops

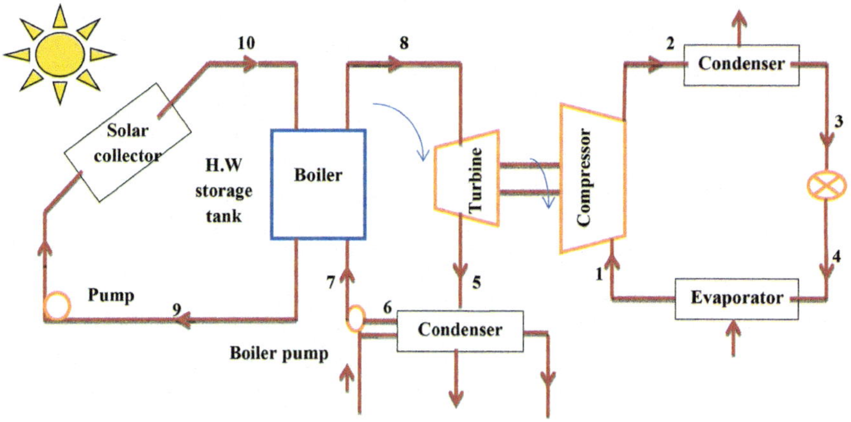

Fig. 5.10 Schematic of Rankine cycle cooling system

across the vapour generator between the solar collector fluid and the Rankine cycle fluid. It is also possible to vapourise the fluid in the collector itself, thus eliminating the need for a separate vapour generator. However, this scheme may introduce the problems of control instability, possible fluid over temperature and decomposition, pressurize collectors and high duct pressure losses associated with the vapour flowing from the collector to the expander.

There are some desirable thermodynamic properties of working fluids used for Rankine cycle/turbine subsystems:

- The refrigerant vapour should possess high latent heat of vapourisation. The critical temperature and density should also be high for the refrigerant vapour.
- The fluid should be chemically stable at all temperatures to which the fluid is exposed in contact with all metals.
- High molecular weight.
- Highest boiling point temperature in the heat input portion of the cycle should be at least 30 °C below the critical temperature.
- Low vapour pressure and higher revolution per minute (RPM) of the turbo compressor are desirable.
- A most important factor in the choice of fluid is the slope of the vapour line on $T - S$ diagram. It should be positive, i.e., $dT/dS > 0$

  This would enable the perfect isentropic expansion through the turbine.

- The heat transfer coefficient and thermal conductivity should be as high as possible. The low viscosity is also a desirable characteristic feature of the working fluid.
- Low toxicity, low flammability and the low cost are additional features of the working fluids. Several freons, e.g., R113, R114 etc. have been used for the power cycle fluid while the freons like R11, R12, R22 and R21 have been used for the refrigeration cycle.

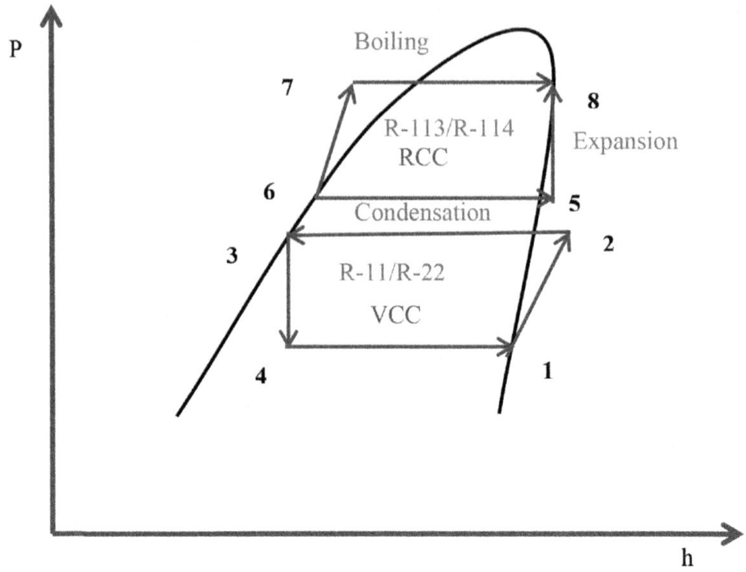

Fig. 5.11 *P–h* diagram of solar-assisted Rankine cycle

It is desirable to predict the thermal performance of the system and evaluating the sizing of the system components. A representation of the solar Rankine cycle air conditioning system is shown in *P–h* diagram (Fig. 5.11). It can be seen that the cycle 1–2–3–4 represents the refrigeration cycle and 5–6–7–8 represents the heat engine cycle and thus both compression and expansion process leads to the super-heated vapour.

Solar Driven Combined Power and Cooling System

An integrated system used for combined power and cooling is shown in Fig. 5.12. The system consists of two subsystems, namely, Rankine power cycle and vapour compression cooling cycle. Two different working fluids having low GWP can be used to drive two cycles, for example, R113 for Rankine cycle and R22 for air conditioning system. The turbine, heat exchanger and a condenser are the components of the first sub system. The succeeding subsystem consists of compressor, condenser, expansion device and evaporator.

Rankine cycle heat engine produces mechanical power and is used to run the compressor of the air conditioning system (Sun 1997). A working fluid such as R113 used to drive a turbine gains its heat of vapourisation from input solar energy. The mechanical energy is produced by expanding the fluid in the turbine. The solar collector provides warm water, exchanges its heat with the Rankine fluid circulated through the heat exchanger. The fluid after expanding in the turbine subjected to condensation in the Rankine cycle condenser and the liquid is then pumped back to the heat exchanger placed in-between the condenser and turbine. Due to the heat transfer

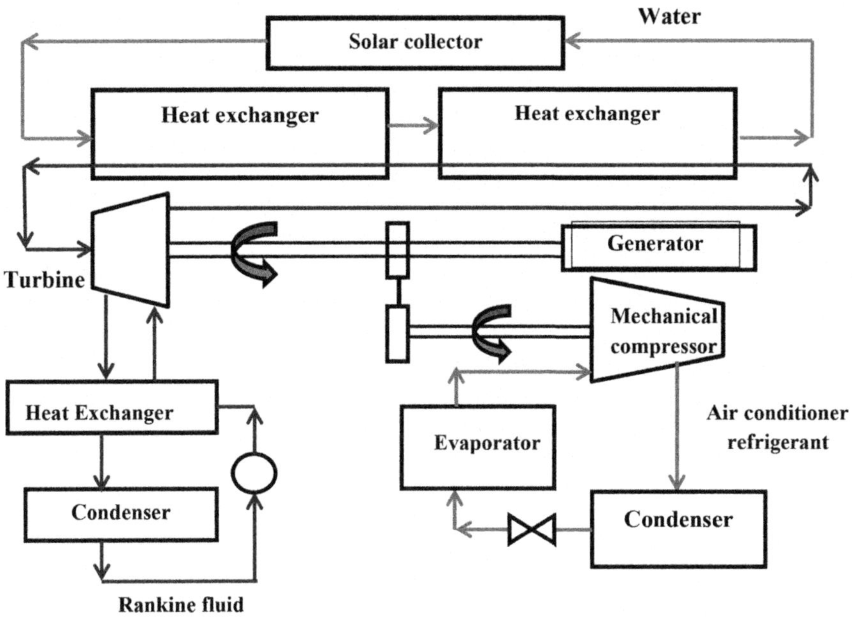

Fig. 5.12 Solar heated Rankine combined power/cooling system

between the expanded fluid in the turbine and condensed liquid, the temperature of the liquid increases and hence the input energy in the solar collector can be reduced.

The second subsystem works as similar to the conventional vapour compression system. The difference in the above conventional Rankine vapour compression system and integrated system is that the current system can act as only a powerful engine or can be used to drive the airconditioner.

The system uses different fluids in the Rankine and airconditioning cycle, allowing improved overall efficiency. The COP of the cooling cycle is as high as that of vapour compression systems, but the efficiency of the heat engine cycle is about 10% only. The coefficient of performance of the system reported is about 0.71. The overall system is an alternative to thermally activated cooling technologies with the advantage of providing mechanical power, or electricity when cooling is not required.

Fuel Assisted Steam Regenerative Brayton Cycle Solar Cooling System

A conventional vapour compression cooling system can be driven by a Brayton cycle rather than the Rankine cycle in a solar/fuel assisted power generating loop. The Steam Regenerative Brayton Cycle (SRBC) is similar to that of the conventional air standard gas turbine cycle. However, the working fluid in the SRBC is steam rather than air. The schematic of the system combined with the solar collector/storage loop together with the vapour compression system is shown in Fig. 5.13.

The system consists of solar collectors, storage tank, flashing chamber, compressor, a regenerator, superheater, turbine, a condenser and evaporator (Fig. 5.13). The system is having the advantage that it only needs a non-pressurised

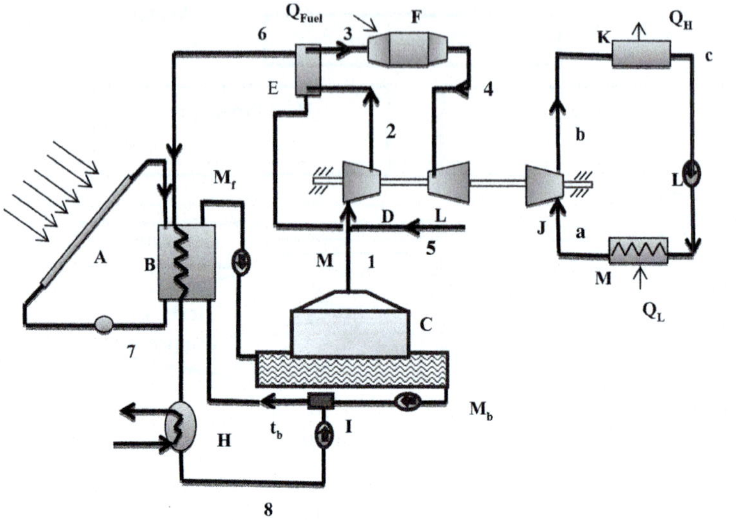

A- Solar Collector	D- Vapour Compressor	G- Turbine	J- Compressor
B- Storage Tank	E- Regenerator	H- Condenser	K- Condenser
C- Flashing Chamber	F- Super Heater	I-Mixing Box	L- Expansion Valve M- Evaporator

Fig. 5.13 Schematic of the fuel assisted Brayton cycle

standard flat plate collector with water as the working fluid and maximum temperature of 95 °C. The hot water is subjected to evaporation in a flash chamber and also it is compressed to 100 kPa. The saturated steam generated at this temperature is compressed to higher pressure via a mechanically driven compressor. A regenerator is placed between the low-pressure turbine and a compressor. The compressed steam is sensibly heated in a regenerator and it is superheated further in a gas-fired superheater. The superheated steam expanded to low-pressure in the low-pressure turbine (LP). The vapour compression unit is driven by the shaft work available at the turbine. The turbine exhaust is condensed in the condenser. The condensate is pumped into a mixing chamber, where it is mixed with the efflux from a flashing chamber. The combined flow is then pumped back to the thermal storage tank.

5.7.2 *Direct Conversion Solar VCR Systems*

Solar energy can be used for the refrigeration through Photovoltaic Electric Conversion (PVEC), thereby producing electricity which can run the compressor of VCR cooling system. The word 'photovoltaic' combines two terms—'photo' means light and 'voltaic' means voltage. Photovoltaic (PV) systems are used to convert sunlight

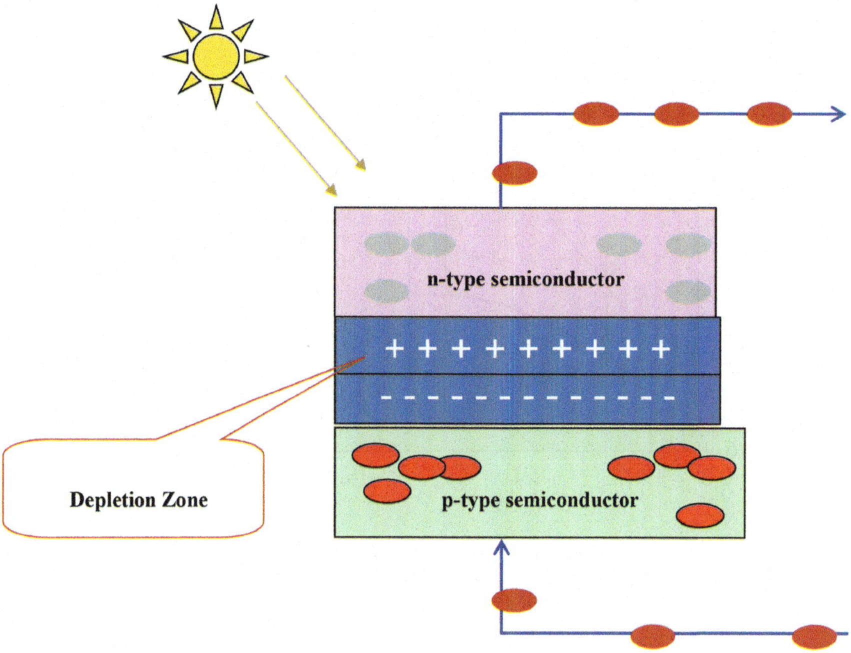

Fig. 5.14 Physics of photovoltaic generation

into electricity. The photovoltaic effect can be defined as the generation of potential when solar radiation ionizes the region in or near the built-in potential barrier of a semiconductor. A potential barrier can be formed in a semiconductor by the following methods (Fig. 5.14).

(i) In the metal–semiconductor or barrier formed by the deposition of a suitable transparent metallic film by the evaporation or sputtering on to a semiconductor.
(ii) The suitable impurity is introduced into the opposite impurity type semiconductor. The *p–n* junction is formed based on the above process.

Thus, the photovoltaic (PV) power generation refers to the process of direct conversion of solar energy into electricity using a device known as solar cell, which absorbs the incident photon to produce photo generated, electron–hole pairs and separates them for the generation of photovoltage and photocurrent by means of an inbuilt electric field as in a *p–n* junction. To absorb the solar radiation (photons) effectively, the bandgap of the semiconductor (absorber) should be as small as possible. This will result in a large part of the spectrum being absorbed. Since larger is the number of photons absorbed, larger will be the carrier (electron–hole) pair's generation and hence larger will be the current. However, if the band gap is smaller, a larger number of thermally generated carriers are produced which compensates the effect due to the photogenerated carriers. This results in the lowering of voltage. Thus, for the higher voltages, the bandgap of the semiconductor (absorber) should be as high as possible.

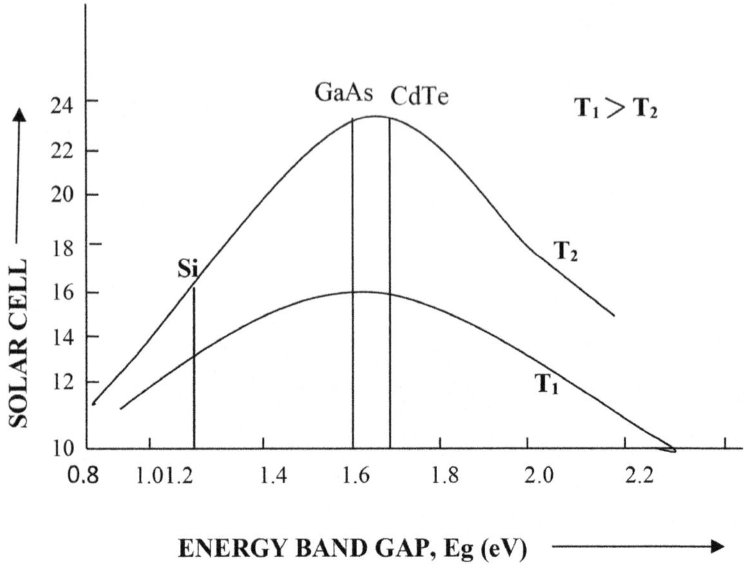

Fig. 5.15 Solar cell efficiency as a function of the energy band

This affects the current since a smaller portion of the spectrum is absorbed. The power output and hence the efficiency of a solar cell depends upon the current and voltage generated, and it follows that there is an optimum bandgap (1.4–1.5 eV) for maximum efficiency of the solar cells as shown in Fig. 5.15 for different materials. It is seen that the optimum bandgap and the maximum efficiency decreases as the temperature increases. The materials having high and close to the bandgap values are CdTe and GaAs. Although the silicon is not that efficient absorber, its fabrication technology is well established and so it is still favourable material for solar cells. A simple sketch of Si solar cell and its dimension are shown in Fig. 5.16.

Semiconductor material, typically silicon, is used in thin wafers or ribbons in most commercially available cells. The semiconductor material has a positive charge on one side and the other side is negatively charged. An electric current is generated when the negative side is activated by sunlight hitting the positive side of the cell. Photovoltaic (PV) cells are made up of at least two semiconductor layers. One layer containing a positive charge, the other a negative charge. Sunlight consists of little particles of solar energy called photons.

As a PV cell is exposed to this sunlight many of the photons are reflected. These photons are absorbed by the solar cell. The electrons are freed from the negative side when enough photons are absorbed by the negative layer. These free electrons are naturally migrated to the positive layer creating a voltage differential similar to a household battery. These two layers are connected to an external load, the electrons flow through the circuit creating electricity. The solar energy cell can produce electricity in Watts. The power output is enhanced by combining different cells in a weather-tight package called a solar module. These modules (from one to several

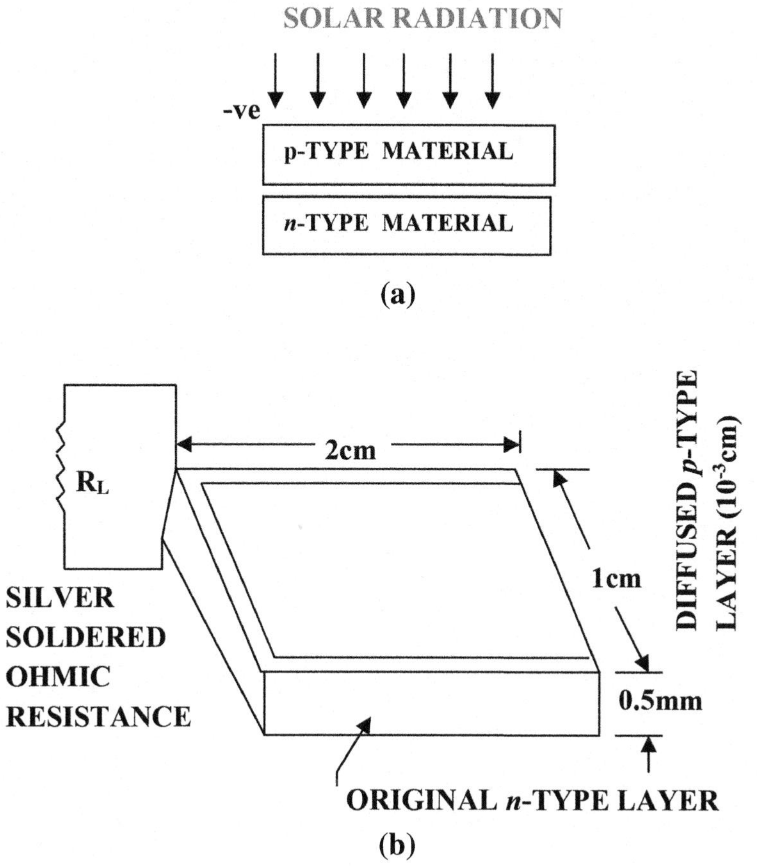

Fig. 5.16 **a** Sketch of a solar cell and **b** dimensions of a solar cell

thousand) are then wired up in serial and/or parallel with one another, into what's called a solar array, to create the desired voltage and amperage output required by the given project.

The electrical characteristics of a solar cell can be understood from a simplified equivalent circuit shown in Fig. 5.17, which consists of a constant current generator, nonlinear junction impedance and load resistance. Light causes a current to flow in the load, the magnitude of this current is the difference between the photon generated short circuit current I_s and junction current I_j which is given by the equation:

$$I_j = I_o\left(e^{\lambda V} - 1\right) \tag{5.36}$$

where I_o is the reverse saturation current, V the voltage applied to the junction and $\lambda = e/kT$ (k = Boltzmann's constant and T the absolute temperature), so if

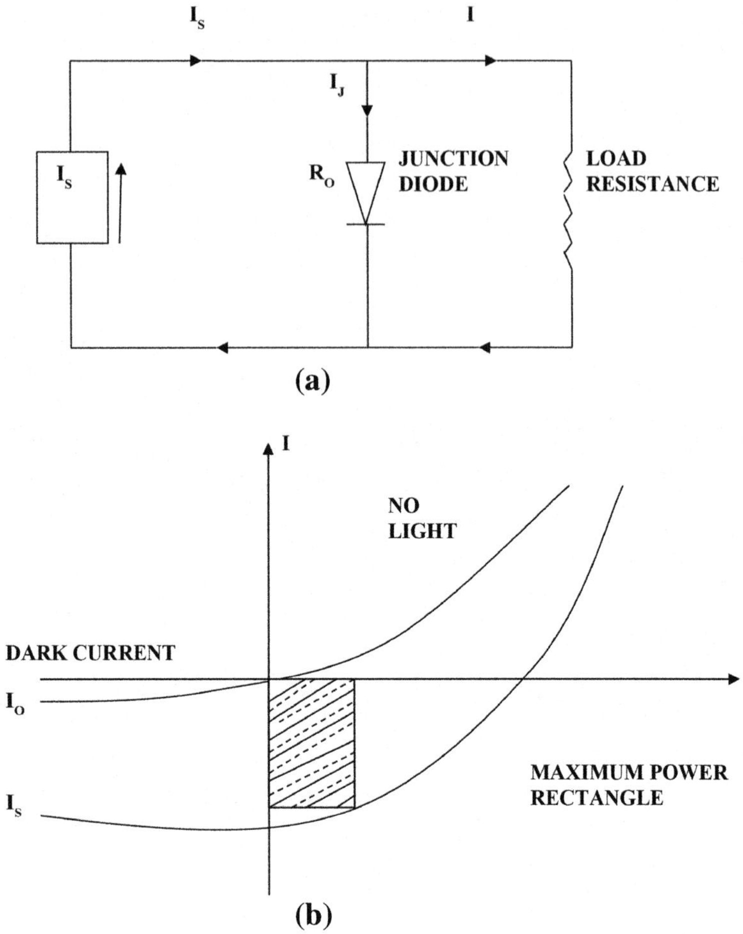

Fig. 5.17 **a** Equivalent circuit of a solar cell and **b** I-V characteristics of the PV cell

$$I = I_s - I_o\left(e^{\lambda V} - 1\right) \tag{5.37}$$

and the maximum voltage V_m occurs when $I = 0$, i.e., open-circuit voltage

$$V_m = \frac{1}{\lambda}\ln\left(\frac{I_s}{I_o} + 1\right) \tag{5.38}$$

While for the short circuit condition $V = 0$ and $I_{so} = I_s$, thus, the output power and solar cell efficiency can be evaluated for a given input power incident on the solar cell. Typical values of the efficiency are 10–15%. Figure 5.18 shows the I–V characteristics of a solar cell.

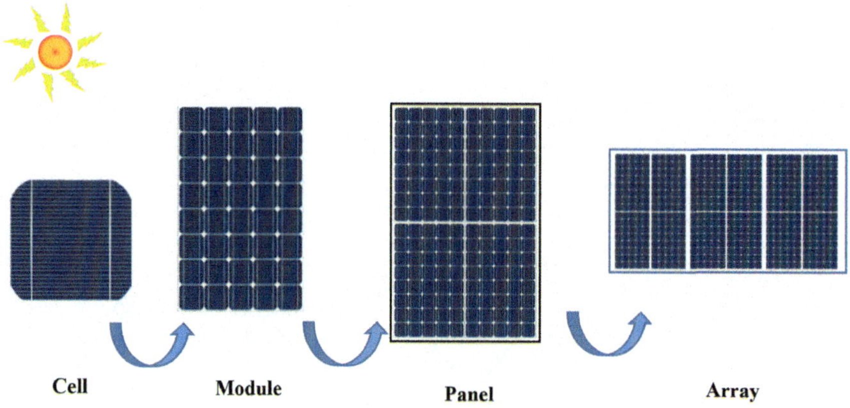

Fig. 5.18 The PV array

The voltage and current generated by a single-crystalline Si solar cell are only around 0.5 V and 30 mA/m^2 and, therefore, a string of such cells is made by connecting them in series to get voltage added. A solar cell module consists of such strings of cells sandwiched between the two glass sheets for the protection from the environment. These modules can be externally connected in series to increase the voltage and in parallel to increase the current. Thus, one can design a PV power generation system by taking account of an approximate number of modules and connecting them in series or parallel combination to meet the load. However, since the Sun's position varies during the day thereby changing the angle of incidence (θ) on the solar panels, the power output is given by Dutta (1987),

$$P_{out} = P_{in}.\cos\theta.\eta.F.A \tag{5.39}$$

where P_{in} is the insolation, η the solar cell efficiency, F the array design factor, A the array area and θ the angle of incidence on the normal to the cell surface.

Thus, P_{out} is maximum when $\theta = 0$, around noon. Hence, the automatic tracking must be used to maintain $\theta = 0$, throughout the day thereby increasing the power generation by 35% in this case.

Although the solar cell seems to be a relatively simple device, an efficient cell is to be finely tuned between the optical and electrical material properties. The major limitations on the conversion efficiency of photovoltaic solar energy converters are:

(i) Reflection losses from the surfaces.
(ii) The utilisation of only a part of the energy of the absorbed photon.
(iii) Incomplete absorption of photons available in the solar spectrum.
(iv) An incomplete collection of electron–hole pairs generation from photon absorption.
(v) A voltage factor is given by the ratio of open-circuit voltage to the equivalent energy gap potential difference.

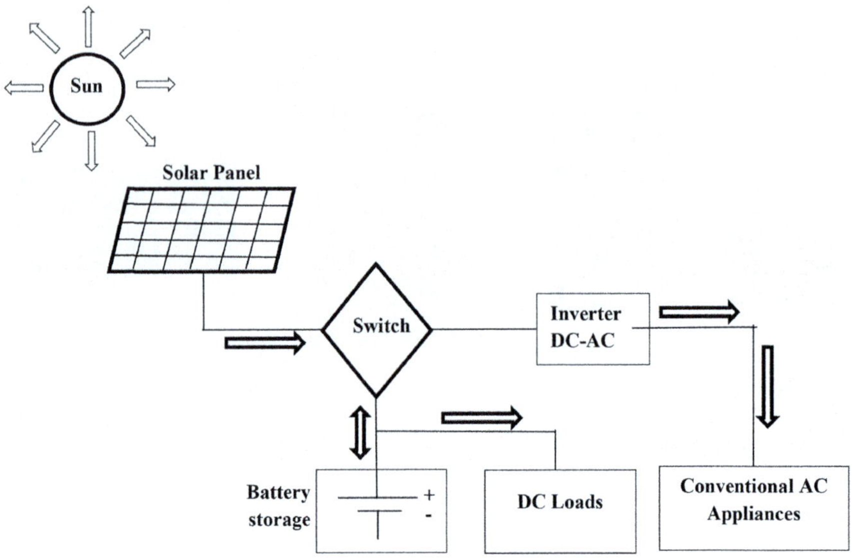

Fig. 5.19 Schematic diagram of a standalone PV system (Alobaid et al. 2017)

(vi) Additional degradation of current and voltage due to the series resistance, shunt resistance, nonohmic contacts, depletion layer effects and surface effects etc.

While the above basic factors are being investigated, the efforts are also being made for new design options resulting in gradually increasing conversion efficiencies (Alobaid et al. 2017). The silicon solar cell is still the main candidate in flat array *PV* market.

Photovoltaic Vapour Compression Cooling System

Solar energy can be used for the refrigeration through PVEC, thereby producing electricity which can run the compressor of the VCR system as shown in Fig. 5.19.

PV Modules: The PV cells converts light energy from the Sun into electricity.

Battery: The battery stores DC voltage when the sunlight is available and discharges in the absence of daylight.

Inverter: The inverter converts the DC electrical power into AC and then delivers the electrical energy to the AC loads.

AC unit: The vapour compression AC unit is a conventional cooling or refrigeration system that is run by the power received from the inverter.

Figure 5.20 represents a PV powered vapour compression cooling system. The system consists of a VCR system and a photovoltaic panel. A DC motor is directly coupled to a compressor and a refrigeration cabinet with thick insulation is used to store extra cold storage in the form of ice or other eutectic material for the time when the compressor is not working. A motor-driven compressor is used to prevent

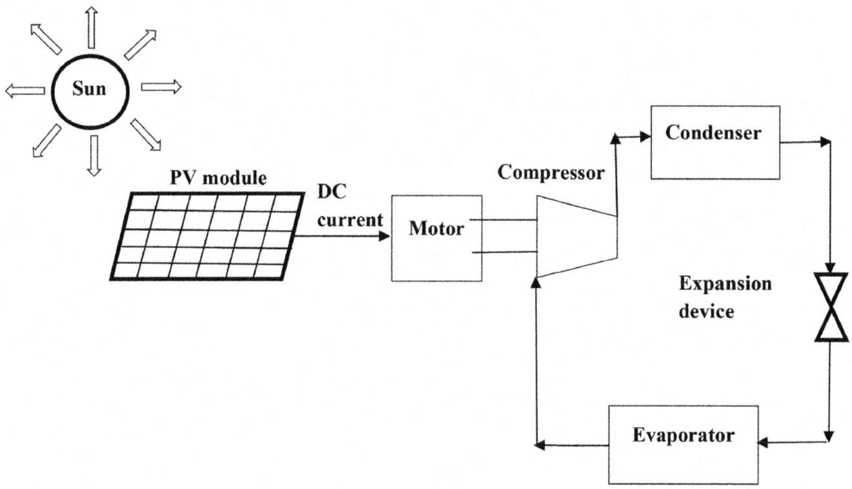

Fig. 5.20 Schematic diagram showing PV powered VCR (Askalany et al. 2012)

the over-discharge of the battery. Thus, the success of photovoltaic conversion base system is governed by the performance and cost-effective design of the photovoltaic converter.

The system is still quite expensive due to high installation cost and low efficiency of the photovoltaic panel. It is, however, effective and simple in an area far from an electricity grid if the required cooling capacity is low.

References

Alobaid, M., Hughes, B., Calautit, J. K., O'Connor, D. & Heyes, A. (2017). A review of solar driven absorption cooling with photovoltaic thermal systems. *Renewable and sustainable energy reviews, 76,* 728–742.

Arora, C. P. (2000). Refrigeration and Air Conditioning, second ed., *Tata McGraw-Hill Publishing Company Limited,* New Delhi, India.

Askalany, A. A., Saha, B. B., Kariya, K., Ismail, I. M., Salem, M., Ali, A. H. & Morsy, M. G. (2012). Hybrid adsorption cooling systems–An overview. *Renewable and Sustainable Energy Reviews, 16*(8), 5787–5801.

Calm, J. M. & Hourahan, G. C. (2001). Refrigerant Data Summary. *Engineered Systems, 18*(11), 74–77.

Dutta, V. (1987). Photovoltaic power systems, Summer school on advanced energy systems, Sponsored by ISTE, June 1–12, CES, IIT Delhi.

Mohanraj, M., Jayaraj, S., Muraleedharan, C. & Chandrasekar, P. (2009). Experimental investigation of R290/R600a mixture as an alternative to R134a in a domestic refrigerator. *International Journal of Thermal Sciences, 48*(5), 1036–1042.

Mohanraj, M., Muraleedharan, C. & Jayaraj, S. (2011). A review on recent developments in new refrigerant mixtures for vapour compression-based refrigeration, air-conditioning and heat pump units. *International Journal of Energy Research, 35*(8), 647–669.

Sun, D. W. (1997). Solar powered combined ejector-vapour compression cycle for air conditioning and refrigeration. *Energy Conversion and Management, 38*(5), 479–491.

Chapter 6
Vapour Absorption Cooling Systems

Absorption cooling is one of the oldest cooling processes known in literature and it has long been realised that the absorption cooling machines could be employed to produce cooling from solar energy. The absorption refrigeration machine was originated in 1824 by Faraday who achieved cooling by vapourisation of ammonia. However, the earliest successful attempt using solar energy was made by Mouchot in 1878. An absorption cooling system can be operated with a low-grade heat generated by conventional flat plate solar collectors and/or using low temperature available from industry.

6.1 Basic Absorption Process

A survey of the vapour absorption cooling systems (VAB) has been presented by Kaushik and Chandra (1982). The basic absorption process consists of the utilisation of difference in the binding energy of a fluid (known as refrigerant) in two vessels at different temperatures, the fluid can be bounded by a liquid or solid solvent called as the absorbent. The main feature of this process is given in Fig. 6.1. Vessel I contains the refrigerant (R) bound to the absorbent (A) of the vessel, while vessel II is pure refrigerant. When two vessels are at the same temperature and the valve between them is closed, the vapour pressure in vessel II will be higher than that in vessel I, thus $P_{\text{I}} < P_{\text{II}}$, when $T_{\text{I}} = T_{\text{II}}$. When the valve is opened, the refrigerant gas will flow from vessel II to vessel I due to the existing pressure difference. The vapour pressure in vessel I exceeds the equilibrium vapour pressure and the vapour condenses in it. Due to the heat of condensation, the temperature of vessel I rises. The reverse process takes place in vessel II thereby its temperature gets decreased. Finally, the vapour pressure tends to equalise in two vessels, because of mass transfer which causes a temperature difference between the vessels.

S. C. Kaushik et al., *Solar Cooling*, https://doi.org/10.1007/978-3-031-42410-6_6

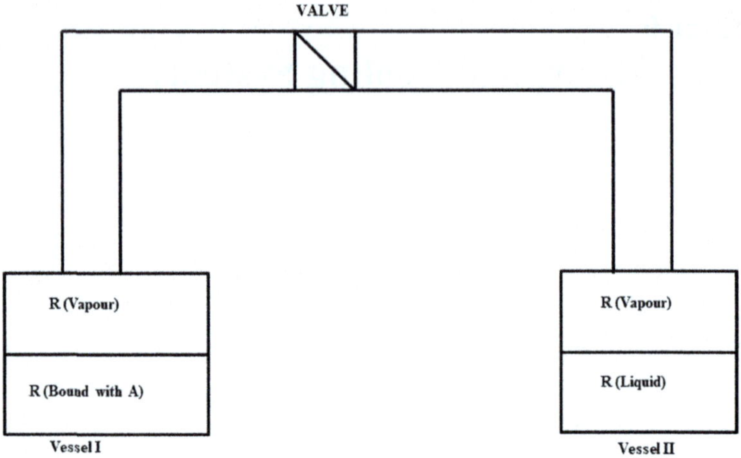

Fig. 6.1 Basic vapour absorption process (Kaushik et al. 1988)

Thus,

$$P = P \text{ and } T > T \tag{6.1}$$

Further, when the heat is applied to the cold vessel, the new pressure equilibrium is disturbed and the vapour pressure in vessel II exceeds the vapour pressure in vessel I, causing mass transfer and the consequent flow of heat from vessel II to vessel I. Thus, heat gets pumped from a lower temperature to higher temperature. The absorption process can be used for cooling (when cold vessel takes heat from the surroundings) or heating (when heat developed in the warm vessel is utilised). The heat pumping ability of the device thus gets consumed and to restore it, the system must be regenerated by reversing the mass transfer and supplying the heat to vessel I while withdrawing heat from vessel II. During the time between phases of heat pumping and regeneration, energy can be stored. Thus, an absorption process has three main characteristics namely, cooling, heating and energy storage as discussed by Kaushik and Chandra (1982).

6.2 Basic Absorption Cycle

A conventional cooling system employs a compressor to pressurise and to elevate the temperature of the refrigerant since the refrigerant is in the vapour state and therefore it takes a lot of energy to compress a gas. But in case of a pump, the energy consumption for pressurizing a liquid requires very less power as compared to a compressor. In a vapour absorption system, the compressor is replaced by a pump and a heating unit known as thermal compressor to save power.

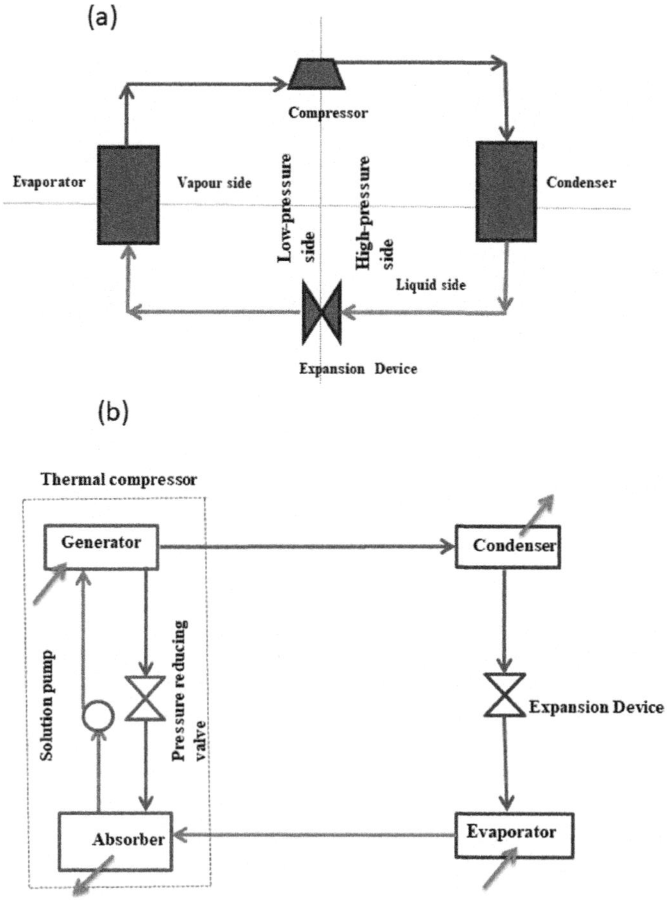

Fig. 6.2 A vapour **a** compression and **b** absorption cooling cycle

A basic vapour absorption cooling (VAB) cycle is a heat operated cycle as illustrated in Fig. 6.2b. The generator, a condenser, an evaporator, an absorber, a pump and expansion device are the main components of an absorption cooling system. By comparing it with a vapour compression unit as shown in Fig. 6.2a, we notice that the absorption system (Fig. 6.2b) looks very much like the vapour compression system, except that the mechanical compressor (work operated) is replaced by a complex absorption mechanism consisting of an absorber, pump and generator. The absorber-generator assembly called the thermal compressor (being operated by low-grade energy, heat) is devolving on increasing the pressure of the absorbent (refrigerant) from the low evaporator pressure to the condenser pressure level.

As compared with the conventional compressors, the thermal compressor in an absorption cooling system has the advantage that liquid is compressed instead of vapour. Since the steady flow work is proportional to the specific volume and hence,

Table 6.1 Differences between VCR and VAB cooling systems

Compression system	Absorption system
High grade energy or work operated device	Heat operated
COP is higher	Low COP
Lot of heat loss	Less heat loss
Regular maintenance	Very low maintenance
Higher noise	Lower noise
More moving parts	Less moving parts
Higher cost due to electricity as energy input	Cheap fuel (such as waste heat or biomass) is used as the energy input

the work input for the absorption unit is very small as compared to the vapour compression cooling system. To achieve the cooling process in an absorption cooling system, it is essential to produce a means of removing the refrigerant vapour as fast as it is formed. In this case, an evaporator is connected to another vessel containing a substance capable of absorbing the refrigerant vapour called absorber (Fig. 6.2b). The substance used to absorb the refrigerant is known as absorbent. Next stage is the release of refrigerant vapour at a convenient pressure for its subsequent liquefaction or condensation in a condenser. This is accomplished by pumping the solution to a high-pressure, the solution enters the generator, where heat is added to the refrigerant-absorbent combination, and the refrigerant is driven off from the solution as vapour. The absorbent solution after releasing refrigerant vapour is returned to the absorber through a pressure reducing valve. The refrigerant-absorbent mixture is thus thermally compressed without using the compressor shaft work as demanded by compression cooling system. Vapour compression (VCR) and vapour absorption cooling system are differentiated and is given in Table 6.1.

6.3 Selection of Working Fluids

The success of an absorption cooling system depends on the selection of suitable working pairs of refrigerant and absorbent. This aspect has been considered in reasonable detail by Kaushik et al. (1988). In VAB systems, only a few working pairs are suitable for its application. Although the physical and chemical properties of the pure refrigerant and pure absorbent are important, thermodynamic properties of their solutions are critical and useful in evaluating the solubility of new fluids for improved absorption refrigeration systems. Thus, the thermophysical and thermodynamic properties also determine the cooling performance of absorption cooling systems.

In general, the selection is based on the following parameters:

(1) Physical and chemical properties of the fluids
(2) Value of certain thermophysical and thermodynamical properties of the fluids.

The desirable properties of working fluid used in VAB are:

(i) Refrigerant should exhibit high solubility with the solution in the absorber, that is, it should exhibit negative deviation from Raoult's law at absorber.
(ii) There should be a large difference in the boiling points of refrigerant and absorbent (greater than 200 °C) so that only refrigerant is boiled in the generator. This ensures that only pure refrigerant circulates through refrigerant circuit (condenser-expansion valve-evaporator) leading to isothermal heat transfer in evaporator and condenser.
(iii) It should exhibit small heat of mixing so that a high COP can be achieved. However, this contradicts requirement (i). Hence, in practice, a trade-off is required between solubility and heat of mixing.
(iv) The high thermal conductivity of working pair is desirable for better performance. The refrigerant-absorbent mixture should also possess very low level of viscosity.
(v) Stability of the working fluid is also important. Crystallization of the refrigerant-absorbent pair is not feasible.
(vi) Ease of availability, chemically stable low cost are the other desirable features of the mixture.

The choice of a working fluid is an important factor in the design of an absorption cooling system. Many combinations are proposed but most commonly used refrigerant absorbent pairs commercial systems are water–lithium bromide (H_2O–LiBr) and ammonia–water (NH_3–H_2O). The water–lithium bromide is limited for above 0 °C applications such as airconditioning. Water is the refrigerant and lithium bromide the absorbent. In ammonia–water system ammonia is used as refrigerant and water as absorbent. Combinations of both natural and synthetic refrigerants are also tried in absorption cooling systems to overcome the limitations of currently employed working pairs such water–lithium bromide and ammonia–water.

The water–lithium bromide is limited for above 0 °C applications such as airconditioning. Water is the refrigerant and lithium bromide the absorbent. In ammonia–water system ammonia is used as refrigerant and water as absorbent. Combinations of both natural and synthetic refrigerants are also tried in absorption cooling systems to overcome the limitations of currently employed working pairs such as water–lithium bromide and ammonia–water. The NH_3–H_2O pair is one of the oldest combinations successfully employed for industrial refrigeration and airconditioning. Ammonia has certain favourable thermodynamic characteristics, i.e., large negative deviation from Raoults law, low molecular weight and hence large heat of vapourisation of the refrigerant. H_2O is a proper absorbent due to its low cost, availability and non-toxicity. Both the absorbent and refrigerant meet the requirement for hydrogen bonding extremely well, and are highly stable. In general, quite compatible with ordinary materials for construction. However, because of high affinity between NH_3 and H_2O, although their boiling points are spread apart by 133 °C, some water is always vapourised from the generator with ammonia vapour and a complex rectifier system is needed.

Another potential working fluid required for the VAB system operation is NH_3–$LiNO_3$, which even does not require any rectifier for its operation and can be used for low evaporator temperature. NH_3 gets absorbed by $LiNO_3$ to form a clear syrupy liquid and problems such as absorbent migration, which are associated with the other solid absorbents for NH_3 (e.g., $CaCl_2$) do not arise since NH_3–$LiNO_3$ is in liquid form. The NH_3–NaSCN combination also carries appreciable properties, such as inexpensiveness, non-explosive character, chemical stability, safety and non-corrosiveness in steel containers. The solubility of NaSCN in ammonia is so large that it is possible to have their solution at room temperature with less than one atmospheric pressure, while the vapour pressure of liquid ammonia is ten atmospheres. Because of the high viscosity, this combination is only used for intermittent absorption cooling systems.

The H_2O–LiBr and NH_3–$LiNO_3$ pairs do not require rectification of the refrigerant vapour in the generator of the vapour absorption system.

Further, precooler could also be avoided unlike in case of NH_3–H_2O combination. However, the H_2O–LiBr system suffers from two disadvantages at low-temperature and high concentration of the absorbent, the absorbent crystallises and since water is the refrigerant, the evaporator cannot operate at a temperature below 4.4 °C. The usual lower limit of temperature is 2 °C in such an absorption chiller. Therefore, although a vapour absorption cooling system operating on an H_2O–LiBr cycle gives better performance than the aqua-ammonia and NH_3–$LiNO_3$ systems, its use is limited to air conditioning since water is the refrigerant. Both NH_3–H_2O and NH_3–$LiNO_3$ systems are recommended for cooling applications (Amaris et al. 2018).

Recently some alcohol–salt mixtures (methanol–LiBr–ZnBr, ethanol–LiBr–$ZnBr_2$) and water-multicomponent salt combinations are also proposed as suitable working fluids for the vapour absorption cooling system. The choice of the working fluid for open-cycle absorption cooling is very selective and water has to be the refrigerant. Further work on new working fluids suitable for the vapour absorption system is also being undertaken and pressure–temperature–concentration (P–T–X) measurements for refrigerant-absorbent pairs are desirable to assess the potential of these fluids. Fluorocarbon refrigerant-based working fluids such as R22 or R21 have been studied. These refrigerants have solubility with the number of organic solvents. Two solvents investigated so far are dimethylformamide (DMF) and dimethylether of tetraethylene glycol (DMETEG). The combination of LiBr–H_2O or NaOH–H_2O and an inorganic salt mixture may be the most preferable working pair for the VAB cooling system. The main issue of the proposed combination is its stability at very high temperature and high concentration.

6.4 Solar Operation of VAB System

A satisfactory operation of solar absorption cooling system depends on the design of the solar collection system. Two solar collector designs can be considered as representative of the range of collector types suitable for solar cooling. First, a conventional flat plate collector having two glass cover plates and another collector, which has highest efficiency, consists of absorbing surface and piping housed in an evacuate glass cylinder. Comparative efficiency curves of various solar collectors that is preferred for solar absorption cooling systems are shown in Fig. 6.3.

It can be seen that the evacuated tube and concentrating parabolic collectors yield reasonable collection efficiency even at higher collection temperature. This feature further prompted the use of these collectors to supply high-temperature input heat to the generator of absorption cooling systems. Specifically, the advanced absorption cooling systems which require high-temperature heat source can be operated satisfactorily with the recent development of high efficiency and high-temperature collectors.

First, a conventional flat plate collector can be modelled with the performance prediction equation (Wilber and Mancinbi 1976)

$$Q_R = 0.83A_p\left(0.6I_s\cos\phi - 0.0031\left(t_p - t_{\text{air}}\right)\right) \quad (6.2)$$

$$Q_R = 0.748A_p(0.8I_s\cos\phi - v(t_{\text{in}} - t_{\text{air}})) \quad (6.3)$$

$$v = 0.00103 + 3.3\times10^{-6}(t_{\text{air}} - 2.8) + 3.9\times10^{-6}\left(t_p - 5.6\right)$$

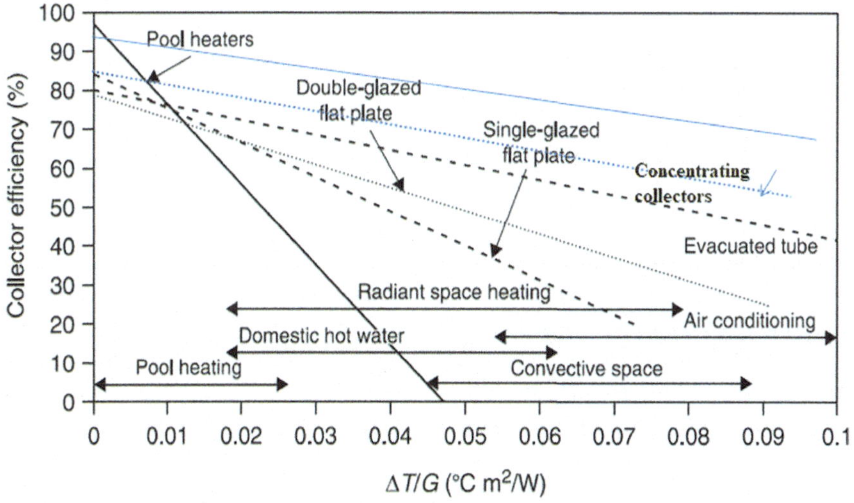

Fig. 6.3 Comparative efficiency curves for various solar collectors

where Q_R is the heat flow rate from the collector (kW), A_p the area of the collector (m^2), I_s the incident solar radiation (kW/m^2), ϕ the angle between Sun and normal to the collector, t_p the average collector temperature (°C) and t_{air} the ambient air temperature (°C).

Thus, knowing the useful heat available from solar collectors, solar collection efficiency can be evaluated. It is also of interest to know the relative collection efficiencies of several solar collectors. The collection efficiency is defined as

$$\eta = \frac{Q_R}{A_p I_s} \tag{6.4}$$

Normally, collection efficiency is determined by a field test or a solar simulator and is usually expressed as a function of the variable,

$$\frac{\Delta t}{I_s}, \text{ where } \Delta t = \left(\frac{t_{\text{in}} + t_o}{2} t_A \right) \tag{6.5}$$

where t_o be the outlet fluid temperature (°C).

The use of solar heat for absorption cooling system has certain limitations in practice as compared to fuel-fired system (viz. intermittency, low energy concentration, hitch of control etc.) and thus solar operated cooling device has to be specially designed to suit the characteristics of solar heat collectors. Solar collection efficiency decreases with the increase of solar collection temperature. Thus, in general, a solar cooling device can be less efficient but never more efficient than the fuel operated device. A solar cooling device is thus similar to any other heat operated cooling device except for the added limitations imposed by the solar source of heat. Two important facts that emerge from the thermodynamic considerations of solar operation of cooling systems are:

(i) The magnitude of cold produced is always considerably less than the heat supplied at the hot end of it.
(ii) Temperature lifts $T_G - T_A$ in an ideal system is close to the temperature depression $T_C - T_E$. In real systems, the lift is always greater than the depression.

These facts apply to a single-stage absorption system. Multi stage cooling system shall produce other results, i.e., high COP can be attained at the expense of high elevated temperature. The first one shows the minimum quantity of heat supplied to the collector for a given refrigerating effect, while the second one shows the operating temperature of the generator and hence the collector.

In determining an absorption system to be driven by solar energy, we should select the operating conditions, which result in higher COP value to minimise the size and cost of the system. Although there are many parameters which affect the performance of the system, the designer is not free to select all of them. For absorption system contains two working fluids (absorbent and refrigerant) and two phases (liquid and vapour) the minimum number of degrees of freedom is two. Thus, if we chose the evaporation temperature and heat rejection sink temperature, the other conditions such as the generator temperature, pressure and equilibrium solution concentration are also fixed.

To study the cost-effectiveness of the solar cooling system, it is necessary to know the daily cooling COP. Hence, cooling is produced during sunshine hours. When the heat input in the generator of absorption cycle is taken from a solar collector, available operating generator temperature imposes restrictions on the thermodynamic processes involved in an absorption cooling system. If the input heat is time-dependent, the output of the absorption cooling system will also be time-dependent. Moreover, the amount of refrigerant generated is dependent on heat transfer to the generator from the solar collector and hence on the solar insolation. Thus, the mass of refrigerant generated tends to reach a peak about noon when the solar flux is highest. Since the solar energy input and ambient air temperature are time-dependent, the generator heat input can also be expressed as a function of time and in terms of the collector parameters.

Simulated generator heat input from a solar collector is derived as follows:

Performance equation for the solar collector is

$$q_u = \left[(\alpha T)I_s(\theta) - U_p(t_i - t_a(\theta))\right]F_n \tag{6.6}$$

where U_p is the overall heat transfer coefficient of the collector and t_i the inlet water temperature to the collector. Useful heat energy is used to heat water, according to the energy balance, we have

$$A_p q_u = \dot{m}_p C_p (t_o - t_i) \tag{6.7}$$

where t_o is the outlet collector temperature. Assume the heat losses between the solar collector and heat exchanger are negligibly small, then, the useful heat collected will be fully delivered to the generator unit. The assumed condition is valid for heat exchanger in the generator having infinite heat transfer area. Further, assume that the temperature of generator is 5 °C less than the outlet collector temperature, i.e., the outlet temperature $t_o = t_G + 5$ °C.

So,

$$t_i = t_G + 5 - Aq_u/m_p c_p \tag{6.8}$$

Hence, the heat flux in the generator (heat transfer based on the per unit collector area) is determined from

$$q_G = q_u = F_p\left\{(\alpha T)I_s(\theta) - U_p(t_G + 5t_a(\theta))\right\} + \left(\frac{A_p U_p F_p}{\dot{m}_p c_p}\right) q_u$$

$$q_G = \frac{F_p\{\alpha T\}I_s(\theta) - U_p(t_G + 5 - t_a(\theta))}{1 - \frac{F_p A U_p}{\dot{m}_p c_p}} \tag{6.9}$$

For an ordinary collector parameter

$$F_p = 0.901; (\alpha\tau) = (0.925 \times 0.913) = 0.8448$$

$$\dot{m}_{\mathrm{p}}/A_{\mathrm{p}} = 0.015\,\mathrm{kg\,m^2s}$$

Specific heat,

$$C_{\mathrm{P}} = 4.187\,\mathrm{kJ/kg\,°C},$$

The overall heat transfer coefficient, $U_{\mathrm{P}} = 0.0049\,\mathrm{kW/m^2\,°C}$.

For optimum conditions assuming, mean $t_G = 85$ °C so that the absorption system is operating nearer to its maximum performance, then

$$q_G = \frac{1}{0.93}[0.761 I_s(\theta) - 0.0044(t_G + 5 - t_a(\theta))] \tag{6.10}$$

Therefore, heat input to the generator per unit collector area can be expressed as

$$q_G = 0.8182 I_s(\theta) - 0.0047 t_a(\theta) - 0.4258 \tag{6.11}$$

Thus, solar operation of an absorption system can be studied and daily cooling capacity and COP can be evaluated. This would enable us to assess the solar energy fraction achieved by the absorption system.

6.5 Types of Solar VAB System

Based on the above cycles and processes there are three types of absorption cooling systems, viz. open cycle, continuous closed cycle, and intermittent cycle.

An absorption cooling system is economically attractive when there is a source of inexpensive thermal energy at a temperature of 100–200 °C. In the solar absorption cooling system, heat is supplied to a device called a generator. The generator temperature depends on the refrigerant-absorbent pair. Electricity is normally required for circulating the refrigerant-absorbent pair. The coefficient of performance (COP of these refrigeration cycles is higher than the COP of other thermally operating cycles. It can be used in both refrigeration and air conditioning applications. The absorption cooling cycle can be designed in different configurations like, the basic single effect cycle, the multi-stage absorption cycle, and the Platen Munters cycle which can be operated without a pump or the open cycle type.

6.5.1 Open Cycle Absorption Solar Cooling System

The most promising cooling system is based on the open cycle absorption process. An open-cycle absorption cooling system (Fig. 6.4) phases out the associated heat exchangers of the closed cycle combining the process and heat transfer loop.

Thus, in an open cycle absorption cooling system, flow of heat and mass takes place from the system. The condenser is eliminated in the open cycle system. It is replaced with infinite reservoirs of heat and mass, the absorption process changes the composition of the flow of matter. These systems have the advantage of direct utilisation of solar energy for the regeneration process.

6.5.1.1 Basic System Operation

The physical process of an open cycle VAB system is described as follows (Fig. 6.5).

Initially, the refrigerant, water gets evaporated at low-temperature. The evaporation takes place at low-temperature corresponding to the low-pressure created in the absorber evaporator assembly. The refrigerant vapour absorbs latent heat from the space to be cooled. A strong solution is then pumped to the regenerator. In the regenerator, the refrigerant vapour is driven out through the supply of heat at a higher partial pressure and the weak solution returns back to the absorber. These processes

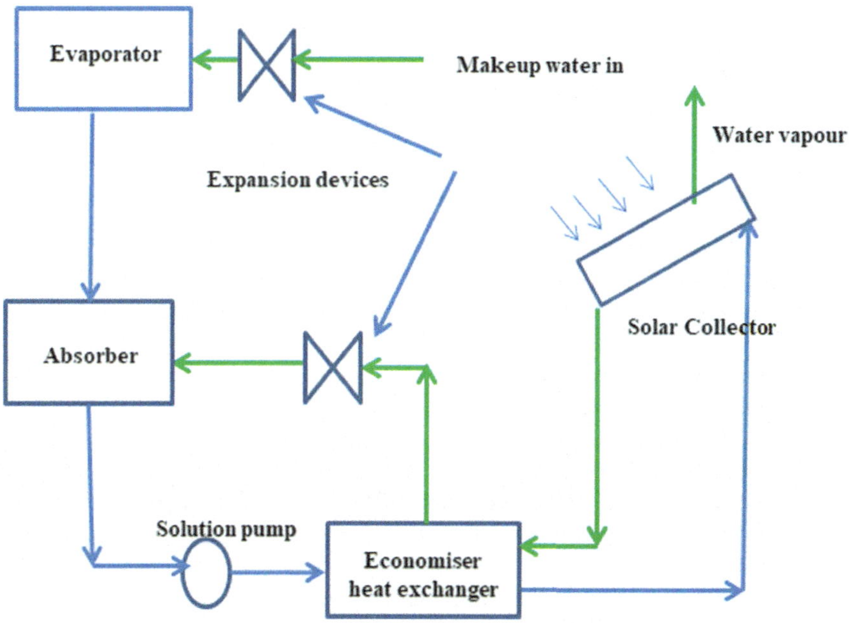

Fig. 6.4 An open-cycle solar VAB cooling system

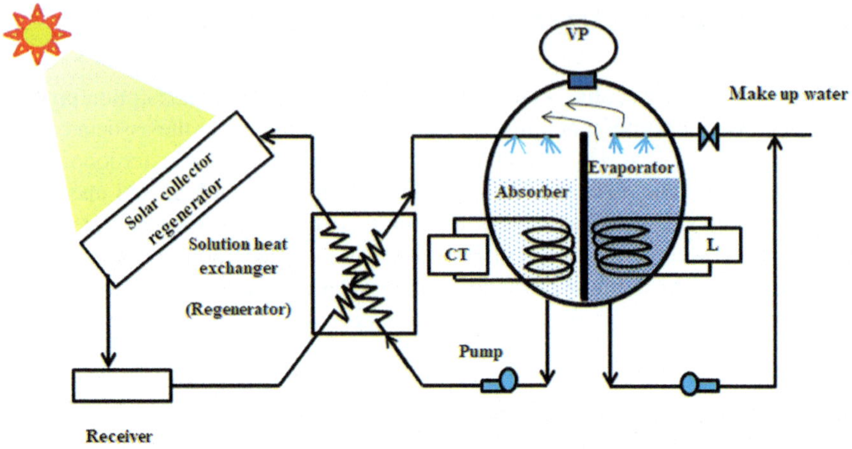

Fig. 6.5 An open cycle VAB system. A—Absorber; E—Evaporator; CT—Cooling tower; P—Pump; VP—Vacuum pump; L—latent heat

result in the transformation of low-pressure vapour to high-pressure vapour. Thus, the weak solution regenerates by losing refrigerant vapour to the atmosphere instead of delivering it to the condenser. In this system, the cooling effect is produced by the evaporation of refrigerant added externally in the evaporator.

6.5.1.2 Open Cycle Absorption Airconditioning: Solar-Collector-Regenerator System

Types of Solar-Collector-Regenerator

In the open cycle absorption cooling system we can have two possible options:

(i) **Open Evaporative Cooling Process**
In this process, room humid air is passed through the absorber. Thus, dry air is passed through the water-filled evaporator, where evaporation takes place and cooled air is processed in the space to be cooled. Ultimately, the system is open to ambient from the evaporator part. The system option has been widely studied by Sheridan (1961) (Fig. 6.6).

(ii) **Open Regeneration Absorption Cooling Process**
The primary system components are solar collector/regenerator, absorber and the evaporator. The condenser is eliminated in this configuration. The absorption solution flows over the regenerator surface thereby losing refrigerant to ambient air freely. The open cycle absorption system becomes very feasible whenever the refrigerant is water since the weak solution regenerates by losing the refrigerant to the atmosphere (Figs. 6.7 and 6.8).

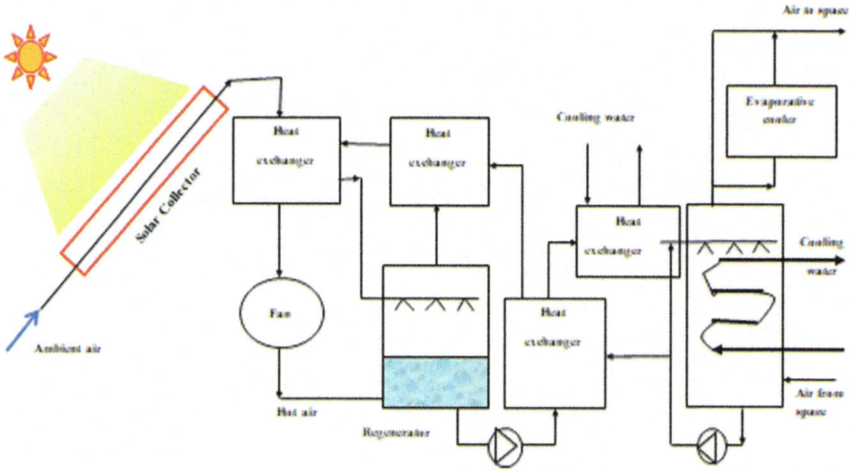

Fig. 6.6 Absorption-dehumidifier solar air conditioning system (Sheridan 1961)

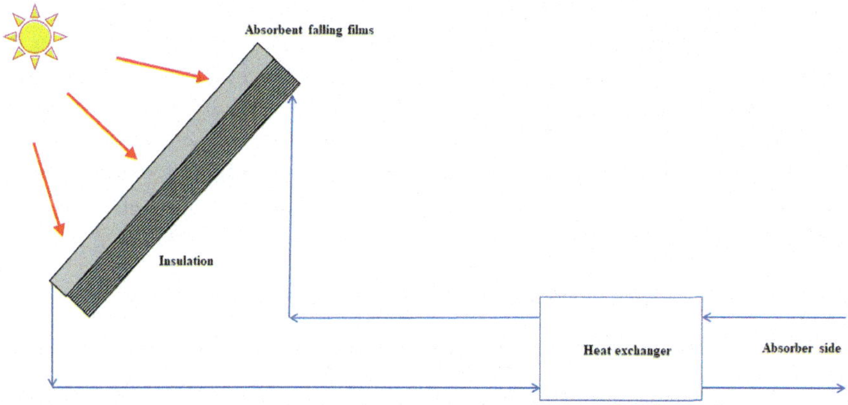

Fig. 6.7 An open surface solution regenerator

A number of designs for the desiccant regenerators have been proposed by various researchers (Gandhidasan 1983; Hollands 1963; Mullick and Gupta 1974; Peng and Howell 1982). Hollands investigated the regeneration of LiCl solution in a roof type single basin solar still regenerator. Since the vapour pressure of a concentrated LiCl solution is lower than the vapour pressure of water at the same temperature, for the useful evaporation to take place the vapour pressure of the solution should be greater than the vapour pressure of water at the glass temperature. Mullick and Gupta (1974) proposed and developed a natural convection type solar-collector-cum-regenerator where moisture is removed by the air flowing between the glass cover and the absorbing plate.

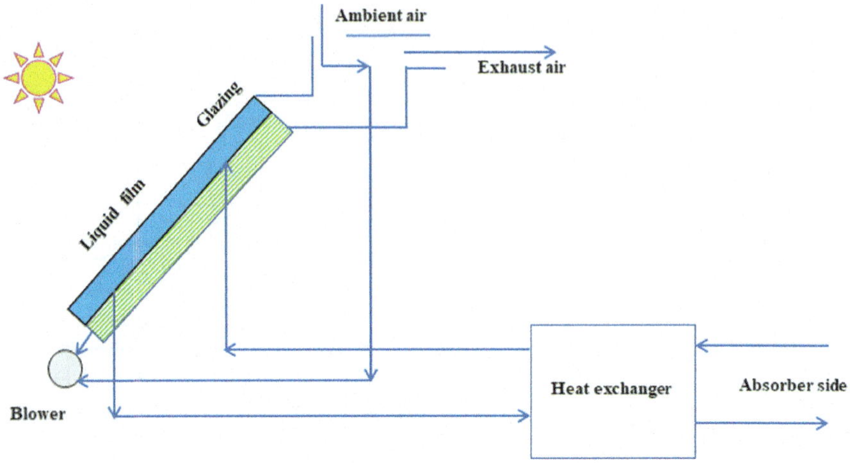

Fig. 6.8 Sloping surface air circulation solar regenerator

In both regenerators, the desiccant solution is made to flow as a thin film over the inclined metallic blackened surface. The absorption of solar radiation raises the temperature and pressure of the solution flowing over the surface, thereby increasing the driving potential, i.e., vapour pressure difference for water desorption from the solution surface. To design both the solar collector and regenerator system for a certain cooling capacity of an open cycle absorption cooling system, the heat and mass transfer models are developed to predict the systems regeneration efficiency at steady-state conditions.

Design and Performance Studies on Solar-Collector-Regenerators

1. Hot and Dry Climate: Open Surface and Forced Flow Solar Regenerators

(A) Kaudinya and Kaushik (1986) conducted the theoretical heat transfer models for the configuration. Assumptions taken are:

(i) Parameters like refrigerant (water) flow rate and solution flow rate are considered per unit width of the regenerator.
(ii) Solution temperature and water mass desorption rate are flow length dependent.
(iii) Glass temperature and water vapour pressure at solution and glass surfaces are on average basis.
(iv) Steady state heat transfer
(v) Heat conduction and the borrow heat loss for both the configurations are negligible.

As in Fig. 6.9a, in the open surface solar regenerator, the evaporated mass of water is lost directly to the ambient. Thus, the rate of water desorption is a function of ambient air velocity (v_a), ambient relative humidity (γ) and ambient temperature

(T_A) as climatic parameters. On the other hand, in a forced airflow regenerator, the air velocity is controlled which may be optimized for maximum water desorption rate. In a forced flow solar regenerator (Fig. 6.9b), the flowing solution is glazed maintaining an air duct through which ambient air is forced to flow in a direction concurrent or counter current to the solution flow.

To study the heat transfer phenomena in both the regenerators, the energy balances for both the regenerators are described below:

(a) Open Surface Solar Regenerator: From the Fig. 6.9a, the energy balance for the solution surface is given by

$$m_s c_s \frac{dt_s}{dz} = \alpha \cdot I_s - U_s(T_s - T_A) - \frac{dm_{ev}}{dz}\lambda_E \tag{6.12}$$

where λ_E is the latent heat of vaporisation and the water mass evaporated from the solution surface is given by

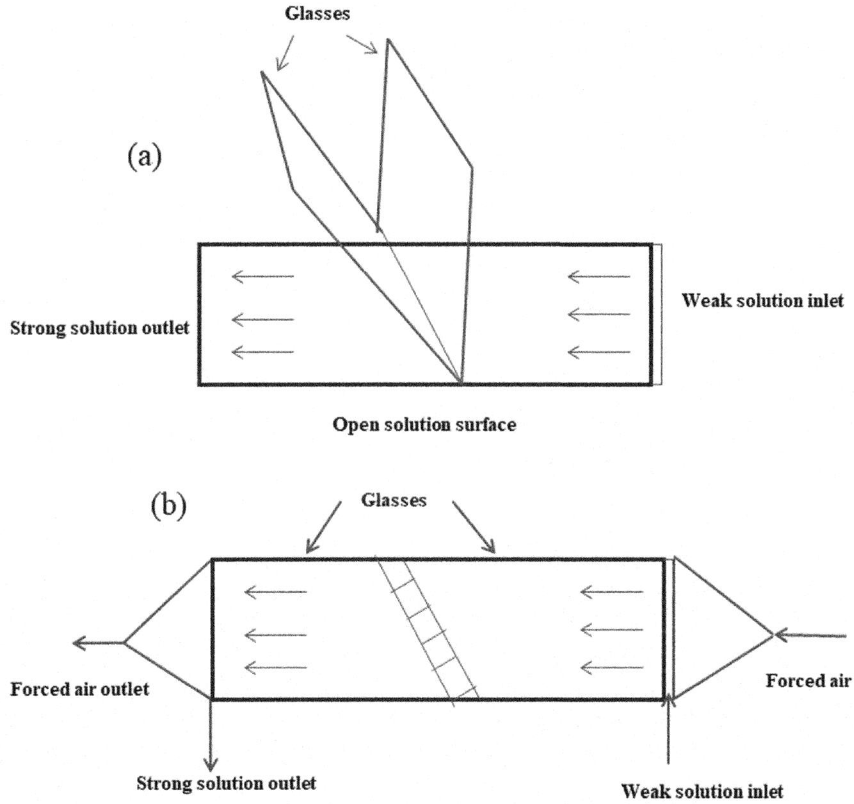

Fig. 6.9 Sketch of **a** open surface solar regenerator and **b** forced flow solar regenerator

$$\frac{dm_{ev}}{dz} = \beta(p_s - p_A) \tag{6.13}$$

The term p_s (solution vapour pressure) is a function of solution temperature/concentration. This is given by

$$p_s = a \cdot T_s + \frac{b}{X_s} + c \tag{6.13a}$$

and ambient air water vapour pressure is

$$P_A = (A_0 + B_0 \cdot T_A) \tag{6.13b}$$

The equation of material balance on the regenerator surface is

$$m_s X_0 = (m_s - m_{ev})X_s \tag{6.14}$$

(b) Forced Flow Solar Regenerator: For a forced flow solar regenerator (See Fig. 6.9b), the equation for the solution surface is given by:

$$M_s \cdot C_s\left(\frac{dt_s}{dz}\right) = (\alpha\tau)(1-\gamma) \cdot L_s - U_s(T_s - T_B) - h_s(T_s - T_a) - \left(\frac{dm_{ev}}{dz}\right)\lambda_e \tag{6.15}$$

Now, consider the equation for air stream

$$2m_a C_a(T_a - T_{a1}) = h_s(T_s - T_a) - U_A(T_a - T_A) \tag{6.16}$$

The positive vapour pressure difference between two streams renders the water mass desorption in the air stream.

The mass flow rate and mass transfer coefficient are related together and it is given by Kakabaev and Khandurdyev (1969) in the following form

$$\frac{dm_{ev}}{dz}(P_S - P_A) = \frac{1244 m_a}{p}(P_a - P_{ai}) \tag{6.17}$$

where P_S, P_A and P_{ai} are as given in Eqs. (6.13a) and (6.13b) for T_s, T_a or T_{ai}, respectively.

With the help of Eqs. (6.11) to (6.14) in open surface and Eqs. (6.15) and (6.17) and other common equation in a forced flow solar regenerator respectively, a common second-order differential equation for refrigerant desorption rate may be achieved as

$$A\frac{d^2 m_{ev}}{dz^2} + B\frac{dm_{ev}}{dz} + C \cdot m_{ev} + D = 0 \tag{6.18}$$

The solution of the Eq. (6.18) are given by

$$m_{ev} = \frac{1}{c\left(y_1 \cdot e^{k1 \cdot z} + y_2 \cdot e^{k2 \cdot z} - D\right)} \quad (6.19)$$

$$K_{1,2} = \frac{-B \pm B^2 - 4AC}{2A}$$

C, D are known constants, y_1 and y_2 can be found out from the two boundary conditions common for both the configuration as

$$m_{ev|z=0} = 0 \quad (6.20a)$$

$$\frac{dm_{ev}}{dz|z = 0} = \beta(P_{s0} - P_A) \quad (6.20b)$$

From the numerical studies, it can be concluded that for a specified length of the regenerator, there is an optimum flow of the solution after which saturation of the water mass desorption rate exists in an open surface solar regenerator. Initially, with the increase of solution flow, the average solution temperature, as well as pressure, increases yielding an increase in water desorption rate from the solution surface. The mass desorption rate increases with the solution flow rate for both working fluids up to a certain flow ratio (flow rate of the solution to the flow rate of the forced air stream) and decreases with the further increase of the flow ratio. The study also shows that LiBr–H_2O gives a higher desorption rate over that of LiCl–H_2O due to higher partial water vapour pressure of earlier as compared to the latter.

If the open roof of the building to be cooled is utilised as a solar regenerator, the important and additional advantage is that it reduces the solar flux in coming to the room through the roof surface. Moreover, the heat capacity of the concrete surface also significantly helps in desorption even during off sunshine hours. The regenerator works fairly in hot and dry climates.

(c) Experimental Studies: An experiment on open and forced flow solar regenerator has been performed with water–lithium bromide and water–lithium chloride solution as the working fluids (each of 35% initial concentration) to observe the cooling effect, over a hot summer day.

The effects of ambient temperature, ambient relative humidity, ambient/forced air velocity and solar intensity on the refrigerant (water) desorption rate during regeneration of the two liquid desiccants, have been experimentally observed for both the regenerator configurations in one and the same unit at each hour of the day. The earlier theoretical results for the water desorption rate from the same configuration with the same working fluid and other identical conditions have been validated experimentally.

Figure 6.10 shows the schematic arrangement of the experimental apparatus considered for the regeneration study. The regenerator is (1.9×0.85) m^2 area of

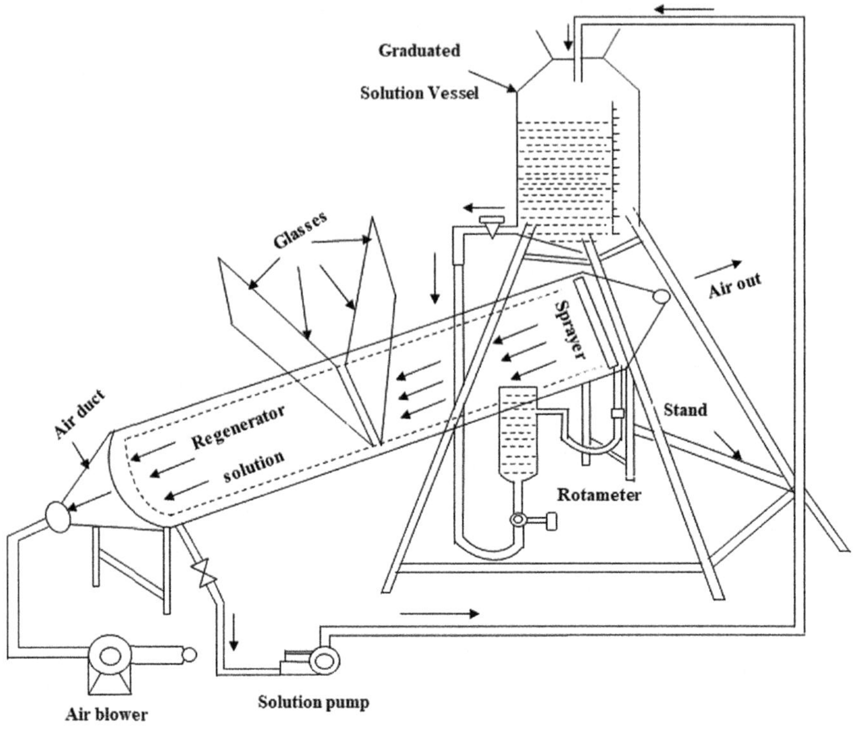

Fig. 6.10 Experimental setup of open and forced flow solar regenerator (Kaudinya and Kaushik 1986)

corrugated aluminium, a blackened sheet with a movable glass cover to make the system desirably open for forced flow solar regenerator. The solution container set up at a certain height is a graduated glass vessel with an outlet control valve to ensure the constant flow rate of the absorbent solution. The solution is sprayed on the regenerator uniformly with the help of a solution distributor as a regenerator header. The solution, at each hour, is collected in the graduated solution vessel connected by a solution pump to observe the reduced solution column in the solution vessel keeping the outlet of the vessel off during the measurement. An air blower is integrated to the system to force the air through the duct of the closed regenerator. The air is forced in a parallel or crossflow mode to the solution flow, to see the comparative effects of the air stream on the regenerator performance. The common parameters affecting the regenerator performance are initial solution concentration (known), flow rate (controlled), solution temperature (measured at the inlet, outlet and intermediate locations), airstream velocity (measured at inlet and outlet) and ambient temperature, ambient relative humidity, ambient air velocity and solar intensity (all these parameters are observed by appropriate devices at the time of measurements). The direct measurement of water desorption rate, however, is necessary which has been the main consideration of the present experiment at the time of observation.

First, the outlet of the solution vessel is kept off; each drop of the solution from the outlet pipe of the regenerator is pumped to the solution vessel to see the exact level of the solution column in the container. On the other hand, in a forced flow solar regenerator, the ratio of the inlet solution mass flow rate to the forced air mass flow rate is also optimized for the existing ambient conditions and then the maximum water desorption rate from theoretical and experimental studies are compared.

2. Hot and Humid Climate: Brine Still Solar Regenerator

Regenerator of the desiccant solution gets impeded by humid climates as laid in the coastal regions where both temperature and humidity are high in summer. So, dehumidification/cooling are essential for a major part of the year and a simple evaporative cooler is not so effective. However, the regenerative cooling system may adequately meet the demand of the cooling load in open cycle absorption cooling systems (OCACS), the regeneration process is dependent on ambient air parameters, i.e., temperature, velocity and relative humidity. At lower and higher values of the first two parameters, the system may work but the regenerative system hardly compromises with higher values of the third parameter. Thus, there is a wide scope for such a regenerator system which may work irrespective of relative humidity. Undoubtedly such a system is a brine still solar regenerator.

A brine still solar regenerator shown in Fig. 6.11 is commonly utilised for regenerating the liquid desiccant. The liquid desiccant solution of given concentration gets heated by solar energy, which increases the partial vapour pressure of the solution. The water vapour desorbs when the driving potential (i.e., the vapour pressure difference between the solution and internal glass surface) is positive. Condensation of desorbed water at the internal glass surface takes place and distillation is taken out. This process is independent of the ambient relative humidity. Hence, a substantial water adsorbate may achieve from the brine still solar regenerator even in hot and humid climates.

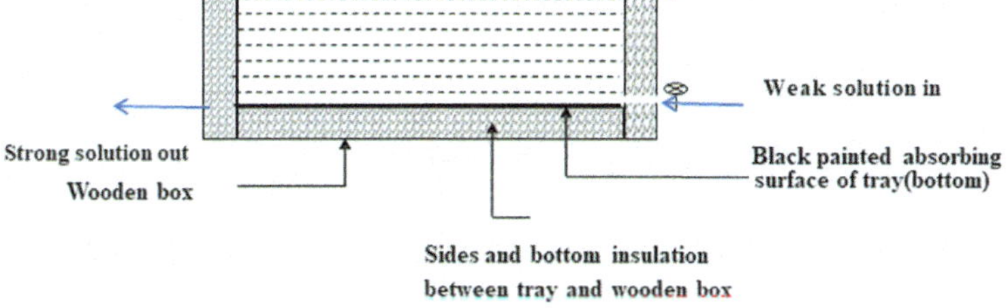

Fig. 6.11 A brine still solar regenerator

6.5.2 *Closed Cycle Continuous Solar Cooling System*

A simple closed-cycle absorption system is a heat actuated unit using a refrigerant which is alternatively absorbed and liberated at certain temperature and pressure. The closed cycle involves two processes loops. The primary loop is for cooling process. The secondary loop is employed to enable heat transfer from the refrigeration load. The primary components essential in this system (Fig. 6.12) are absorber, generator, condenser and evaporator. Thermal energy is applied to the generator to drive off refrigerant vapour. Refrigerant vapour is then subjected to condensation in the condenser. In evaporator, the refrigerant absorbs heat from the space to be refrigerated. In a continuous absorption cooling system, an external source of power is necessary to operate the pump and control systems and these systems are more preferred for comfortable cooling rather than refrigeration.

6.5.2.1 Thermodynamic Modelling of Closed Cycle Absorption Cooling System

A continuous VAB is well established and has been extensively studied in the literature for refrigeration and comfort cooling applications. The performance of these absorption cooling systems mainly depends on the choice of working fluid which is a combination of refrigerant and suitable absorbent. Many combinations of working fluids have been proposed in the literature but most widely used is ammonia as a refrigerant with liquid/solid absorbent (viz. $H_2O/LiNO_3/SrCl_2/CaCl_2$ etc.).

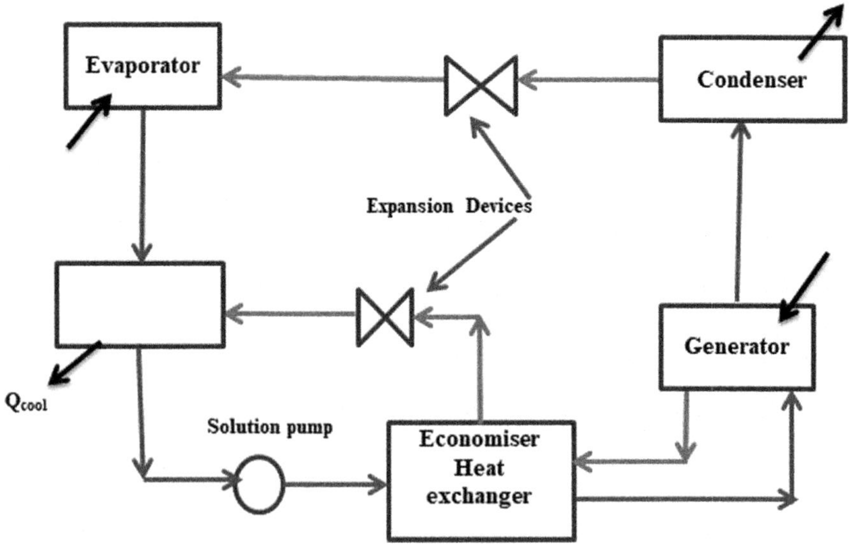

Fig. 6.12 A closed cycle continuous solar cooling system

In this section, the modelling procedure for continuous cycle absorption cooling system and the results for H_2O–LiBr, NH_3–H_2O and NH_3–$LiNO_3$ as the working fluids are presented.

The overall heat balance for the absorption cycle is given by

$$Q_G + Q_E = Q_A + Q_C \tag{6.21}$$

where the solution pump work is neglected as compared to the other heat transfer rates and the cooling COP is given by

$$\mathrm{COP} = Q_E/Q_G \tag{6.22}$$

Another important parameter for the absorption cooling system is the recirculation ratio defined as the ratio of strong solution flow rate to the refrigerant flow rate, i.e.,

$$\text{Recirculation ratio} = \frac{M_s}{M_r} = (X_R - X_S)/(X_S - X_W) \tag{6.23}$$

To evaluate the actual COP of the system, mass, heat and material balance equations around each component of the system are solved along with the thermodynamic properties state equations.

The basic system equations are in a general form given by

$$\sum M_i = 0, \sum m_i X_i = 0, \sum M_i h_i = 0 \tag{6.24}$$

where M_i is the mass flow rate of the solution, X_i is the solution concentration, h_i is the enthalpy. The state equation for potential working fluids has already been known and presented in Appendix A. An iterative procedure is required to solve the nonlinear state equation if necessary.

The input conditions assumed for the system modelling are the system components: temperature, the effectiveness of the preheater and subcooled heat exchanger, refrigerant concentration as well as the unit mass flow rate of the vapour leaving the generator. The refrigerant mass flow rate can also be fixed and evaluated by the ratio of known cooling capacity and the latent heat of vapourisation of the refrigerant vapour at the evaporator temperature. The refrigerant concentration leaving the generator is unity for H_2O–LiBr and NH_3–$LiNO_3$ and is assumed to be fixed for NH_3–H_2O to account for the fixed effectiveness of the rectifier heat exchanger. This concentration is assumed to be 0.99 for NH_3–H_2O system.

6.5.2.2 *Single Stage Water–Lithium Bromide Vapour Absorption Cooling System*

A water–lithium bromide single-stage absorption cycle (Fig. 6.13) consists of a generator, condenser, evaporator, an absorber, expansion device and heat exchanger.

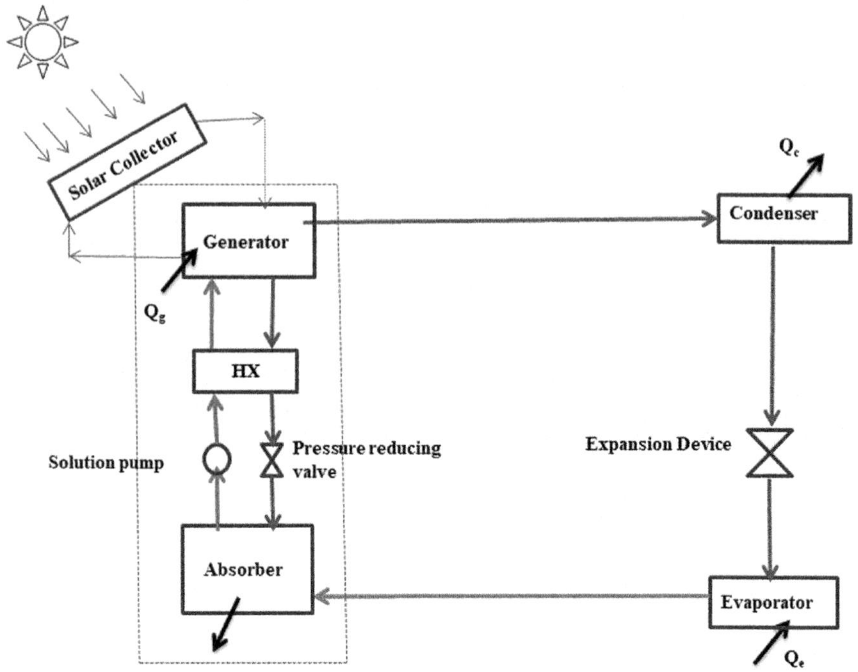

Fig. 6.13 Single-stage H_2O–LiBr—vapour solar absorption cycle for solar air conditioning

In this system, water is the refrigerant and LiBr is the absorbent. The system is powered by solar energy. The solar heat input to the generator drives the system.

An expansion valve is placed in-between condenser and evaporator to expand the liquid refrigerant from the condenser pressure level to the evaporator pressure. Another valve is connected between the generator and absorber to reduce the pressure of a week solution to absorber pressure. The refinements of the cycle can be made by adding liquid–liquid heat exchanger which is used to recover heat from the weak solution and transfer it to the strong solution entering the generator. Another liquid–vapour heat exchanger can be added between the condenser and expansion valve to precool the condensed liquid refrigerant before subjected to expansion.

The subcooling will increase the COP of the system. Addition of solution heat exchanger between generator and absorber will reduce the heat requirement in the generator which improves the COP but it also reduces the cooling requirement in the absorber which is beneficial for the cooling system.

A pressure–temperature–concentration (*P*–*T*–*X*) equilibrium diagram for the water–LiBr cycle with various thermodynamic processes are shown in Fig. 6.14.

In the generation process, the solution concentration (in terms of weight of LiBr) increases from 55 to 66% whereas the equilibrium solution temperature increases from 72 to 82 °C at the high-pressure prevailing in the system (~48 mm Hg). On the other hand, in the absorber, the solution concentration decreases from 60 to 55%

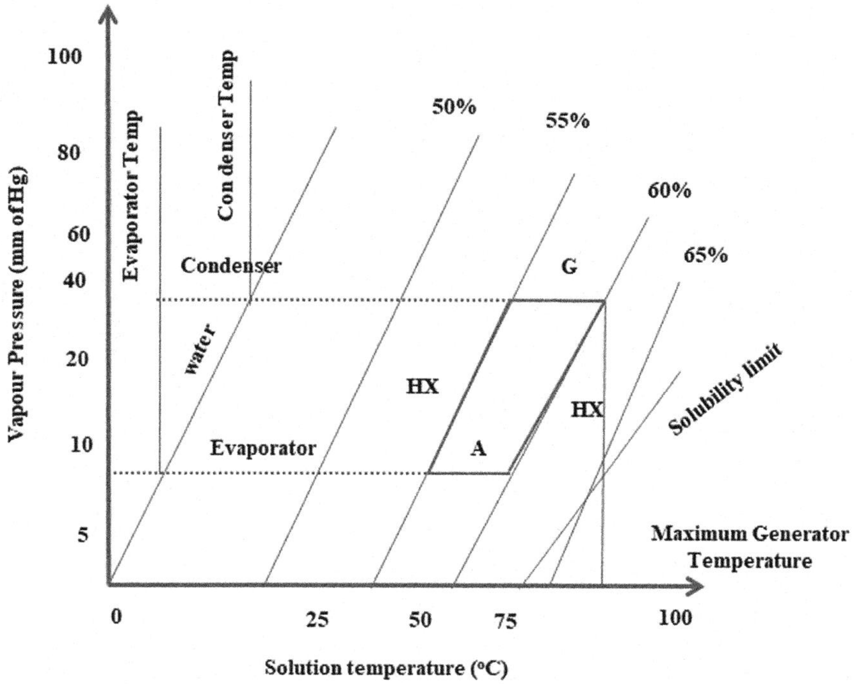

Fig. 6.14 *P–T–X* diagram for H_2O–LiBr cycle (Duffie et al. 1980)

as the solution temperature drops from 48 to 38 °C at the low-pressure prevailing in the system (~8 mm Hg). The pressure difference between the low- and high-pressure sides of LiBr–H_2O system is small enough so that a vapour lift pump and thermosyphon action can be used for solution circulation from absorber to the generator. However, there are problems of water cooling of the absorber/condenser components, restrictions on generator temperature and crystallisation of salt during system operation.

6.5.2.3 *NH_3–H_2O Solar VAB System*

The NH_3–H_2O refrigeration system was first patented by the Frenchman Ferdinand Carre in 1859. In this system, the ammonia serves as the refrigerant and water as the transport medium (absorbent). Low-pressure ammonia vapour leaves from the evaporator enters the thermal compressor unit, where it is subjected to pressurisation by the application of low-grade energy source. In the absorber unit of the thermal compressor, the NH_3 vapour dissolves and reacts with water to form a strong solution of NH_3–H_2O (enriched with ammonia) (Fig. 6.15).

This is an exothermic reaction, and hence a large heat is rejected during this cycle. Since the absorption is inversely proportional to the temperature of the absorber unit,

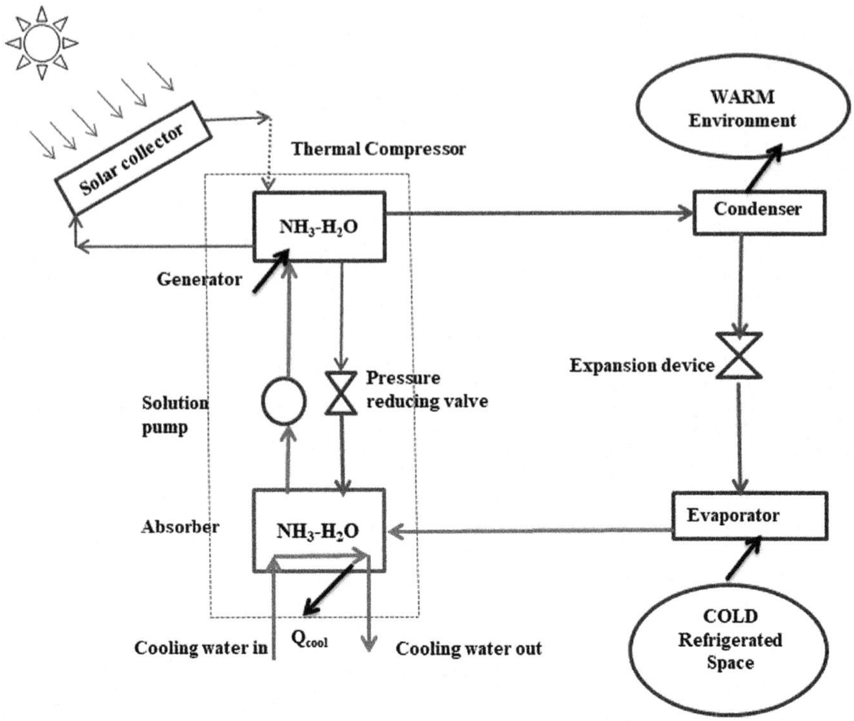

Fig. 6.15 Schematic diagram of solar-powered single effect NH_3–H_2O vapour absorption cooling system (Kaushik 1989)

it is necessary to cool the absorber unit to maintain its temperature as low as possible to maximise the ammonia dissolved in water. The liquid solution can then be pumped in to the required pressure with very much less input work. The liquid solution is then heated in the regenerator from a source (solar heat). The solubility of the refrigerant in the absorber is very low at this temperature. As a result, the refrigerant vapour is then driven off from the solution. The rest of weak solution then flows back to the absorber through the expansion device. Once the pressure of ammonia is raised by the thermal compressor unit, it is cooled and condensed in the condenser by rejecting heat to the surroundings, and then it is throttled to the evaporator pressure. The low-pressure ammonia refrigerant flows to the evaporator, where it absorbs the heat from space to be refrigerated.

6.5.3 Actual VAB Cooling Systems

There are two drawbacks in the simple vapour absorption cooling systems:

(1) The low-temperature of the rich solution entering the generator and the high-temperature of the weak solution of the absorber.
(2) Presence of absorbent in the vapour leaving the generator and going to the refrigerant circuit.

The actual practical system consists of a regenerative heat exchanger incorporated in-between the weak and strong solution. This will reduce the amount of heat added in the generator and decreases the amount of heat rejected in the absorber. The size of the absorber and generator can be reduced. The heat exchanger is called a liquid–liquid heat exchanger.

Figure 6.16 shows actual vapour absorption cooling system. The ammonia vapour (refrigerant) is driven off from the strong solution of NH_3 by an external heating source. A rectifier unit is placed to remove the water vapour carried with ammonia. High-pressure ammonia vapour gets condensed in the condenser. The liquid ammonia is passed to the expansion device, where it is expanded to the evaporator pressure level. The low-pressure refrigerant enters the evaporator, absorbs heat from the space to be refrigerated and leaves the evaporator as a saturated vapour. The weak solution (aqua-ammonia) entering the absorber becomes strong solution after absorbing ammonia vapour and then it is pumped back to the regenerator through the heat

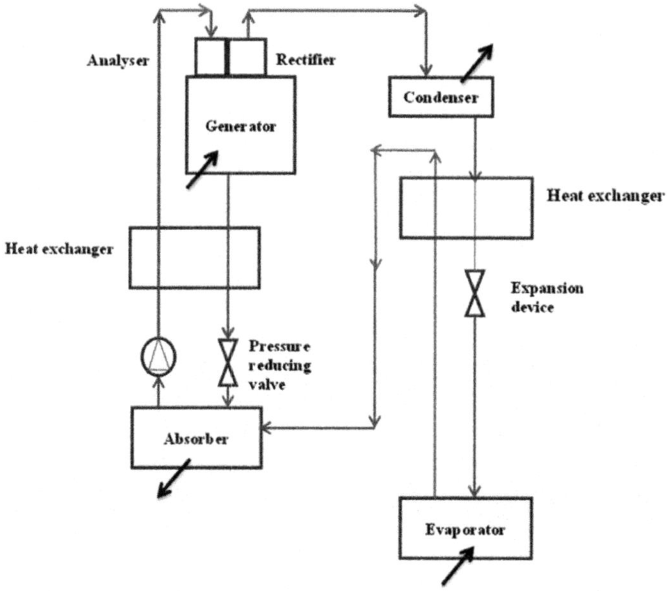

Fig. 6.16 Actual absorption cooling system

Table 6.2 A comparison of different absorption chillers for solar-driven air conditioning systems

Type	Typical COP	Heat source temperature (°C)	Solar heat required per kW of cooling effect (kW)	Type of solar collectors
Single effect	0.7	85	1.43	Flat plate
Double effect	1.2	130	0.83	Flat/ETC
Triple effect	1.7	220	0.59	Evacuated tube collectors (ETC)

exchanger. Pump increases the pressure of a strong solution. The weak solution passes through a throttle valve to reach the pressure level of the absorber. Table 6.2 shows the comparison of different absorption chillers.

6.5.4 Thermodynamic Analysis of VAB Cycle

Performance of an absorption cooling system is expressed in terms of Coefficient of Performance (COP). The effect of thermodynamic parameters on COP in an ideal VAB cycle and non ideal cycle is detailed in this section.

6.5.4.1 Coefficient of Performance (COP)

Case 1: COP of an Ideal Absorption Cycle

An absorption cycle is basically a three-temperature level cycle, viz. high-temperature heat supply, sink and the low-temperature heat supply system and the cooling COP of the ideal absorption cycle is given by

$$\mathrm{COP} = \left(\frac{T_G - T_O}{T_G}\right)\left(\frac{T_E}{T_o - T_E)}\right) \tag{6.25}$$

where T_O be the common sink temperature between the low-temperature heat supply and the higher temperature heat supply in the absorption system. But in practice, the temperature of condenser and absorber are not same, and hence the above equation for the COP is not valid. It is surprising to notice that for different absorber-condenser temperature, different expressions for COP have been presented by earlier workers. In the present section, the authors intend to clarify the basic distinction between the two expressions for COP and support the one with correct thermodynamic basis (Kaushik and Chandra 1982).

The COP of an absorption cycle is defined as the ratio of the heat flow into the evaporator and unit heat flow in the generator. In the VAB system, the total refrigerating effect in the system is dependent on the heat absorbed by the refrigerant in the evaporator. The total energy supplied to the system is the sum of the work done by the pump and the heat supplied in the generator. Therefore, the COP of the system is theoretically given as

$$\text{COP} = \frac{\text{Heat absorbed in the evaporator}}{\text{Workdone by pump} + \text{Heat supplied in the generator}}$$

$$\text{COP} = \frac{Q_e}{Q_g + W_p} \tag{6.26}$$

Neglecting the pump work, we get

$$\text{COP} = \frac{Q_e}{Q_g} \tag{6.27}$$

which is the expression for COP of the system.

The COP for heating is defined as ratio of total heat flow out (from the condenser and absorber) and heat flow into the generator

$$\text{COP}_{\text{heating}} = \frac{Q_A + Q_C}{Q_G} \tag{6.28}$$

However, from the overall heat balance

$$Q_G + Q_E = Q_A + Q_C \tag{6.29}$$

The pump work is assumed to be small as compared to the other quantities in Eq. (6.29).

$$\text{COP}_{\text{heating}} = 1 + \text{COP}_{\text{cooling}} \tag{6.30}$$

With the value of minimum heating COP as unity, the minimum value of cooling COP is zero. The absorption cycle may be thought of combination of a heat engine and a refrigerating machine. The work produced by the heat engine is used to drive the refrigerator in this condition. This conclusion shows that the processes in the condenser and evaporator are thermodynamically reversible, the reduction in the entropy in the condenser equals the gain in entropy in the evaporator.

$$\frac{Q_C/m_r}{T_C} = \frac{Q_E/m_r}{T_E} \tag{6.31}$$

m_r in the Eq. (6.31) represents the refrigerant flow rate. For the complete cycle

$$\frac{Q_G}{T_G} + \frac{Q_E}{T_E} = \frac{Q_A}{T_A} + \frac{Q_C}{T_C} \tag{6.32}$$

Hence, we obtain

$$\frac{Q_G}{T_G} = \frac{Q_A}{T_A} \tag{6.33}$$

Combining Eqs. (6.31) in (6.32) and (6.33) and solving for Q_E/Q_G or $(COP)_{cooling}$

$$COP_{cooling} = \left(\frac{T_G - T_A}{T_G}\right)\left(\frac{T_E}{T_C - T_E}\right) = (COP)_{Ideal} \tag{6.34}$$

Above equation shows that the coefficient of performance of an ideal absorption cycle is equal to the product of thermodynamically reversible heat engine efficiency with the COP of a thermodynamically reversible vapour compression cooling cycle.

Equation (6.34) is similar to that formulated by Stoeker and Reed, and Kaushik and Sheridan. A Carnot heat engine operates between the generator and the absorber thereby agreeing to the fundamental difference between the VCR cycle and a VAB cycle. Thus, the generator-absorber assembly is acting like a compressor in a VCR system. It should also be mentioned that Eq. (6.34) disagrees with that available in ASHRAE hand book and derived by Stanish and Perlmutter where it was found that

$$COP_{cooling} = \left(\frac{T_G - T_C}{T_G}\right)\left(\frac{T_E}{T_A - T_E}\right) \tag{6.35}$$

Equation (6.35) assumes that the ideal VAB comprises of a Carnot heat engine (operates between the T_G and T_C) and a Carnot reversed heat engine, which operates between the evaporator and absorber temperature. Thus, the basis of this equation is the heat rejection of the engine portion is at the condenser and the refrigeration portion is at the absorber, which is not thermodynamically correct and hence the Eq. (6.35) is doubtful.

It is interesting to note that when $T_C = T_A = T_O$, COP reduce to

$$COP_{cooling} = \left(\frac{T_G - T_O}{T_G}\right)\left(\frac{T_E}{T_O - T_E}\right) \tag{6.36}$$

Equation (6.36) is generally used for estimating COP of cooling. Thus, assuming the same sink for heat engine and the refrigeration part, the discrepancy is removed and the earlier workers have just left out the case when T_A is not equal to T_C. The purpose of the study is to emphasis that the Eq. (6.35) for COP is misleading and of doubtful merit.

As the expressions are same for equal T_A and T_C, no one could justify the difference between two cases. The proposed Eq. (6.35) is more logical on thermodynamic basis which derive under the assumption that an ideal absorption cycle is equivalent to a heat engine between a generator and absorber temperature and a heat pump between the evaporator and condenser temperature. Felli and Kumar have further studied the basic thermodynamic equation of an ideal absorption cycle in a more general way.

Eliminating Q_A and Q_C between Eqs. (6.33) and (6.35) respectively, we get

$$\text{COP} = \frac{T_E}{T_c - T_E}\left(\frac{T_G - T_C}{T_G} + \alpha\frac{T_c - T_A}{T_A}\right) \quad (6.37)$$

and

$$\text{COP} = \frac{T_E}{T_A - T_E}\left(\frac{T_G - T_A}{T_G} + \beta\frac{T_A - T_C}{T_C}\right) \quad (6.38)$$

where $\alpha = \frac{Q_A}{Q_G}$ and $\beta = \frac{Q_c}{Q_G}$. Thus, it is seen that an ideal COP does not depend only on component temperatures but also on the heat flow rates associated with the component.

Variation of COP with α and β

(i) If $T_A < T_C$, the quantity $\beta \cdot \frac{T_A - T_C}{T_C}$ in Eq. (6.38) is negative or equal to zero because $\beta \geq 0$. Therefore, COP reaches its maximum when $\beta = 0$ and then

$$\text{COP}_{\max} = \left(\frac{T_E}{T_A - T_E}\right)\left(\frac{T_G - T_A}{T_G}\right) \quad (6.39)$$

(ii) If $T_A > T_C$, the quantity $\alpha \cdot \frac{T_C - T_A}{T_A}$ in Eq. (6.37) is negative or equal to zero as $\alpha \geq 0$. Therefore, COP reaches its maximum value when $\alpha = 0$ and is given by

$$\text{COP}_{\max} = \left(\frac{T_E}{T_C - T_E}\right)\left(\frac{T_G - T_C}{T_G}\right) \quad (6.40)$$

Thus, (i) and (ii) above shows that the COP reaches its maximum when the heat engine operation is improper, because $\alpha = 0$ and $\beta = 0$, i.e., $Q_C = 0$, $Q_A = 0$, which is impossible with the real fluids.

When $T_A = T_C$, it is observed that if $\alpha > 0$, $\beta > 0$ and $T_A \neq T_C$ the quantities $\frac{T_C - T_A}{T_A}$ and $\frac{T_A - T_C}{T_C}$ are different from zero and have the opposite signs. If $T_C > T_A$ (or $T_C < T_A$), then according to the Eqs. (6.37) and (6.38), COP increases when $(T_C - T_A)$ decreases and it reaches its maximum when $T_C = T_A = T_O$ as

$$\text{COP}_{\max} = \left(\frac{T_E}{T_O - T_E}\right) \cdot \left(\frac{T_G - T_O}{T_G}\right) \quad (6.41)$$

This maximum COP is independent of α and β may be interpreted as the Carnot heat engine efficiency between T_G and T_O multiplied by the Carnot COP between T_O and T_E.

Case 2: COP of Non-ideal Absorption Cycle

The Carnot cycle is an ideal reversible one and hence it achieves the maximum theoretical COP. A non-ideal absorption cycle employing any working fluid cannot achieve the Carnot performance due to the irreversibilities, temperature and pressure drops and the working fluid properties etc.

In an actual VAB cycle (Fig. 6.17), the compressor work is reduced by dissolving the refrigerant vapour in a suitable absorbent in the absorber. The liquid solution can then pump to the required pressure with very much less pumping work.

In the cooling operation (refrigeration and airconditioning), input heat is transferred into the refrigerant and absorbent in the generator and into the refrigerant in the evaporator. The process in absorber and condenser are exothermic, and hence heat must be removed from the absorbent refrigerant pair. In actual system operation, the heat quantities associated with the solution streams flowing into and out of the generator cannot be ignored. The refrigerant leaves the generator at relatively higher temperature, hence it must be cooled to the absorber temperature before it exits the absorber unit of the configuration. Also, it must be reheated to the exit generator temperature before it leaves the generator. The actual COP of an absorption unit is given by

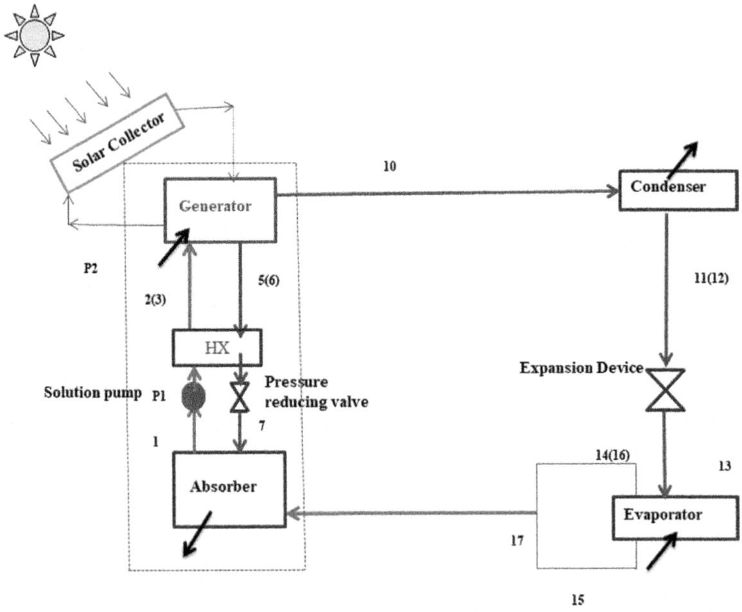

Fig. 6.17 Solar operated actual VAB cooling system

$$\mathrm{COP}_{\text{cooling}} = \frac{Q_{\mathrm{E}}}{Q_{\mathrm{G}}} = \text{Heat extracted by the evaporator/Heat input to the generator}$$

$$\mathrm{COP}_{\text{cooling}} = \frac{\Delta H_{\mathrm{V}}}{[\Delta H_{\mathrm{d}}(RR + 1) + \Delta H_{\mathrm{V}}]} \tag{6.42}$$

where ΔH_V is the heat of vapourisation of refrigerant per unit mass flow rate of the refrigerant, ΔH_d the heat of dilution (or mixing) per unit mass flow rate of the solution and RR the recirculation ratio being equal to the mass flow rate of the weak solution leaving the generator per unit mass flow rate of the refrigerant. Thus, it can be seen that the value of $\mathrm{COP}_{\text{cooling}}$ is always less than 1.0 unless the value of ΔH_d is zero or negative. The heat of dilution of an ideal solution is zero. Therefore, theoretical $\mathrm{COP}_{\text{cooling}}$ of the absorption cycle using conventional working fluids such as NH_3–H_2O and H_2O–LiBr cannot exceed unity. The practical COP of the commercially available single effect absorption units could range from 0.5 to 0.8 mainly depending upon the absorption cycle employed and air/water cooling of an absorber or the condenser component. Moreover, the success of achieving a higher COP is also governed by the choice of working fluid used.

In the first law of thermodynamics analysis of a refrigerator, the efficiency of the machine is determined as the ratio of energy output to the energy input. No distinction is made between heat and energy transfer. However, in the second law of thermodynamics, it is recognised that heat energy has lower availability than the work energy. This analysis not only allows a comparison between the actual machine and an ideal machine but can be used to determine the degree of efficiency (or irreversibility) associated with the cycle. The irreversibility of a process or cycle is defined as the difference between the reversible process (viz. compression, condensation or evaporation that could theoretically be produced and the actual process that is possible by control volume under investigation.

An equation for any actual process produced by the control volume is obtained by applying the first law of thermodynamics while an equation for the theoretically maximum reversible process is obtained by applying both first and second laws of thermodynamics. These equations are derived in terms of entropy values corresponding to different state points and used to determine the overall irreversibility of the cycle or the thermodynamic limit on the COP of the cycle. An understanding of these limitations can be used as an aid in the reduction of overall irreversibility for any refrigerant through the modification of the basic cycle and hence can also be used for establishing the thermodynamic basis for the choice of working fluid.

For carrying out the thermodynamic analysis of the proposed vapour absorption cooling system, the following assumptions are made which corresponds to the various state points referred in the Fig. 6.17 (without preheater and sub-cooler heat exchanger):

1. Component temperatures are uniform throughout the length and are independent of the time for steady-state conditions.
2. System high-pressure is the equilibrium saturation pressure corresponding to the condenser temperature. High-pressure in the generator is the same as in condenser.
3. No pressure changes except through the flow pump.
4. The system low-pressure is the saturation pressure corresponding to the evaporator temperature, both evaporator and absorber are at its low-pressure.
5. Heat transfer to the surroundings, other than unavoidable, does not exist.
6. Thermodynamic properties of the equilibrium states are equal to the corresponding equilibrium states at the same temperature and concentration.
7. Refrigerant vapour leaving the evaporator at state point 14 is the only saturated vapour.
8. Refrigerant vapour leaving the generator at the state point 10 is in equilibrium at the high-pressure and the generator temperature.
9. Refrigerant vapour leaving the condenser (at state point 11) is saturated at the condenser temperature.
10. Weak solution leaving the generator (at state point 5) is at the equilibrium temperature and the final solution concentration in the generator.
11. Strong solution leaving the absorber (state point 1) is a saturated liquid at the temperature of final concentration in the absorber.
12. The refrigerant concentration in the vapour entering the condenser is assumed to be known and constant (being equal to 1 for H_2O–LiBr and less than 1 for NH_3–H_2O depending upon the effectiveness of the rectification process).
13. Pumping is isentropic.
14. Assume a weak solution contain more percentage of refrigerant and less of absorbent, and the strong solution contains more percentage of absorbent and less of refrigerant.

The above assumptions are used to formulate the mass, material and energy balance equations.

$$\sum_i m_\mathrm{i} = 0, \quad \sum_j m_\mathrm{j} X_i = 0 \tag{6.43}$$

$$\sum_i m_\mathrm{i} h_\mathrm{i} \pm Q_\mathrm{i} = 0$$

Referring to Fig. 6.4 without preheater and subcooler heat exchangers, the following relations exist between various heat transfer rates and the properties of working fluid in steady-state cycle:

$$\begin{aligned}
Q_{\mathrm{G}} &= m_{10}h_{10} + m_5h_5 - m_2h_2 = m_{\mathrm{r}}(h_{10} - h_5) + m_{\mathrm{a}}(h_5 - h_2) \\
Q_{\mathrm{C}} &= m_{10}h_{10} - m_{11}h_{11} = m_{\mathrm{r}}(h_{10} - h_{11}) \\
Q_{\mathrm{E}} &= m_{12}h_{12} - m_{15}h_{15} = m_{\mathrm{r}}(h_{12} - h_{15}) \\
Q_{\mathrm{A}} &= m_{14}h_{14} + m_7h_7 - m_1h_1 = m_{\mathrm{r}}(h_{14} - h_7) + m_{\mathrm{a}}(h_7 - h_1)
\end{aligned} \tag{6.44}$$

where $m_{\mathrm{r}} = m_{10} = m_{11} = m_{12}$ is the refrigerant flow rate and $m_{\mathrm{s}} = m_1 = m_2$ is the strong solution mass flow rate given by the mass and material balance equation, viz.

$$m_{\mathrm{s}} = m_{\mathrm{r}} + m_{\mathrm{w}}$$

$$m_{\mathrm{a}}X_{\mathrm{s}} = m_{\mathrm{r}}X_{\mathrm{r}} + m_{\mathrm{w}}X_{\mathrm{w}} \tag{6.45}$$

Thus,

$$\frac{m_{\mathrm{a}}}{m_{\mathrm{r}}} = \frac{X_{\mathrm{R}} - X_{\mathrm{w}}}{X_{\mathrm{s}} - X_{\mathrm{w}}} \tag{6.46}$$

Equations (6.45) and (6.46) are valid for NH_3–H_2O fluid cycle as shown in Fig. 6.4 if the generator-rectifier assembly is considered as one component and the solution heat exchanger is assumed to be eliminated.

Thus, the actual COP of the absorption cycle is defined as

$$\mathrm{COP}_{\mathrm{actual}} = \frac{Q_{\mathrm{E}}}{Q_{\mathrm{G}}} = \frac{h_{12} - h_{14}}{\left(h_{10} - h_5 + \frac{(X_{\mathrm{R}} - X_{\mathrm{w}})}{(X_{\mathrm{S}} - X_{\mathrm{w}})} \cdot (h_5 - h_2)\right)} \tag{6.47}$$

Applying the Second law of thermodynamics to relate thermodynamic properties and cycle characteristics. Apply the seady state flow conditions viz.

$$\begin{aligned}
Q_G &\leq T_G[m_r(\phi_{10} - \phi_5) + m_s(\phi_5 - \phi_2)] \\
Q_C &\geq T_C \cdot m_r(\phi_{10} - \phi_{11}) \\
Q_E &\leq T_E \cdot m_r(\phi_{12} - \phi_{14}) \\
Q_A &\geq T_A[m_s(\phi_7 - \phi_1) + m_r(\phi_{14} - \phi_7)]
\end{aligned} \tag{6.48}$$

It must be noticed that the equality sign is for the reversible cycle and the inequality sign is for a reversible process. Thus, the overall energy balance of the absorption cycle is given by (neglecting the pump work)

$$Q_G + Q_E = Q_A + Q_C \tag{6.49}$$

Thus, we obtain the condition

$$[T_A\{m_s(\phi_7 - \phi_1) + m_r(\phi_{14} - \phi_7)\} + T_C \cdot m_r(\phi_{10} - \phi_{11})] \leq (Q_G + Q_E) \leq [T_G\{m_r(\phi_{10} - \phi_5) + m_s(\phi_5 - \phi_2)\} + T_E \cdot m_r(\phi_{12} - \phi_{14})] \quad (6.50)$$

Dividing Eq. (6.50) by Q_G and introducing the basic definition of Q_G and COP, We get

$$COP_{LL} \leq COP_{actual} \leq COP_{UL} \quad (6.51)$$

where the lower limit (LL) and the upper limit (UL) of COP are given by the equations

$$COP_{LL} = \frac{[T_A\{m_s(\phi_7 - \phi_1) + m_r(\phi_{14} - \phi_7)\} + T_C m_r(\phi_{10} - \phi_{11})]}{[m_r(h_{10} - h_5) + m_s(h_5 - h_2)]} - 1 \quad (6.52)$$

and

$$COP_{UL} = \frac{[T_G\{m_r(\phi_{10} - \phi_5) + m_s(\phi_5 - \phi_2)\} + T_E m_r(\phi_{12} - \phi_{14})]}{[m_r(h_{10} - h_5) + m_s(h_5 - h_2)]} - 1 \quad (6.53)$$

It must be mentioned that to evaluate these COP values the thermodynamic properties (viz. entropy and enthalpy) of the working fluids should be known in the form of state equations. The lack of data on entropy values for the binary mixtures possesses a problem. Methods of evaluation of entropy of mixtures need to be developed especially for H_2O–LiBr, where LiBr is in solid form in its pure state.

It can be further mentioned that these upper and lower limit COP, besides being dependent upon the environmental temperature of the components of the cycle also depend on thermodynamic properties of the refrigerant-absorbent mixture. The actual COP of the cycle will not go out of the bounds specified by the limit. Also, with the understanding the ideal Carnot COP gives the uppermost limit regardless of the working fluids, the following inequalities can be written as

$$COP_{LL} \leq COP_{actual} \leq COP_{UL} \leq COP_{ideal} \quad (6.54)$$

Thus, the evaluation of lower and upper limits of COPs can also be used for the comparative study of different working fluids.

6.5.5 General Analysis for COP of Vapour Absorption Cooling System

The energy balance for the cycle can be written as (Fig. 6.17)

$$Q_G + Q_E + W_p = Q_C + Q_A + Q_p \quad (6.55)$$

$$\frac{Q_{\mathrm{G}}}{T_{\mathrm{G}}}+\frac{Q_{\mathrm{E}}}{T_{\mathrm{E}}}+\Delta S_o-\left(\frac{Q_{\mathrm{A}}+Q_{\mathrm{C}}+Q_{\mathrm{P}}}{T_{\mathrm{o}}}\right)=0 \tag{6.56}$$

where we have assumed $T_A = T_C = T_o$.

Now, putting the value of $Q_G + Q_E + W_p = Q_C + Q_A + Q_P$.

Equation (6.56) from Eq. (6.55)

We get

$$\frac{Q_{\mathrm{G}}}{T_{\mathrm{G}}}+\frac{Q_{\mathrm{E}}}{T_{\mathrm{E}}}+\Delta S_o-\left(\frac{Q_{\mathrm{G}}+Q_{\mathrm{E}}+W_{\mathrm{P}}}{T_o}\right)=0 \tag{6.57}$$

Now dividing Eq. (6.57) by $Q_G + W_P$ we get

$$\left(\frac{Q_{\mathrm{G}}}{Q_{\mathrm{G}}+W_{\mathrm{P}}}\right)\frac{1}{T_{\mathrm{G}}}+\left(\frac{Q_{\mathrm{E}}}{Q_{\mathrm{G}}+W_{\mathrm{P}}}\right)\frac{1}{T_{\mathrm{E}}}+\frac{\Delta So}{(Q_{\mathrm{G}}+W_{\mathrm{P}})}=\left(\frac{Q_{\mathrm{G}}+Q_{\mathrm{E}}+W_{\mathrm{P}}}{T_o}\right)\frac{1}{T_o}$$

$$\left(\frac{Q_{\mathrm{G}}}{Q_{\mathrm{G}}+W_{\mathrm{P}}}\right)\frac{1}{T_{\mathrm{G}}}+\left(\frac{Q_{\mathrm{E}}}{Q_{\mathrm{G}}+W_{\mathrm{P}}}\right)\frac{1}{T_{\mathrm{E}}}+\frac{\Delta S_o}{(Q_{\mathrm{G}}+W_{\mathrm{P}})}=\left(1+\frac{Q_{\mathrm{E}}}{Q_{\mathrm{G}}+W_{\mathrm{P}}}\right)\frac{1}{T_o} \tag{6.58}$$

We know that

$$\mathrm{COP}=\frac{Q_{\mathrm{E}}}{Q_{\mathrm{G}}+W_{\mathrm{P}}}$$

Putting this value in Eq. (6.58)

$$\left(\frac{Q_{\mathrm{G}}}{Q_{\mathrm{G}}+W_{\mathrm{P}}}\right)\frac{1}{T_{\mathrm{G}}}+\mathrm{COP}\frac{1}{T_{\mathrm{E}}}+\frac{\Delta S_o}{(Q_{\mathrm{G}}+W_{\mathrm{P}})}=\frac{1}{T_o}(1+\mathrm{COP})$$

or

$$\mathrm{COP}\left(\frac{1}{T_{\mathrm{E}}}-\frac{1}{T_o}\right)=\frac{1}{T_o}-\frac{1}{T_{\mathrm{G}}}\left(\frac{Q_{\mathrm{G}}}{Q_{\mathrm{G}}+W_{\mathrm{P}}}\right)-\frac{\Delta S_o}{(Q_{\mathrm{G}}+W_{\mathrm{P}})} \tag{6.59}$$

Now adding and subtracting $1/T_G$ in Eq. (6.59), we get (in the right hand side)

$$\mathrm{COP}\left(\frac{1}{T_E}-\frac{1}{T_o}\right)=\left(\frac{1}{T_o}-\frac{1}{T_G}\right)+\frac{1}{T_G}-\left(\frac{Q_G}{Q_G+W_P}\right)\frac{1}{T_G}-\frac{\Delta S_o}{(Q_G+W_P)}$$

$$\mathrm{COP}\left(\frac{1}{T}-\frac{1}{T_o}\right)=\left(\frac{1}{T_{\mathrm{o}}}-\frac{1}{T_{\mathrm{G}}}\right)+\frac{1}{T_G}\left(\frac{W}{Q_{\mathrm{G}}+W_{\mathrm{P}}}\right)-\frac{\Delta S_o}{(Q_{\mathrm{G}}+W_{\mathrm{P}})} \tag{6.60}$$

Dividing Eq. (6.60) by $\frac{1}{T_E}-\frac{1}{T_o}$.

We get

$$\mathrm{COP} = \frac{\left(\frac{1}{T_o} - \frac{1}{T_G}\right)}{\left(\frac{1}{T_E} - \frac{1}{T_o}\right)} + \frac{W_P/T_G}{(Q_G + W_P)\left(\frac{1}{T_E} - \frac{1}{T_o}\right)} - \frac{\Delta S_o}{(Q_G + W_P)\left(\frac{1}{T_E} - \frac{1}{T_o}\right)}$$

But we know that

$$\mathrm{COP_C} = \frac{\left(\frac{1}{T_o} - \frac{1}{T_G}\right)}{\left(\frac{1}{T_E} - \frac{1}{T_o}\right)}$$

$$\mathrm{COP_C} = \left(\frac{T_G - T_O}{T_G}\right)\left(\frac{T_E}{T_O - T_E}\right) \tag{6.61}$$

This is the ideal Carnot COP for vapour absorption cooling system.

$$\mathrm{COP} = \mathrm{COPc} + \frac{W_P/T_G}{(Q_G + W_P)\left(\frac{1}{T_E} - \frac{1}{T_O}\right)} - \frac{\Delta S_o}{(Q_G + W_P)\left(\frac{1}{T_E} - \frac{1}{T_O}\right)}$$

Equation (6.61) is the same as Eq. (6.25)

$$\begin{aligned}\mathrm{COP} = {} & \mathrm{COP_C} + (T_o/T_G)\left(\frac{T_E}{T_O - T_E}\right)\left(\frac{W_P}{(Q_G + W_P)}\right) \\ & - \left(\frac{T_E}{T_O - T_E}\right)\left(\frac{T_o \Delta S_o}{(Q_G + W_P)}\right)\end{aligned} \tag{6.62}$$

Putting the value of $\mathrm{COP_C}$ from the Eq. (6.61) in (6.62), we get

$$\begin{aligned}\mathrm{COP} = {} & \left(\frac{T_G - T_O}{T_G}\right)\left(\frac{T_E}{T_O - T_E}\right) + \frac{W_P(T_o/T_G)}{(Q_G + W_P)}\left(\frac{T_E}{T_O - T_E}\right) \\ & - \left(\frac{T_E}{T_O - T_E}\right)\frac{T_o \Delta S_o}{(Q_G + W_P)}\end{aligned}$$

$$\mathrm{COP} = \frac{T_E}{T_O - T_E}\left(\frac{T_G - T_O}{T_G} + \frac{W_P(T_o/T_G)}{(Q_G + W_P)} - \frac{T_o \Delta S_o}{(Q_G + W_P)}\right)$$

But $\mathrm{COP_{VCR}} = \frac{T_E}{T_O - T_E}$, this is, the ideal Carnot COP of the vapour compression cycle working between T_E and T_C. So

$$\mathrm{COP} = \mathrm{COP_{VCR}}\left(\frac{T_G - T_O}{T_G}\right) + \frac{W_P(T_o/T_G)}{(Q_G + W_P)}\mathrm{COP_{VCR}} - \mathrm{COP_{VCR}}\frac{T_o\Delta S_o}{(Q_G + W_P)} \tag{6.63}$$

Now

$$\mathrm{COP_{VCR}}\left(\frac{T_\mathrm{G} - T_\mathrm{O}}{T_\mathrm{G}}\right) = \mathrm{COP_{VAB}}$$

$$\mathrm{COP} = \mathrm{COP_{VAB}} + \frac{W_P(T_o/T_G)}{(Q_G + W_P)}\mathrm{COP_{VCR}} - \mathrm{COP_{VCR}}\frac{T_o\Delta S_o}{(Q_G + W_P)} \tag{6.64}$$

Now

$$\mathrm{COP_R} = \mathrm{COP_{VAB}} + \frac{W_P(T_o/T_G)}{(Q_G + W_P)}\mathrm{COP_{VCR}} \tag{6.65}$$

Equation (6.65) is defined as the reversible COP of the vapour absorption cooling system.

$$\mathrm{COP} = \mathrm{COP_R} - \mathrm{COP_{VCR}}\frac{T_o\Delta S_o}{(Q_G + W_P)} \tag{6.66}$$

$\mathrm{COP_C}$ or $\mathrm{COP_{VAB}}$ = COP vapour absorption cooling system

$\mathrm{COP_{VCR}}$ = COP vapour compression cooling system.

6.5.6 An Intermittent Absorption Cooling System

An intermittent system alternates between two modes as shown in Fig. 6.18a and b.

In the cooling mode, the refrigerant is allowed to vapourise in the evaporator, thereby extracting heat from the space to be cooled. The absorber absorbs the refrigerant leaving from the evaporator. Heat of absorption is removed by water or air. Next is the regeneration mode. The operation of the evaporator or the condenser, and the absorber or generator are exchanged. Heat is supplied to the generator to desorb refrigerant vapour, which is cooled in the condenser by water or air in the condenser. Thus, intermittent cooling operation is in two-phase regeneration and refrigeration. Former is the regeneration of refrigerant vapour by heating the working fluid and successive condensation. The refrigerant takes the latent heat from the space to be cooled and the refrigerant vapour is reabsorbed by the absorbent. Thus, the energy storage is within the intermittent absorption cycle in the form of separated refrigerant and absorbent mass.

Because of the intermittent nature of solar energy, an intermittent absorption cooling cycle has been considered as the most logical and simple approach for the

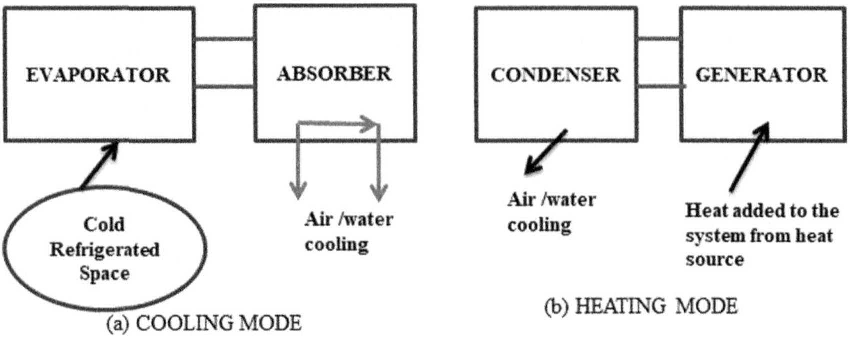

Fig. 6.18 An intermittent absorption solar cooling system, refrigeration and regeneration phases (Kaushik and Chandra 1982)

solar cooling. The continuous system requires electrically driven solution pump and auxiliary heat source and/or associated energy storage devices for their continuous system operation while the intermittent cooling systems are free from these requirements and could be used in remote areas where electric power is not readily available and if available, it is more expensive. These intermittent systems have been largely used for food preservation while the continuous systems have been used for the comfort cooling. In an intermittent cycle, distillation of the refrigerant from the absorbent solution occurs during the refrigeration and thus the refrigerant is condensed and stored. During refrigeration process of the cycle, the refrigerant is evaporated and reabsorbed. It must be mentioned here that these intermittent systems are more viable. During the intermittent operation, the thermodynamic processes are far from reversible and are dependent upon the properties of the refrigerant absorbent pair used and the amount of refrigerant present at the commencement of regeneration.

6.5.7 Solar Electrolux Refrigerator

The schematic of the solar electrolux refrigerator shown in Fig. 6.19 is also called as a no-moving part and no-auxiliary energy supply system. An inert gas is used to maintain a constant total pressure in the whole system. The refrigerant partial pressure can be low in the evaporator, obtaining the refrigeration effect. The three fluid system uses ammonia as the refrigerant, water as absorbent and hydrogen as the inert gas (Valizadeh and Ashrafi et al. 1996). Principle of the operation of Electrolux system is similar to the single-stage absorption cycle. The deviation between the two is in the electrolux system, the total pressure is the same in the entire system. To compensate the pressure difference that may occur between the high-pressure and low-pressure side of the system, hydrogen gas is circulated between the evaporator and the absorber. Ammonia vapour in the generator is condensed in the condenser, by rejecting its latent heat to a condensing medium.

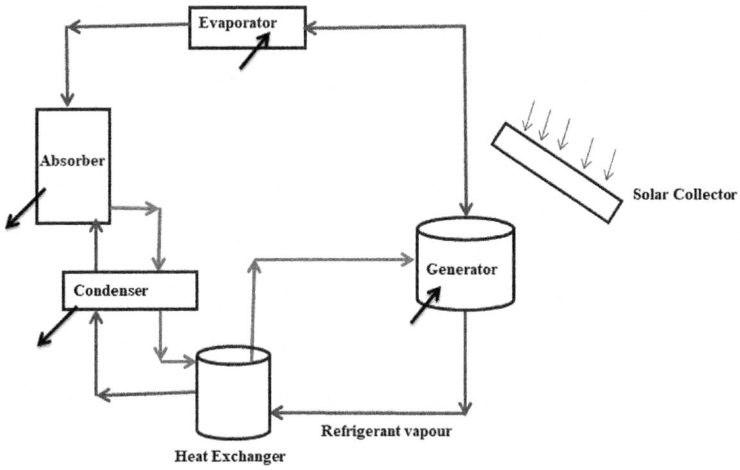

Fig. 6.19 Schematic of a solar electrolux refrigerator

The refrigerant liquid-ammonia that enters in the evaporator is exposed to the hydrogen atmosphere, and a cooling effect is achieved when the refrigerant evaporates due to its low partial pressure. The ammonia–hydrogen mixture then enters in the absorber, where the ammonia gets absorbed in the water solution. Hydrogen returns to the heat exchanger and the evaporator, while the aqueous ammonia solution is pumped to the generator by a thermosyphon pump. Weak solution goes back to the absorber by gravitational flow. The generator is connected to a solar collector where the refrigerant ammonia is driven off from the strong solution by receiving heat from the solar collector.

6.6 Vapour Absorption Cooling System with Storage

Since solar energy is intermittent in nature for the continuous working of any solar operated absorption system the energy storage is essential. It adjusts the temporal mismatches between the variable load and an intermittent energy source, thereby improving the system capability and permitting increased solar contribution to the total energy demand. In continuous operation of these absorption cooling systems, heating/cooling is accomplished by using low-grade heat available from the solar collection and the auxiliary source is also needed during peak off-hours or when cloudy. Thermal energy storage is also more critical in VAB systems for sizing the heat operated devices more in relation to average load than the peak load and providing the reserve for peakloads. Possible energy storage options for absorption systems are shown in Fig. 6.20. Conventionally, solar thermal energy can be stored as sensible heat in water or rocks.

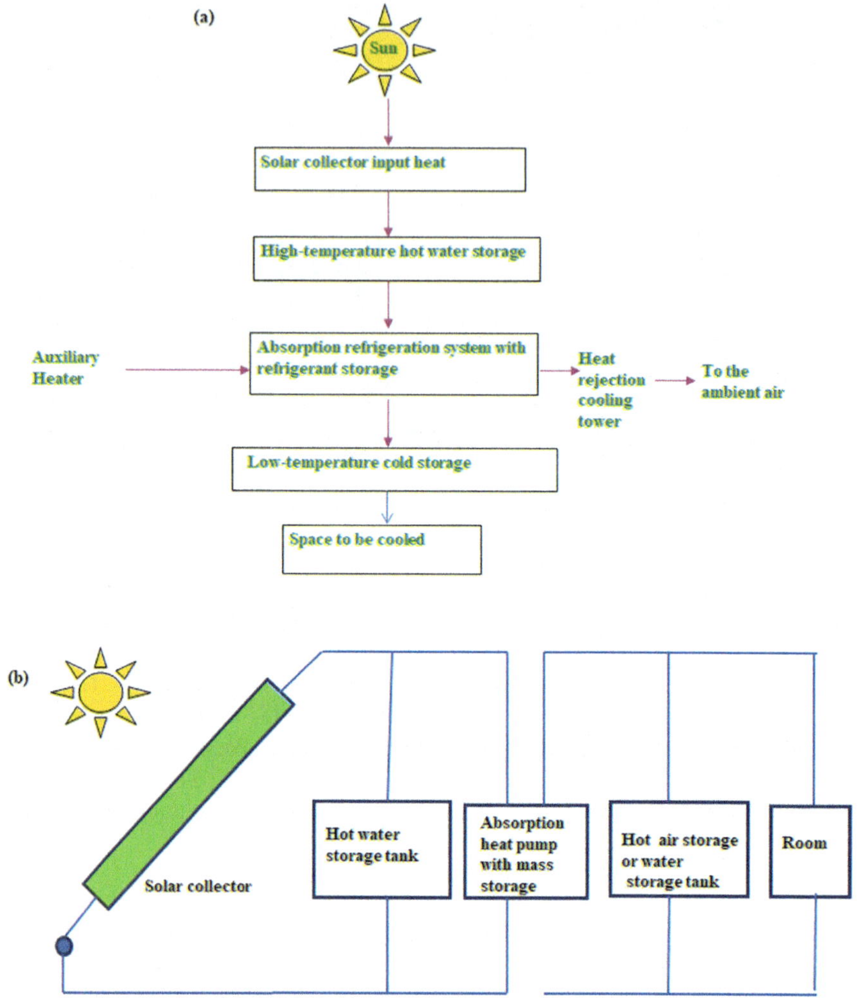

Fig. 6.20 Solar space **a** cooling system with energy storage options and **b** heating with energy storage options

Heat can be stored in sensible/latent form, and by thermochemical techniques. These types of storage systems are well explained in Chap. 4. Possible configurations of thermal energy storage used in absorption cycle systems are shown in Fig. 6.20a and b either on input side as hot storage between the low-grade heat source and the generator of the cycle or on output side as cold/ heat storage between the output component and the space to be cooled/heated.

Phase change material storage is another option for thermal energy storage in solar vapour absorption cooling systems. The latent heat storage is based on heat of absorption or release when a phase change material (PCM) undergoes a phase change.

The latent heat storage by PCM in comparison with sensible heat storage (SHS) possesses a greater density of stored energy and operates in a narrower operational temperature range.

Section 6.6.1 deals with the absorption space conditioning system with the solar energy storage in the form of refrigerant storage within the cycle itself. The refrigerant storage possesses the advantages such as high energy storage per unit volume, low energy losses, low storage pressure. Hence, strength of the storage vessel can be reduced.

6.6.1 *Basic Operations of Absorption Cycle with Refrigerant Storage*

6.6.1.1 An Absorption Airconditioning System

The concept of the refrigerant storage for solar cooling system was suggested by Shriedan (1961). Solar air conditioning with the refrigerant storage has been investigated in detail and shown to be technically feasible. The vapour absorption cooling system with refrigerant storage considered is shown in Fig. 6.21. The system consists of a generator, rectifier, condenser, evaporator and an absorber along with the preheater or subcoooling heat exchangers, pressure reducing valves and pumps. Basically, for the refrigerant storage in absorption cooling system, a storage volume is provided in association with the condenser to store the refrigerant during the hours of excess generation (i.e., hours of high insolation). In solar absorption cooling system, storage is also needed in the absorber to accommodate not only the refrigerant but also the sufficient absorbent to keep the concentration within the available limit. The condenser store is assumed to have a regulating valve at its outlet to control refrigerant flow. Heat input to the generator is supplied through a flat plate solar collector. The heat of condensation in the condenser and the heat of solution at the absorber are rejected to the absorber by a cooling tower from which water circulates through the absorber store, absorber, condenser store and condenser. Room air is circulated through the evaporator at constant inlet temperature.

6.6.1.2 Solar Absorption Heat Pump System with Refrigerant Storage

The basic principle of an absorption cycle heat pump is similar to the absorption cooling cycle. The concept of absorption cycle heat pump with refrigerant storage, proposed by Dixit et al. (2015), is given in Fig. 6.22.

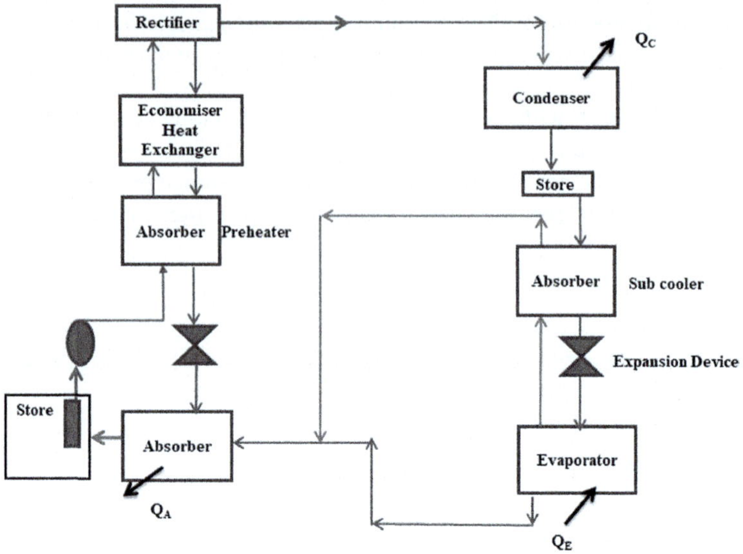

Fig. 6.21 A vapour absorption heat pump with integral refrigerant storage

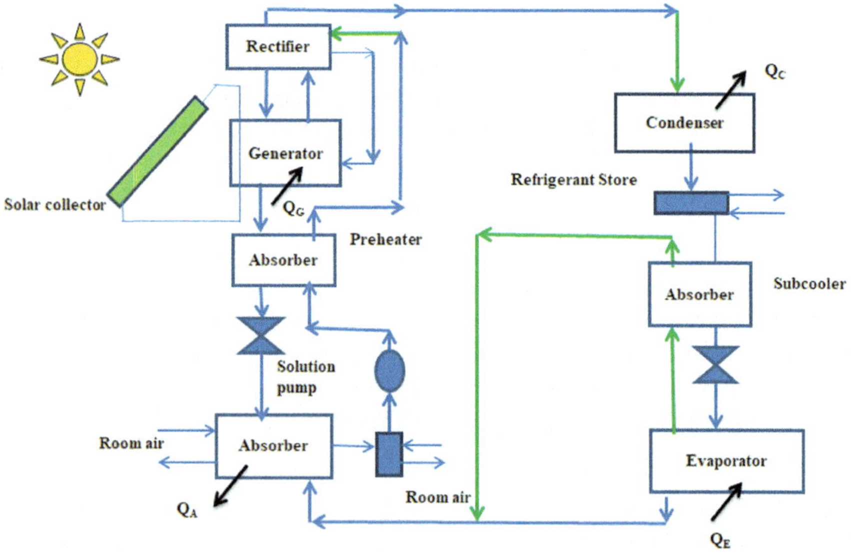

Fig. 6.22 A solar assisted vapour absorption heat pump with integral refrigerant storage

6.6.2 Basic System Equations and System Modelling

The modelling of the VAB system with refrigerant storage is explained in this section (Kaushik et al. 1988).

Following assumptions are made for the analysis of the cooling systems.

Common assumptions are also valid in the modelling of solar absorption cooling systems.

(1) Storage rate remains constant over the time interval under consideration and the total storage time for the internal can be found by first-order approximation.
(2) Mass of the refrigerant at any time in components other than the stores is negligible. Modelling of different components of the system is discussed below.

(a) Solar Collector: A flat plate collector is modelled by the following equation

$$Q_{\mathrm{p}} = F_{\mathrm{R}}\left[(\alpha\tau)I_s(\theta) - U_{\mathrm{p}}(t_i - t_a(\theta))\right]Ap \tag{6.67}$$

This Q_{p} is also equal to the heat taken by the circulating heat transfer fluid

$$Q_{\mathrm{p}} = \dot{m}_{\mathrm{p}}C_{\mathrm{w}}(t_{\mathrm{o}} - t_{\mathrm{i}}) \tag{6.68}$$

The useful heat from the collector is transferred to the generator through the heat exchanger relation. Therefore, we have

$$Q_{\mathrm{p}} = Q_{\mathrm{G}} = (UA)_{\mathrm{G}}\frac{[(t_o - t_{\mathrm{G}}) - (t_{\mathrm{i}} - t_{\mathrm{G}})]}{\ln\left(\frac{t_o - t_{\mathrm{G}}}{t_i - t_{\mathrm{G}}}\right)} \tag{6.69}$$

(b) Generator: The generator is assumed to be a heat exchanger with collector heat transfer fluid flowing on one side of the heat exchanger and a two-phase flow on the other side. Since a larger part of the generator heat is utilised for boiling of generator solution, the lower side of heat exchanger is assumed to be at the outlet temperature. The residential time in the generator is too small for equilibrium conditions to be set up. Therefore, an equilibrium factor is defined as the ratio of the difference in concentrations of ammonia solution leaving the generator and absorber

$$\in = \frac{x_s - x_w}{x_s - x_w} \tag{6.70}$$

$\in =$ 1 corresponds to the full equilibrium state. Here x_w is the equilibrium concentration of the solution in the generator at the pressure and temperature of the generator.

The basic energy balance equation for the generator is

$$Q_G + \dot{M}_5h_5 + \dot{M}_{10}h_{10} = \dot{M}_6h_6 + \dot{M}_9h_9 \tag{6.71}$$

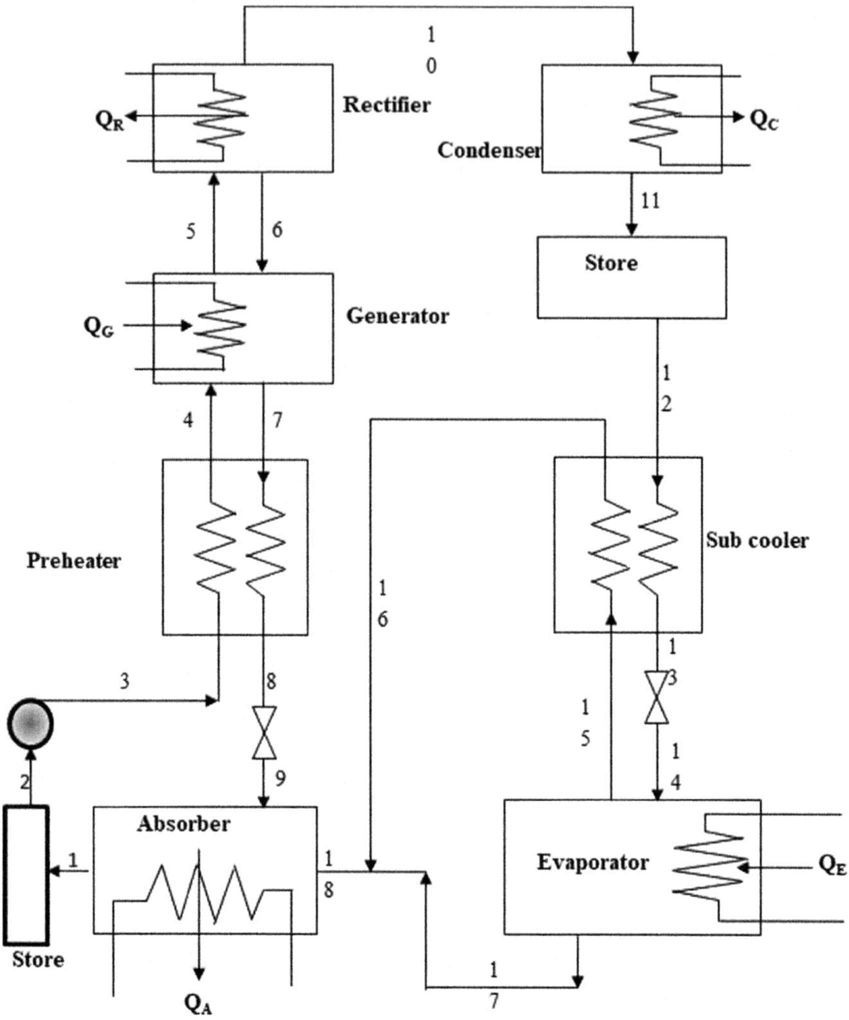

Fig. 6.23 A vapour absorption system with refrigerant storage (Kaushik 1989)

where the state points correspond to Fig. 6.23.

(c) Rectifier: Rectifier is also a heat exchanger. It is assumed that the temperature throughout the rectifier is the same.

Following are the basic energy balance equations around the rectifier:

$$\dot{M}_9 h_9 + \dot{M}_{10} h_{10} = \dot{M}_{11} h_{11} + Q_R \tag{6.72}$$

$$Q_R = \dot{M}_4 h_4 - \dot{M}_5 h_5 = (UA)_R \frac{[(t_R - t_5) - (t_R - t_4)]}{\ln\left(\frac{t_R - t_5}{t_R - t_4}\right)} \tag{6.73}$$

where Q_R is the heat evolved at the rectifier.

(d) Condenser: The condenser is also analysed based on the same expressions of a heat exchanger. However, the process in the condenser is different from that of the other heat exchangers of the system because of the condensation of the refrigerant. UA product of condenser heat exchanger becomes a function of refrigerant flow rate as the heat transfer due to condensation varies over a wide range with flow rate and hence with the thickness of condensate.

An equation is used to find out the actual UA product of the condenser

$$(UA)_c = \frac{3.77}{2.4 + 55M_{12}^{\frac{1}{3}}} \tag{6.74}$$

where $(UA)_N$ is the nominal rating of a condenser. The above equation holds good only for small variations over 3 kW/s of $(UA)_N$.

For the condenser, we can write the following energy balance equations:

$$\dot{M}_{11} h_{11} = \dot{M}_{12} h_{12} + Q_C \tag{6.75}$$

$$Q_C = M_a C_w \left(\dot{t}_{c2} - t_{c1}\right) \tag{6.76}$$

$$Q_C = (UA)_c \frac{[(t_c - t_{c1}) - (t_c - t_{c2}]}{\ln\left[\frac{t_c - t_{c1}}{t_c - t_{c2}}\right]}$$

(e) Condenser Store: Mass and energy balance on condenser store are given by the two first-order differentials

$$\frac{dM_{cs}}{d\theta} = \dot{M}_{12} - \dot{M}_{13}$$

$$\dot{M}_{12} h_{12} = \dot{M}_{13} h_{13} + \frac{dE_{cs}}{d\theta} + Q_{cs}$$

$$\frac{dE_{cs}}{d\theta} = \frac{dM_{cs}}{d\theta} U_{cs} + M_{cs} \frac{dU_{cs}}{d\theta} \tag{6.77}$$

where U is the specific internal energy and obtained from the state equation, $U = h - Pv$, (v is the specific volume), and

$$Q_{cs} = (UA)_{cs} [\frac{(t_{cs} - t_{cs1}) - (t_{cs} - t_{cs2})}{\ln\left[\frac{(t_{cs} - t_{cs1})}{t_{cs} - t_{cs2})}\right]} \tag{6.78}$$

Concentration of the refrigerant leaving the condenser store is calculated from the mass and ammonia mass balance conditions and is given by

$$X_{13}^{m} = \frac{M_{12}^{n} X_{12}^{n} + M_{cs}^{n-1} X_{13}^{n-1}}{M_{12}^{n} + M_{cs}^{n-1}} \tag{6.79}$$

(f) Subcooler: Subcooler is a liquid vapour heat exchanger and is characterised by the effectiveness given by

$$\eta_{11} = \frac{h_{16} - h_{18}}{h_{16} - h_{18.13}} \tag{6.80}$$

where $h_{18.13}$ is the enthalpy of the vapour at the concentration of state 18 and temperature of the state 13.

Energy balance around the subcooler is given by

$$\dot{M}_{13} h_{13} + M_{16} h_{16} = \dot{M}_{14} h_{14} + \dot{M}_{18} h_{18} \tag{6.81}$$

(g) Evaporator: Energy balance equations for the evaporator are given by

$$Q_E + \dot{M}_{15} h_{15} = \dot{M}_{16} h_{16} + \dot{M}_{17} h_{17}$$

$$Q_E = \dot{M}_{16} C_w (t_{E1} - t_{E2}) = (UA)_E \frac{(t_{E1} - t_E) - (t_{E2} - t_E)}{\ln\left[\frac{t_{E1} - t_E}{t_{E2} - t_E}\right]} \tag{6.82}$$

(h) Absorber: The absorber is also modelled as a heat exchanger. The equilibrium conditions are considered to exist in the absorber heat exchanger. The low side pressure is determined from the strong solution concentration at state point 1 and the temperature of the absorber

$$Q_A + \dot{M}_1 h_1 = \dot{M}_8 h_8 + \dot{M}_{19} h_{19}$$

$$Q_A = \dot{M}_a C_a (t_{A1} - t_{A2}) = (UA)_A \frac{(t_A - t_{A1}) - (t_A - t_{A2})}{\ln\left[\frac{t_A - t_{A1}}{t_A - t_{A2}}\right]} \tag{6.83}$$

(i) Absorber store: Absorber store is similar to condenser store. Mass and energy balance in absorber store are first order differential equations which can be solved by first order approximation.

$$\frac{dM_{AS}}{d\theta} = \dot{M} - \dot{M}_2$$

$$\dot{M}_1 h_1 = \dot{M}_3 h_2 + \frac{dE_{As}}{d\theta}$$

$$\frac{dE_{AS}}{d\theta} = \frac{dM_{AS}}{d\theta} U_{AS} + \frac{dU_{AS}}{d\theta} M_{AS} \quad (6.84)$$

The concentration of the strong solution in the absorber store is calculated by mass and ammonia balances with the assumption that the mass accumulated in the components other than the condenser and absorber stores is negligible.

$$X_s = \frac{M_1 X_1 - M_{cs} X_{13}}{M_{AS}} \quad (6.85)$$

(j) Solution Heat Exchanger: The solution heat exchanger is modelled as a liquid–liquid heat exchanger whose effectiveness is given by

$$\eta = \frac{h_6 - h_7}{h_6 - h_{7.3}} \quad (6.86)$$

where $h_{7.3}$ is the enthalpy determined at x_7 and t_3. The energy balance around this heat exchanger can be written as

$$\dot{M}_3 h_3 + \dot{M}_6 h_6 = \dot{M}_4 h_4 + \dot{M}_7 h_7 \quad (6.87)$$

(k) Cooling Load: A single house cooling load varies sinusoidally with a peak variation of 3 kW about a mean load of 7 kW is considered. This load corresponds to a load of a house with approximately 250 m^2 living space and heat loss coefficient of $U_B = 1.5$ W/m^2 °C. The analytical model of the cooling load is considered as

$$Q_{EDES} = 7.0 - 3.0 \cos[\Pi(\theta - 2)/12] \text{ in kW} \quad (6.88)$$

(l) Cooling Tower Model: In cooling tower model, it is assumed that the outlet water temperature can be expressed empirically as a quadratic function of both the inlet water temperature and wet bulb temperature

$$t_{w2} = k_1 + k_{2,twb} + k_3 t_{wb}^2 + k_4 t_{w1} + k_5 t_{w1}^2 + k_6 t_{wb} t_{w1} + k_7 t_{wb}^2 \cdot t_{w1} + k_8 t_{wb} \cdot t_{w1}^2 + k_9 t_{wb}^2 \cdot t_{w1}^2 \quad (6.89)$$

The constants k_1 to k_9 are evaluated by determining the cold water temperature t_{w2}, for three hot water temperatures t_{w1}, at each one of three wet bulb temperature t_{wb}. There are, thus, nine values of the function tw_2 ($t_{wb}.t_{w1}$) at independent values of the variables t_{wb} and t_{w1}. Hence, k_1 to k_9 are calculated by using the simple inversion matrix.

6.7 Advanced Absorption Cooling Cycles

6.7.1 *Double Effect NH_3–H_2O Vapour Absorption Cooling System*

Figure 6.24 shows a double effect absorption cooling system with ammonia–water as working fluid.

The system consists of two generators, viz. generator 1 and generator 2, two heat exchangers, condenser, an expansion device, an absorber and an evaporator. The main feature of the system is that the NH_3–H_2O strong solution pumped from the absorber is divided at the exit of the second solution heat exchanger HXII and sent separately to two generators. Generator 1 is connected to a solar collector and the refrigerant, NH_3 vapour gets heated up and driven off from the solution. The storage tank is added to the system, which acts as a sensible heat storage device. The hot water available in the tank can be used as a driving heat source at the period of low solar availabilities. Generator 2 is driven by the heat obtained from generator 1. The refrigerant desorbed from generator 2 condenses inside the condenser, subjected to expansion in expansion device and finally, it passes to the evaporator where the cooling occurs. The refrigerant liquid from the condenser, after expansion, continues to the evaporator, and it is evaporated at low-pressure, extracting the heat of vapourisation from the space to be cooled. Then the fluid passes through the heat-exchangers (HX-1 and HX-II) from

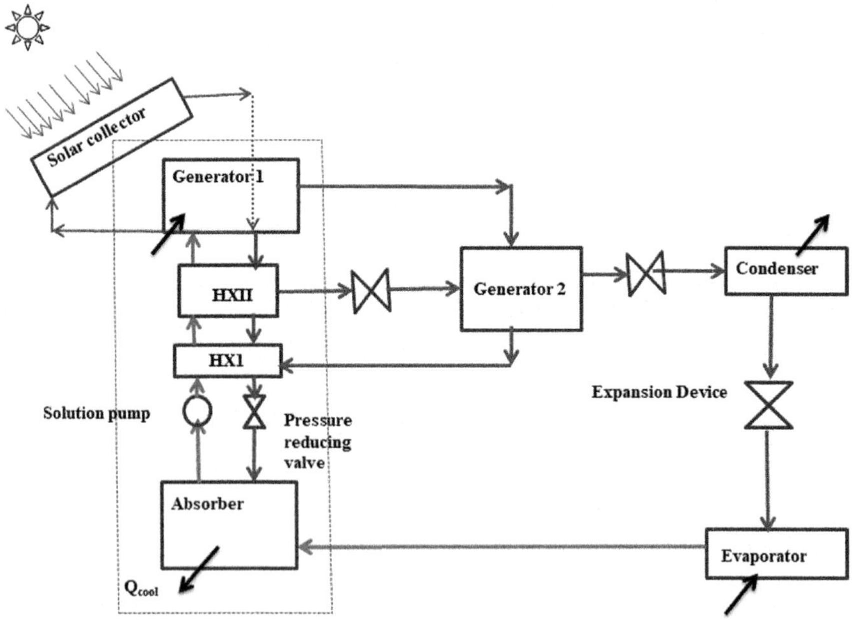

Fig. 6.24 A double effect NH_3–H_2O solar vapour absorption cooling system (Grossman 2002)

the absorber to generator 1 using a pump. The system is subjected to three different pressure levels from high, medium and low. Thus, the heat input to the double effect cycle is less than the single effect system. Hence, a double effect system has a higher COP. Two single-effect systems effectively form a double-effect absorption cooling system; therefore, the COP of a double-effect system is almost twice that of the single-effect absorption system.

6.7.2 Two-Stage Dual Fluid Cycle

Kaushik (1989) introduced a two-stage dual fluid absorption refrigeration system. The system uses LiBr–H_2O and H_2O–NH_3 as working fluids at the first and second stage, respectively. The schematic of the device is given in Fig. 6.25. In this system, the first stage operates through LiBr–H_2O working fluid combination, while the second stage uses H_2O–NH_3 combination. Therefore, a rectifier is avoided at the first stage and is needed only in the second stage. The condenser and absorber are maintained at a temperature close to ambient during the operation. The evaporator is assumed to be operated at a temperature of 5 °C. In the second stage, the absorber is assumed to be maintained at 5 °C by the evaporator of the first stage. The working in the second stage is similar to the first stage except that a rectifier is required in the second stage of the working. Also, an evaporator temperature as low as –20 °C can easily be produced at the second stage. The system analysis indicates that at the second stage, low condenser temperature can yield a better performance at lower generator temperature but at higher generator temperature, a high value of COP

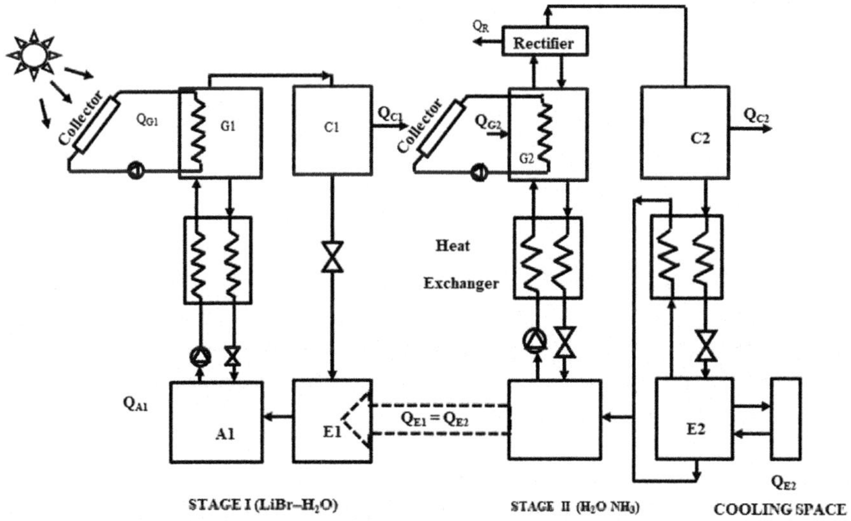

Fig. 6.25 Two-stage dual fluid system (Kaushik 1989)

is obtained at a higher condenser temperature. The study rendered the need of maintaining high generator and low absorber temperature when the system is working at low evaporator temperature. An ordinary flat plate collector is recommended at the first stage of operation and an evacuated tube collector is a better choice for the second stage to produce very low evaporator temperature.

6.7.3 *Multi-effect H_2O–LiBr Solar VAB Cooling System*

Different types of multi effect VAB cooling systems such as half effect, double effect and tripple effect are illustrated in this section.

6.7.3.1 *Half-Effect Water–Lithium Bromide Vapour Absorption Cooling System*

Figure 6.26 shows schematic of a half-effect water–lithium–bromide system. The half-effect absorption system was introduced for an application with a relatively low-temperature heat source. The configuration of the system is exactly similar to that of the NH_3–water double effect system. The high-temperature heat from an external source transfers to both generators. Both absorbers reject heat out to the surrounding. The operating pressure of absorber II and generator I are at intermediate level. The coefficient of performance (COP) of the half effect system is slightly low as compared to the single effect system. The research indicates that a half effect absorption cooling system can reject around 50% more heat than the single effect system does (Kaushik and Chandra 1982). However, it has the advantage of being operated with a relatively low-temperature heat source.

6.7.3.2 *Double Effect Water–Lithium Bromide Vapour Absorption Cooling System*

One of the limitations of a single-stage cooling system is its high investment. The capital cost of a single-stage cooling system is very high. The main reason for high investment cost is the cost of a solar collector. To reduce the solar collector cost, the operating temperature of collector is to be lowered. The generator temperature of the LiBr–H_2O system can be lowered by using two-stage LiBr system instead of single-stage system. System shown in Fig. 6.27 is a double-effect LiBr–H_2O absorption chiller, where the weak solution is circulated in series. The system consists of two generators, viz. low-pressure and high-pressure generator, two heat exchangers, condenser, an expansion device and an evaporator. The cycle is divided into a high-pressure stage and low-pressure stage. Diluted LiBr solution (enriched with refrigerant, water) in the high-pressure generator is heated by steam. Then the solution enters the low-pressure generator, which is driven by the solar collector. The driven

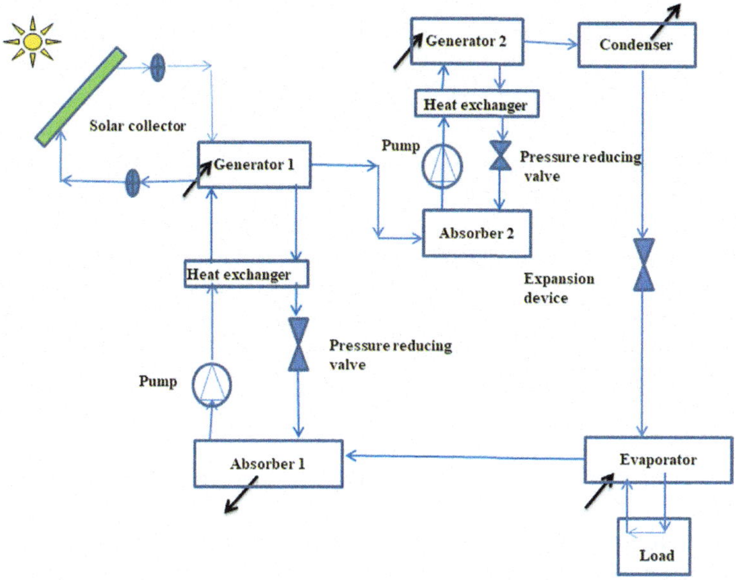

Fig. 6.26 Schematic diagram of half-effect water–lithium bromide vapour absorption system

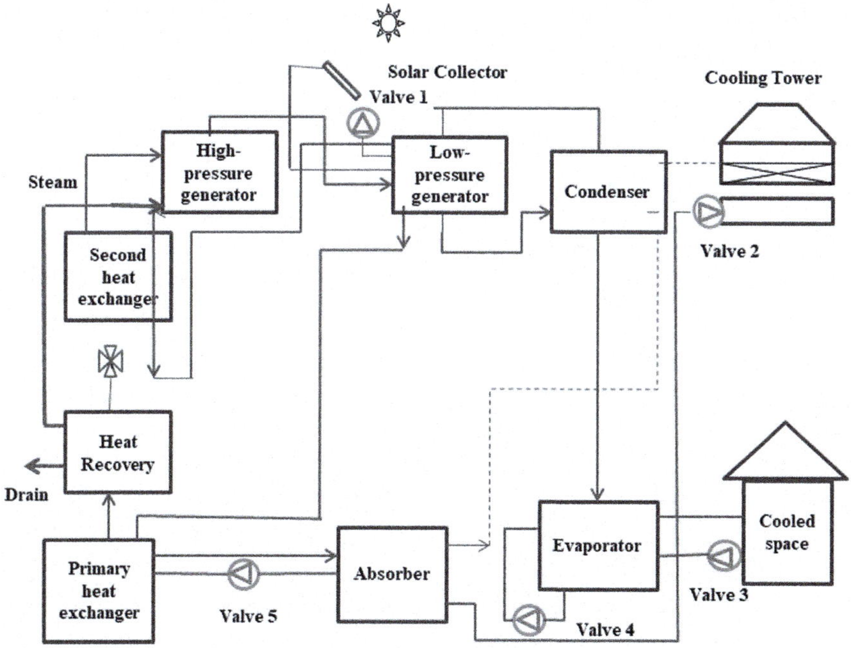

Fig. 6.27 A double-effect H_2O–LiBr absorption system (Mittal et al. 2005a, b)

off water vapour is condensed in the condenser, flows into the evaporator, producing the cooling. The weak solution from the low-pressure generator then transfers to the absorber through a primary heat exchanger and absorbs water vapour generated from the evaporator, thus changing back to a diluted solution. This solution is then pumped back to the high-pressure generator, completing a high-pressure cycle. The functions of the rest of the components are similar to that of single-effect absorption refrigeration system. The increased daily efficiencies of solar collector due to the lower generator inlet temperature are the advantages of refrigeration cycle. Steady operation of the cooling system and the increased daily efficiencies of solar collector due to the lower generator inlet temperature are the advantages of the two-stage system. The drawback of this system is the technical difficulty in the chillers construction and low COP as the normal generator temperature is lesser than the conventional single effect cooling system.

6.7.3.3 *Triple-Effect LiBr–H_2O Solar Vapour Absorption Cooling System*

The world's first triple-effect absorption system using (LiBr–H_2O) was developed in 1978. Figure 6.28 shows parts of the triple-effect VAB system. The system consists of two important parts, namely, solar collector and absorption cooling system.

The generator of the system is operated at three pressure and temperature levels. The high-pressure generator (HPG) is directly connected with a parabolic collector (Fig. 6.28). The triple-effect series absorption cooling system (Fig. 6.28) consists

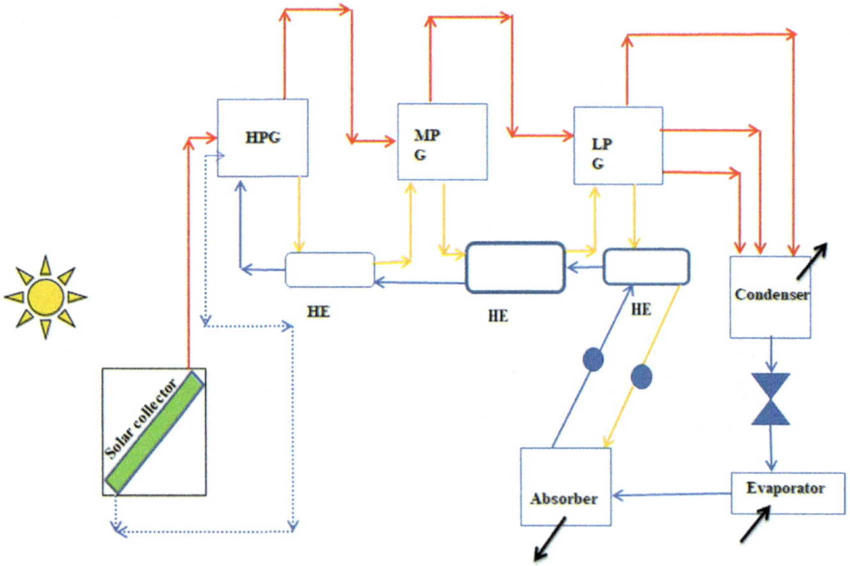

Fig. 6.28 Triple-effect LiBr–H_2O absorption refrigeration system

of an evaporator, an absorber, a condenser, three generators, three heat exchangers, a solution pump, three expansion valves and three reducing valves. As shown in Fig. 6.28, the refrigerant vapour is generated thrice. The first generation is by external heat input at HPG. Second and third generations are accomplished by internal heat exchange at medium pressure generator (MPG) and low-pressure generator (LPG), respectively. Thus, more, the vapour is generated for a unit heat input as compared with the single-effect and double-effect systems.

The system is explained with LiBr–H_2O as absorbent and refrigerant solution. The strong solution (rich in refrigerant) contained in HPG is heated by parabolic trough collector. Water vapour thus generated moves from HPG to MPG. Next, due to high-temperature heated water coming from HPG, converts water of MPG into vapour and moves to LPG. Meanwhile, LiBr from HPG enters the LPG through the heat exchanger. Then all the refrigerant vapour coming from all three generators flows to the condenser, where the vapour gets condensed by rejecting heat and converted to liquid form at a high-pressure. The liquid refrigerant leaving from the condenser then enters the expansion valve and gets evaporated in the evaporator by extracting heat from the surroundings. The LiBr from the heat exchanger is collected in the absorber, and then the water coming from the evaporator mixes with LiBr. The absorbent-refrigerant mixture is pumped back to the HPG to complete one cycle of operation.

6.8 Performance Comparison of Single-Effect, Double-Effect and Triple-Effect Solar VAB Cooling System

Nowadays the majority of absorption systems available are single-effect with solar flat-plate collectors at low-temperature. Recent researches show that there are also double-effect systems available, with thermal COP in the range of 1.0–1.2, and triple-effect systems with thermal COP of 1.7. Multi-effect absorption systems require higher heat supply temperatures, which may be obtained from higher-cost concentrating or evacuated tube collectors.

A comparison between the COP of various systems along with its cost is given in Table 6.3. Comparison indicates that the cost of collector dominates all in the solar vapour absorption cooling systems. Even though the multi-effect absorption systems are having higher COP, the higher cost of the high-temperature collectors makes the total system cost higher than that of the low-temperature systems.

Table 6.3 Comparison of single-effect, double-effect and triple-effect solar vapour absorption cooling systems (Dixit et al. 2015)

Type of systems	Heat source temperature (°C)	Solar energy required per kW of cooling effect (kW)	Type of collector	Collector area (m^2)	Typical collector cost ($\$/m^2$)	Solar system cost ($)	Absorption chiller cost ($)	Total system cost ($)	Typical COP
Single-effect	85	1.43	Flat plate collectors	7.48	110	1234	200	1434	0.7
Double-effect	130	0.84	Flat/CPC	5.08	160	1216	178	1359	1.2
Triple-effect	220	0.59	ETC/CPC	4.49	400	2694	165	2859	1.7

6.9 Absorber Heat Recovery Systems

The recovering heat internally in the basic VAB cycle was first shown by Altenkirch. Kaushik conducted detailed thermodynamic studies in the generator-absorber-heat exchanger (GAX) and hybrid GAX (HGAX) absorption cycle based on the energy and exergy analysis. GAX also sometimes called DAHX, desorber/absorber heat exchanger. Single effect VAB is a good choice as it gives better performance.

The schematic of the GAX system is shown in Fig. 6.29. The duty of a GAX system is to recover heat from the absorber and supply to generator, and hence reduce the heat input at the generator of VAB. The system consists of the generator, absorber, condenser, expansion device, evaporator, rectifier unit, GAXA (GAX absorber) and GAXD (GAX desorber). The saturated strong solution (strong in ammonia) is pumped to high-pressure and then passed through high-temperature part of the absorber to low-temperature part of generator (2) to recover the heat internally. The temperature overlaps between GAXA and GAXD allows internal heat recovery. Next, the generator receives heat from an external heat source integrated to it and desorbs the refrigerant vapour. The remaining weak solution flows back to the absorber unit. The saturated refrigerant liquid leaving the condenser (8) is subcooled by condensate precooler (RHX) while the saturated ammonia vapour (11) is preheated before absorbed in the absorber (12). The subcooled refrigerant (9) is expanded in the expansion device and enters the evaporator, where it takes the latent heat from the space to be cooled and converted to vapour refrigerant. The ammonia vapour at state point 12 is absorbed by the weak solution at absorber to form a strong solution of ammonia and water. Thus, one cycle of operation is completed. A single-effect absorption system can provide as high COP as that for the two-stage-double-effect absorption system by using GAX.

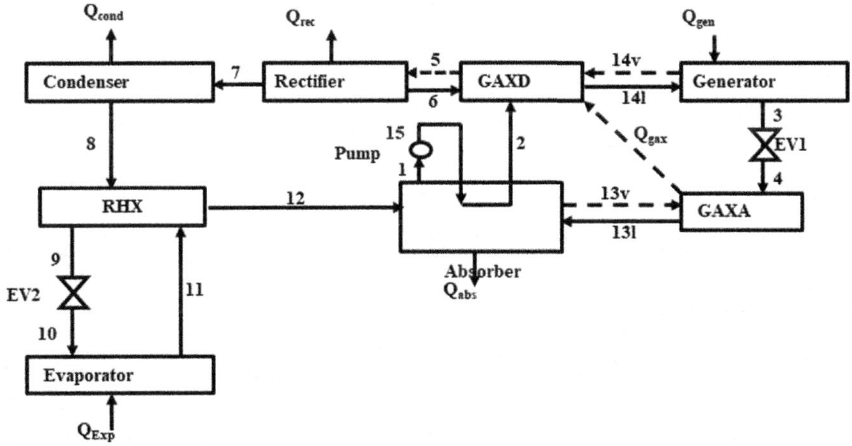

Fig. 6.29 GAX absorption cooling system (Dixit et al. 2015)

6.9.1 Solar GAX Absorption Cooling Cycle

The GAX absorption cooling system integrated with the parabolic trough collector is shown in Fig. 6.30. The system consists of a parabolic trough collector and GAX system.

The main components of the GAX system are the generator, condenser, expansion valve, evaporator and absorber. The PTC consists of a receiver or the heat collection element (HCE) and a trough-shaped reflector. The HCE is located along the focal line of the parabolic trough collector. The heat transfer fluid (HTF), metal tube together with the outer glass envelops constitutes the HCE.

The annular area between the glass cover and the metallic tube is assumed to be a vacuum. The oil can be used as the heat transfer fluid. The oil transfers its heat to the GAX absorption cooling system. Working of the system is similar to that of the simple GAX absorption system explained above in the Sect. 6.9.

6.9.2 Hybrid GAX Absorption Cooling Systems

This hybrid (HGAX) system is a modified version of a simple GAX, where a compressor is also integrated in it. The working principle and construction of both systems are same except in the inclusion of a mechanical compressor present between the condensate precooler and absorber (Fig. 6.31) in the HGAX.

Heat recovery in hybrid system is comparatively at higher side, which allows the absorber to operate at very high pressure and temperature. A low value of the

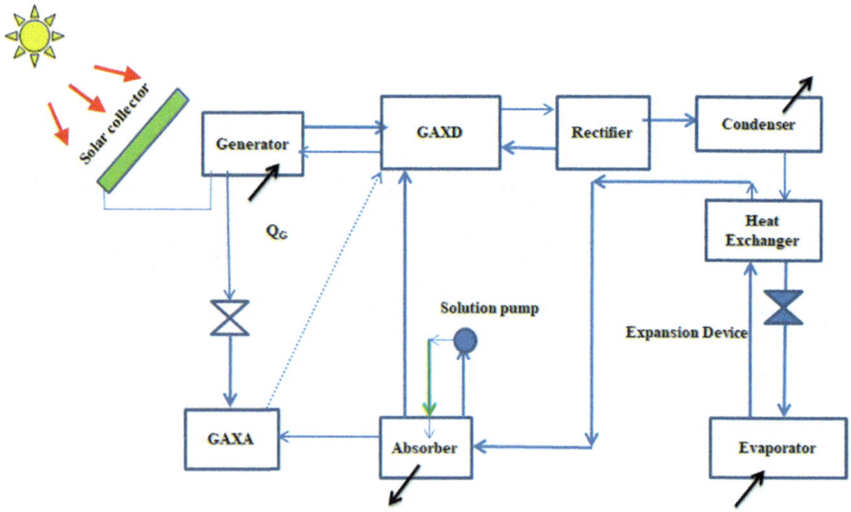

Fig. 6.30 Solar integrated GAX absorption cooling system

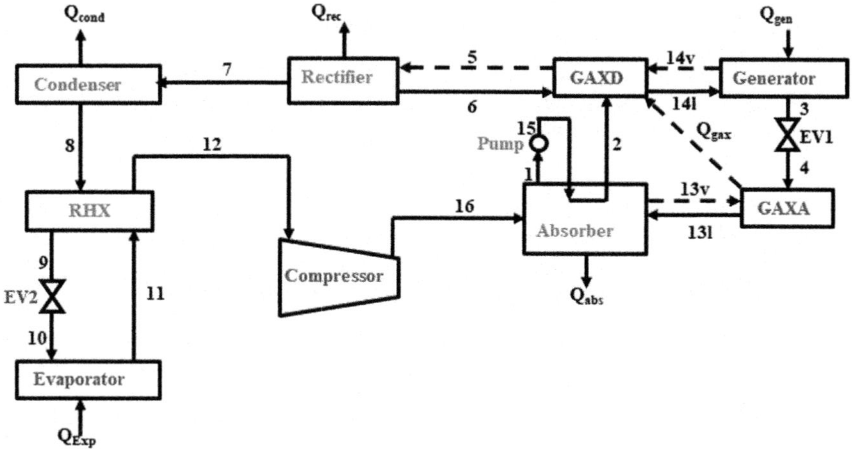

Fig. 6.31 Hybrid GAX absorption refrigeration system (Dixit et al. 2015)

absorber temperature in the absorption cycle is a drawback because the cooling of the absorber will require a cooling medium different from ambient air.

References

Amaris, C., Vallès, M. & Bourouis, M. (2018). Vapour absorption enhancement using passive techniques for absorption cooling/heating technologies: A review. *Applied Energy, 231*, 826–853.

Dixit, M., Arora, A. & Kaushik, S. C. (2015). Thermodynamic analysis of GAX and hybrid GAX aqua-ammonia vapor absorption refrigeration systems. *International Journal of Hydrogen Energy, 40*(46), 16256–16265.

Duffie, J. A., Beckman, W. A. & Worek, W. M. (1980). Solar engineering of thermal processes (Vol. 3). *New York etc.: Wiley*.

Gandhidasan, P. (1983). A simple analysis of an open regeneration system. *Solar Energy, 31*(3), 343–345.

Grossman, G. (2002). Solar-powered systems for cooling, dehumidification and air-conditioning. *Solar energy, 72*(1), 53–62.

Hollands, K. G. T. (1963). The regeneration of lithium chloride brine in a solar still for use in solar air conditioning. *Solar Energy, 7*(2), 39–43.

Kakabaev, A. & Khandurdyev, A. (1969). Absorption solar refrigeration unit with open regeneration of solution. *Appl. Solar Energy (USSR) (Engl. Transl.); (United States), 5*(4).

Kaudinya, J. V. & Kaushik, S. C. (1986). Experimental validation of theoretical studies on open and forced flow solar regenerator. *International Journal of Solar Energy, 4*(1), 13–23.

Kaushik, V. S. (1985). Mass transfer in packed bed reconcentration of lithium bromide for open-cycle absorption cooling (Doctoral dissertation, Colorado State University).

Kaushik, S. C. (1989). Solar Refrigeration and Space Conditioning, Divya Jyoti Prakashan House.

Kaushik S. C. & Chandra, S. (1982). Theoretical analysis of an ammonia water absorption cycle for solar refrigeration and space conditioning applications, *International Journal of Space Research, 6*, 206–226.

Kaushik, S. C. & Kaudinya, J.V. (1986a). Dynamic simulation of aqua ammonia absorption heat pump with refrigerant storage, *Journal of Heat Recovery Systems, 5*(1), 20–26.

Kaushik, S. C., Gadhi, S. M. B., Agarwal, R. S. & Kumari, R. (1988). Modeling and simulation studies on single/double-effect absorption cycle using water-multicomponent salt (MCS) mixture. *Solar energy, 40*(5), 431–441.

Mittal, V., Kasana, K. S. & Thakur, N. S. (2005a). The study of solar absorption air-conditioning systems. *Journal of Energy in Southern Africa, 16*(4), 59–66.

Mittal, V., Kasana, K. & Thakur, N. (2005b). The study of solar absorption air-conditioning systems. *Journal of Energy in Southern Africa, 16*(4), 59–66. https://doi.org/10.17159/2413-3051/2005/v16i4a3103.

Mullick, S. C. & Gupta, M. C. (1974). Solar desorption of absorbent solutions. *Solar energy, 16*(1), 19–24.

Peng, C. P., & Howell, J. R. (1982). Analysis of open inclined surface solar regenerators for absorption cooling applications—Comparison between numerical and analytical models. *Solar Energy, 28*(3), 265–268.

Sheridan, N. R. (1961). Industrial evaporative cooling solar air conditioning, *The International Journal of Engineering*, 47–52.

Valizadeh, H. & Ashrafi, N. (1996). A continuous cycle solar thermal refrigeration system. *Renewable energy, 9*(1–4), 632–640.

Wilber, P. J. and Mancinbi, T. R. (1976). A comparison of solar air conditioning systems, *Solar Energy*, 578.

Chapter 7
Vapour Adsorption Cooling Systems

The vapour adsorption cooling systems are emerging as potential systems for the refrigeration and airconditioning. The genesis of today's technology in vapour adsorption cooling systems was the work reported by a German Physicist, Henrich Kayser in 1880. The basic principle, working of different vapour adsorption cooling (VAD) systems etc. are described in this chapter.

7.1 History of Adsorption Cooling

In the 1920s, GE Hulse proposed first adsorption refrigeration system in which silica gel–SO_2 was used as a working pair for the storage of food in train. 1n 1940–1945, the adsorption refrigeration system with working pair $CaCl_2$–NH_3 was used for the storage of food in the train from London to Liverpool for which the heat source was steam at 100 °C. Since 1930s, new technologies like the discovery of freons and the successful development of the hermetically sealed compressors improved the efficiency of the compression refrigeration system significantly (Anyanwu et al. 2000, 2003; Gurdeep Raj 2002). Because of that, the adsorption refrigeration technology could not compete with the highly efficient chlorofluorocarbons system. But in 1970s, the energy crisis becomes a concern and it offered a great chance for the development of adsorption technology, mainly because that the adsorption refrigeration system is driven by a low-grade heat source like waste heat and solar energy. Adsorption cooling technology is a possible alternative to traditional cooling systems, which absorb significant quantities of electrical energy and causes series problems including natural resource depletion and environmental degradation. In 1990s, the environmental pollution became more serious issue and conventional cooling systems were identified as the cause of ozone depletion potential and global warming potential. Hence, the refrigeration systems driven by thermal compressors (such as adsorbent bed in an adsorption cooling system) became more popular as

S. C. Kaushik et al., *Solar Cooling*, https://doi.org/10.1007/978-3-031-42410-6_7

an alternative to the vapour compression refrigeration and airconditioning systems. Adsorption cooling systems can easily be powered by a low grade energy source like biomass or solar energy. The refrigerants employed are environment friendly and have no moving parts in the system.

7.2 Adsorption

The surface of the solid has a tendency to attract and to retain the molecules of other substance (gas or liquids) with which it comes in contact. This phenomenon of the solid surfaces is termed as adsorption (Gurdeep Raj 2002). Thus, adsorption is a surface phenomenon in which accumulation or concentration of a substance takes place at the surface or interface. The various characteristics of the adsorption phenomenon are given below

- It is a spontaneous process.
- Adsorption is accompanied by a decrease in free energy (ΔG) of the system. When the free energy for adsorption reaches to zero, adsorption equilibrium is said to be established.
- The phenomenon can occur on all surfaces. Four types of interfaces that exist are: liquid–solid, liquid–liquid, solid–solid and gas–solid.
- During adsorption, the entropy of the system decreases as indicated by the Gibbs free energy equation.

From the Gibbs–Helmholtz equation

$$\Delta H = \Delta G + T\Delta S \tag{7.1}$$

where ΔG is the change in free energy, ΔH the change in enthalpy, ΔS the change in entropy and T is the absolute temperature of the system. In adsorption both the entropy and free energy decreases, the value of ΔH decreases.

This decrease in enthalpy is the net heat release during adsorption. Hence, the adsorption process must always be exothermic.

7.3 Types of Adsorption

Based on the nature of forces between the adsorbate and adsorbent, the adsorption phenomenon is classified into two types.

7.3.1 Physical Adsorption

In physical adsorption or physisorption, the physical force of attraction holds the gas/liquid molecules to the solid. The forces of attraction bringing about the physical adsorption are polarisation, permanent dipole moment in the adsorbent molecules, dispersion effects, and short-range repulsive effect. In the case of physisorption, the forces of attraction, which holds the adsorbate molecules to the solid is very weak and usually found to be relatively inert surfaces like adsorption of gases in charcoal and adsorption of nitrogen on mica.

7.3.2 Chemical Adsorption

In chemical adsorption or chemisorption, the chemical forces hold the adsorbate molecules to the surface of adsorbent. In this case, the adsorbate undergoes a strong chemical interaction and hence very high heat of adsorption as compared to the physical adsorption. The physisorption and chemisorption is compared in Table 7.1.

Table 7.1 Comparison between physisorption and chemisorption

Physisorption	Chemisorption
Physisorption involves physical forces	Chemisorption involves the transfer of free electrons between gas and solid
Heat of adsorption is generally less than 40 kJ/mol	High heat of adsorption
Multilayers are possible in physical adsorption	Chemisorption involves monolayer of adsorbate substances
It is appreciable at low temperatures and high pressures	Can occur at high temperatures
Takes place instantaneously	Very slow process as compared to the physisorption
No appreciable activation energy is required	Chemisorption is an activated phenomenon and involves appreciable activation energy
It is a function of coverage of surfaces	The substance is adsorbed at the specific sites of the surface. These sites are called active centres
Not very specific	Often very specific
It is reversible	Not reversible
A general phenomenon which will occur with any gas—solid system at the suitable conditions of pressure and temperature	Chemisorption is possible only at the surfaces where chemical bond is formed

7.4 Absorption and Adsorption—A Comparative Study

The adsorption is a surface phenomenon whereas the absorption is a bulk phenomenon in which the assimilated is uniformly distributed throughout the body of a solid or liquid to form a solution or a compound (Table 7.2). The process of absorption and adsorption are explained in Fig. 7.1 and the main differences between the two are as below:

(i) The absorption involves penetration of substances to a significant level to form another compound. Whereas adsorption takes place only at the surface and not in the body of adsorbent. Thus, absorption is a bulk phenomenon and adsorption a surface phenomenon.

Table 7.2 Comparison between absorption cooling and adsorption cooling systems

Absorption cooling systems	Adsorption cooling systems
In absorption, the substance is distributed uniformly to convert to a solution or a compound	Adsorption takes place only at the surface, not in the body of adsorbent
In absorption cooling system the refrigerant vapour is drawn from the evaporator by absorption by a liquid having high affinity for the refrigerant. Application of heat to the solution raises its temperature level and the refrigerant vapour is driven off so that it can pass to the condenser to get liquefied	In adsorption cooling system adsorbent bed is used to adsorb or desorb the refrigerant vapour according to the changes in the temperature of the adsorbent. The adsorbent bed is heated to drive off the refrigerant vapour and is sent to the condenser
The range of heat source temperature is small	The system can be powered by a heat source having temperature from 50 to 500 °C
The liquid pump is placed in between absorber and generator for increasing the pressure of solution from absorber pressure to generator pressure	It does not need a liquid pump for increasing the pressure of the refrigerant vapour
The liquid absorption cooling systems have the problem of generating some amount of absorbent along with the refrigerant during generation, thus requires a rectifier for its operation when working with NH_3–H_2O	It does not require any such devices for its operation as the possibility of removal of solid adsorbent during desorption stage is less
The chance of corrosion for the components exists	The working pairs commonly used does not produce any corrosion problems for the components
The solid absorbents need special treatment to obtain hard porous granules that increase the first cost of the system	Solid adsorbents used in the adsorption refrigeration do not need special treatments
The coefficient of performance is high as compared to the adsorption cooling systems	The coefficient of performance and specific cooling power will be less

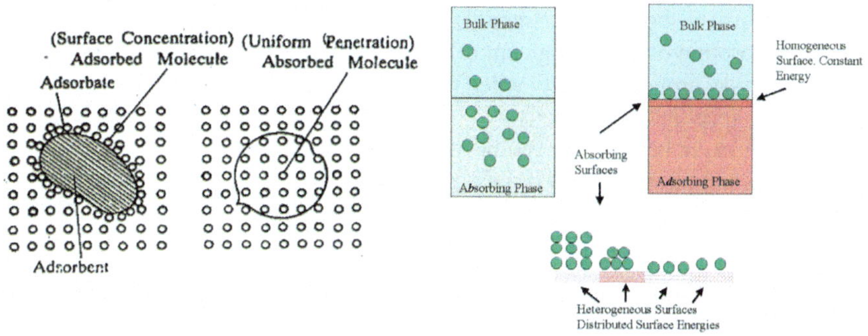

Fig. 7.1 Adsorption and absorption (Gurdeep Raj 2002)

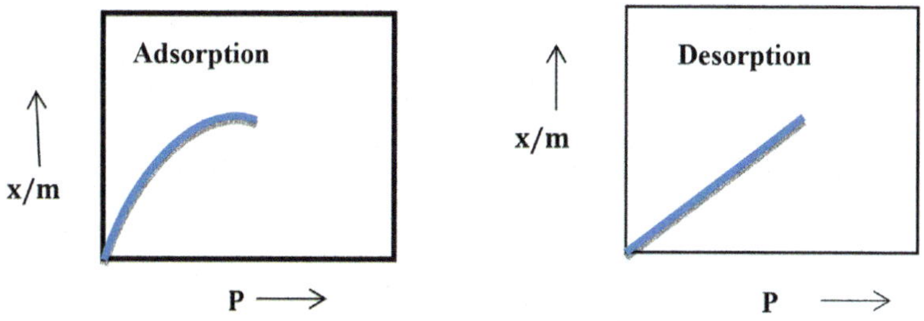

Fig. 7.2 Isotherms of adsorption and desorption

(ii) The adsorbed molecules are entrapped at the surface of the adsorbent than in the adsorbate. But in absorption, the diffusion of molecules into the absorbate takes place

(iii) In adsorption, the equilibrium is achieved in a very short duration. The absorption is a slow process as compared to the adsorption phenomena.

The typical isotherms for both the absorption and adsorption are shown in Fig. 7.2.

7.5 Choice of Adsorbent-Refrigerant Combination

A large number of adsorbents and adsorbates are available for the adsorption cooling systems. Selection of proper adsorbent and adsorbate for the desired application is the preliminary step while designing an adsorption cooling system. Further, the adsorption process involves the separation of a substance from one phase accompanied by its accumulation at the surface of the other. The adsorbing phase is called the adsorption. Performance of adsorbents in the system depends upon its surface area, micro- and macropores, size of granules in powder, while the selection of an adsorbent depends upon the following factors:

(i) Adsorbent should be compatible with the refrigerant selected.
(ii) Adsorption of a large amount of adsorbate at low temperature.
(iii) High desorption capacity when thermal energy is applied.
(iv) Low specific heat capacity.
(v) High latent heat of adsorption compared to sensible heat.
(vi) It should be chemically stable.
(vii) Low cost and easy availability.
(viii) Non-toxic and non-corrosive.

There are mainly three types of adsorbents used in the adsorption systems, namely, physical, chemical and composite adsorbents.

(a) Physical Adsorbents

In physical adsorbents, weak physical forces are responsible to hold adsorbate molecules on the surfaces of adsorbents. Some common adsorbents are given as below:

- Activated carbon
- Activated carbon fibre
- Silica gel
- Zeolites

(b) Chemical Adsorbents

The strong chemical forces are responsible for the adherence of adsorbate molecules on the surface of adsorbent molecules and some of the commonly used are:

- Metal chlorides such as $CaCl_2$, $SrCl_2$, $BaCl_2$, $MgCl_2$ etc.
- Salt and metal hydrides such as LiH, CaH_2 etc.
- Metal oxides such as titanium, zirconium and magnesium oxides.

(c) Composite Adsorbents

The composite adsorbents are the combinations of physical and chemical adsorbents with suitable binders. Composite adsorbents can be prepared by simple mixture, impregnation, and consolidation processes. Based on the affinity towards polar substances (like water) adsorbents are classified into two types (Critoph 1998, 1999; Shabir et al. 2020) as below:

Hydrophilic: These type of adsorbents have affinity towards the polar substances like water. e.g., silica gel, zeolites, active alumina, etc.

Hydrophobic: Adsorbents having attraction towards the oils and gases than for water are termed as hydrophobic, e.g., activated carbon, polymer adsorbents and silicalites.

Fig. 7.3 Silica gel

(i) Hydrophilic Solid Adsorbents

Some of the commonly used hydrophilic solid adsorbents are discussed as below:

Silica Gel: Silica gel ($SiO_2 \times H_2O$) is prepared from pure silica and it retains traces of water at a particular temperature . When it is overheated it loses water and its adsorption capacity. Thus, its application is limited to a temperature of less than 150 °C. Activated silica gel has a poor thermal conductivity of the order of 0.052–0.091 W/mK (Alam et al. 2007).

Silica gel is available in various pore sizes. Silica gel having pore sizes of 2–3 nm are called type A and 3–7 nm are type B. Type A is used for general drying purposes and Type B for relative humidities higher than the 50%. The heat of adsorption of water vapour on silica gel is predominantly due to the condensation of water (Fig. 7.3).

Activated Alumina: Activated alumina is aluminium oxide in the porous form prepared by the dehydration of aluminium hydrates to about 6% moisture level and having surface area ranging from 150 to 500 m^2/g. It is used as a drying agent and adsorbent for polar organic compounds. The activated alumina is commonly used as the desiccant material. This also finds applications in purification of drinking water to remove arsenic and selenium from the drinking water. Additionally, this is used as an adsorbent for the removal of oxygenates and fluoride ions from water (Fig. 7.4).

Zeolites: Zeolites including crystalline zeolites are naturally occurring aluminosilicate minerals (also called molecular sieves). Both natural and synthetic zeolites are used in the adsorption systems. Zeolites are highly porous minerals that can adsorb a large number of water molecules. Most commonly used synthetic zeolites include 3A, 4A, 5A, 10X and 13X. Most of the zeolites have low thermal conductivity and specific cooling power. Type 5A is used to separate cyclic hydrocarbons and some paraffins , while 4A is used for drying. Zeolites are applied in drying of

Fig. 7.4 Activated alumina

process air, removal of CO_2 from natural gas, removal of CO from reforming gas, air separation, catalytic cracking, and catalytic synthesis and reforming (Fig. 7.5). *Calcium Chloride* ($CaCl_2$): $CaCl_2$ is a widely available adsorbent for the most commonly used refrigerants such as methanol and ethanol.

Fig. 7.5 Zeolites

(ii) Hydrophobic Adsorbents

Some of the commonly used hydrophobic adsorbents are discussed as below:

Activated Carbon: Activated carbon is the most widely used adsorbent in adsorption cooling systems. It is easily available. The activated carbon is highly porous with graphics lattice structure. Activated carbon is manufactured by the processes called as carbonization and activation. Coal, peat wood and nutshells (coconut) are the commonly used raw materials for manufacturing activated carbon. One of its main drawbacks is it is combustible. The carbonized particles are subjected to 'activation' to create a porous structure in it. The activation is done in the presence of an oxidising agent like steam or carbondioxide. The pore size mainly depends on the time of exposure to the oxidising agents.

The longer exposure time results in larger pore sizes. Activated carbon is used for adsorption of organic substances and non-polar adsorbates and is also usually used for waste gas (and waste water) treatment. It is the most widely used adsorbent. Activated carbon is available in three different forms, they are, Granular Activated Carbon (GAC) (particle size ranging from 0.2 to 5 mm) Powder Activated Carbon (PAC) (size less than 0.18 mm), Pelleted type of activated carbon (size ranging from 0.8 to 5 mm) as shown in Fig. 7.6.

Preparation of Composite Adsorbents: Composite adsorbents are obtained by combining two or more physical or chemical adsorbents by adding binder in the appropriate proportion (Sha and Baiju 2019). The consolidated adsorbents provides better thermodynamic properties than the base materials commonly used. The composite adsorbent material has an improved adsorbent uptake, thermal diffusivity and porous property than the base adsorbent. This improves the performance of the adsorption cooling system. The method of preparation of the composite involves heating of the base and additive materials around 150 °C for four to five hours to remove the water particles. The binder is mixed with water, and is poured to the material combination. The mixer is then compressed for uniform distribution and is then heated at 120 °C in an oven for four to five

(a) Granular activated carbon (b) Powder activated carbon (c) Pelleted type activated carbon

Fig. 7.6 Forms of activated carbon (Chen et al. 2010)

hours to remove the moisture content fully. The various composite adsorbent-adsorbate used in the adsorption cooling purposes are Maxsorb-III (50%), EG (40%) Binder (10%)-Ethanol, Maxsorb-III (90%) Poly IL [VBTMA] (10%)-Ethanol, SG/LiBr–Ethanol, $CaCl_2$/silica gel–Water, $LiCl_2$/silica–Water, Alumina/Zeolite–Water.

7.5.1 *Adsorbate or Refrigerant*

The substance concentrated or adsorbed at the surface of the adsorbent during adsorption is called the adsorbate. Different types of refrigerants used and its selection is detailed in this section. The following parameters are considered for the proper selection of adsorbate:

(i) Size of the molecule should be small to enable easy adsorption to the adsorbent.
(ii) High latent heat of vaporisation.
(iii) Thermally and chemically stable within the working temperature range
(iv) Non-toxic, non-corrosive and non-inflammable.
(v) It should have a high thermal conductivity.
(vi) Low viscosity.
(vii) Low specific volume.

The commonly used adsorbates (working fluids) for adsorption cooling system are given below:

Water

Water is thermally stable with most of the physical adsorbents. It has a high latent heat of vapourisation 2258 kJ/kg at a pressure of 1.0 bar. Since the freezing temperature of the water is 0 °C it cannot be used for refrigeration applications below 0 °C. It is a good refrigerant when any of natural zeolite, zeolite 13X and silica gel is used as adsorbent.

Ammonia

Ammonia has relatively high enthalpy of vapourisation of about 1368 kJ/kg at atmospheric pressure. It has a low specific volume. Due to the toxicity of ammonia, its application is limited to commercial purposes only. It has a relatively high latent heat of vapourisation of about 1368 kJ/kg at atmospheric pressure. It has a low specific volume. Due to the toxicity of ammonia, its application is limited to commercial purposes only.

Methanol

Methanol has a latent heat of vapourisation 1160 kJ/kg and very low specific volume of $10^{-3} m^3/kg$. But the problem with methanol is that it is highly inflammable. It is the most commonly used refrigerant with activated carbon as adsorbent for low-temperature heat source applications.

Hydrofluorocarbons

Hydrofluorocarbons are currently used in refrigeration industries. They lead to ecological problems like greenhouse effect and global warming.

7.6 Working Pairs for VAD Systems

Several working pairs are available for solid adsorption cooling system. The efficient operation of the system depends upon the careful selection of working pairs. The most commonly used adsorbent-adsorbate combinations are:

(i) Zeolite-water
(ii) Zeolite composites and water systems
(iii) Activated carbon–methanol
(iv) Silica gel-water and silica gel–methanol systems
(v) Activated carbon–ammonia
(vi) $CaCl_2$–methanol systems
(vii) Metal hydrides-hydrogen systems
viii) Complex compound/salts and ammonia or water. The possible combinations of various working pairs suitable for adsorption cooling systems are presented in Table 7.3.

7.7 Characteristics of Working Pairs

Adsorption characteristics of the adsorbent-refrigerant pair are one of the most essential parameters that affect the system performance. The adsorption characteristics such as the adsorption isotherms, adsorption kinetics and heat of adsorption (Table 7.4) needs to be studied in detail to ascertain the properties of employed working pairs. The parameters to be studied are:

(1) Uptake of refrigerant
(2) Heat of adsorption
(3) Heat of desorption
(4) BET (Brunaur-Emmet-Teller) surface area
(5) Adsorption-desorption characteristics
(6) Quantity of refrigerant adsorbed/desorbed at a particular time
(7) Cycle time
(8) Pore size and pore volume

Table 7.3 Various working pairs suitable for adsorption cooling systems (Anyanwu 2000, 2003; Chen et al. 2010; Mahesh and Kaushik 2012; Sha and Baiju 2019; Srinivasan and Dutta 2019)

Adsorbent		Refrigerant	Ozone depletion potential (ODP)	Global warming potential (GWP) (100 years)
Physical adsorbents	Activated carbon	Methanol	0	0
		Ethanol	0	0
		Water	0	0
		Ammonia	0	0
		H_2	0	5.8
		N_2	0	–
		CO_2	0	1
		Diethyl ether	0	–
		SO_2	0	–
		NO	0	–
		R113	0.8	5000
		R141a	0.11	630
		R225cb	0.033	620
		R134a	0	1300
	Maxsorb III	R507	0	3300
		n-Butane	0	4
		Water	0	0
	Zeolites	Ammonia	0	0
		Ethanol	0	0
		Methanol	0	0
		CO_2	0	1
	Silica gel	Water	0	0
Chemical adsorbents	$CaCl_2$	Water	0	0
	$CaCl_2$	Ammonia	0	0
	$MgCl_2$	Ammonia	0	0
	$SrCl_2$	Ammonia	0	0
	$BaCl_2$	Ammonia	0	0
Composite adsorbents	$CaCl_2$/Silica gel	Water	0	0
	AC/Silica gel/ $CaCl_2$	Water	0	0
	Vermiculite/$BaCl_2$	Ammonia	0	0
	Polymer silica-LiCl	Methanol	0	0
	$SrCl_2$/$CaCl_2$/ $BaCl_2$/$MgCl_2$	Ammonia	0	0

(continued)

Table 7.3 (continued)

Adsorbent		Refrigerant	Ozone depletion potential (ODP)	Global warming potential (GWP) (100 years)
	Metal hydrides	H_2	0	5.8
	Graphite/$CaCl_2$	Ammonia	0	0
	Molecular sieve nanoporous carbon (CSG)	R21	0.04	210
		R22	0.055	1700
		R23	–	–
		R30	–	–
		R40	–	–
		N_2	–	–
		R114a	–	–
		R124	0.02	480
		R125	–	–
	$SrCl_2 + CaCl_2$ mixture	Ammonia Methanol Ethanol	0	0
	$SrCl_2 + CaSO_4$ mixture	Ammonia Methanol Ethanol	0	0
	$BaCl_2$		0	0
	$SrCl_2 + CuSO_4$	Ammonia Methanol Ethanol	0	0
	Zeolite/$CaCl_2$	Water Ammonia Methanol Ethanol	0	0

7.7.1 Adsorption Characteristics- Determination Techniques

The adsorption quantity (uptake) needs to be measured to assess the maximum amount of adsorbate that can be adsorbed or desorbed under a certain condition and how long does it take to reach this value. The methods of determining the adsorbed quantity are, volumetric and thermo gravimetric method.

(a) Volumetric method

The volumetric method of estimation of uptake is based on two methods. In the first method, the variation of mass is measured by measuring the change of pressure in a vessel with known volume and in the second method, the change in volume in a

Table 7.4 Potential working pairs

Adsorbent	Refrigerant	ODP	GWP (100 years)	Remarks
Consolidated adsorbents		0	97	Replaces R22
Vermiculite/ $BaCl_2$graphite/$CaCl_2$	R41	0.012	120	
Activated carbon (AC)/ silica gel/$CaCl_2$	R123	0	120	
	R152a	0	12	
	R161	0	20	
Molecular sieve/nano porous carbon (CSG)	R170 R290			Replaces R134a
	Mixed refrigerants R1270/R170 R290/R1270 R290/R170	0	20	
	R290/R600 R290/R600a R290/R600/R600a	0	20	

vessel with known volume is measured by checking the level of liquid. The details of different methods are given below:

(i) Variation of Pressure Method

In this method, a system consisting of a vapour chamber fitted with a pressure gauge, vapour generator as well as an adsorbent stored inside the adsorbent bed is used. The temperature of saturated liquid is controlled by the thermostatic bath located in the vapour generator. The pressure during adsorption and desorption are determined, the uptake is determined by the corresponding pressure. A coil water pipe controls the temperature of the chamber. Another thermostatic bath is used to control the adsorption and desorption temperature.

The volume of the vapour inside adsorber, vapour chamber volume and the mass of adsorbent are usually kept small in order to attain the precise results (Fig. 7.7).

(ii) Liquid Level Measuring Method

The basic principle of this method is the measurement of change in the volume of refrigerant adsorbed or desorbed in the vessel. The level changes can be identified by measuring the changes in volume. More precise devices like magnetostrictive sensors are employed to measure the level changes. The level sensor measures the liquid level in the condenser and evaporator. The area of condenser and evaporator are known. By using the known data along with the density of refrigerant the uptake or adsorption or desorption quantity can be determined (Fig. 7.8).

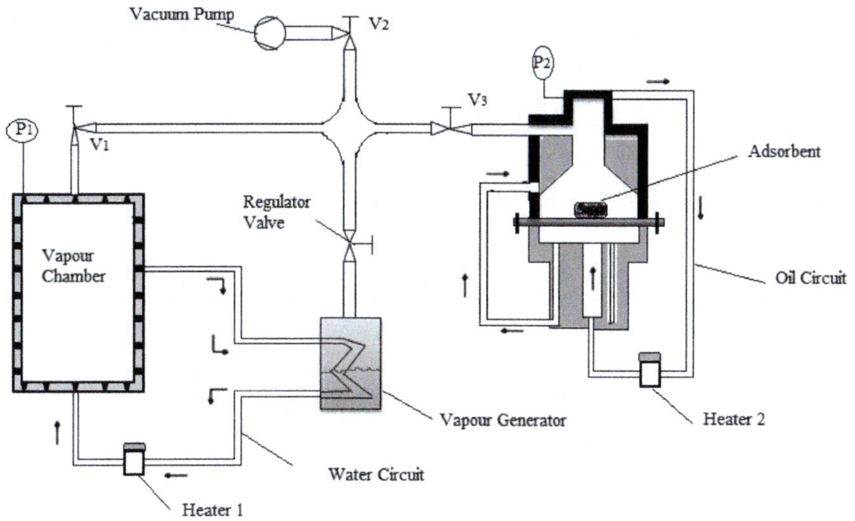

Fig. 7.7 Volumetric method (variation of pressure) (Chen et al. 2010)

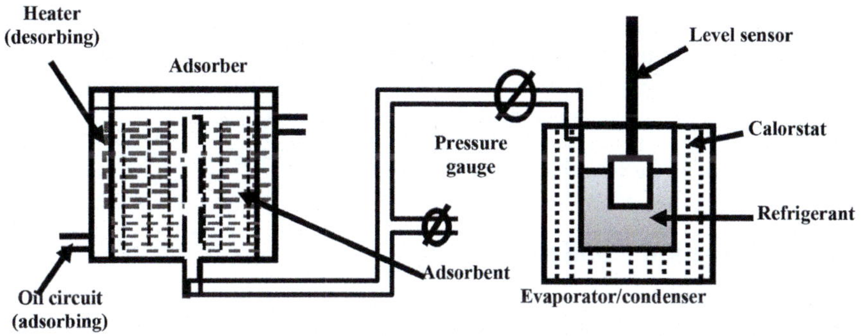

Fig. 7.8 A liquid level method of measurement

(b) Gravimetric Method

In this gravimetric method, changes of mass of the adsorbent during the adsorption and desorption are measured and are used for the uptake determination. In the apparatus, a quartz spring is coupled. The length of the spring, which is connected to an empty basket is known h_1, and the lengths of quartz spring connected to the basket with adsorbent before and after adsorption, respectively, are h_2 and h_3 (Fig. 7.9). Then, the adsorption quantity is given by

$$x = (h_3 - h_2)/(h_2 - h_1) \tag{7.2}$$

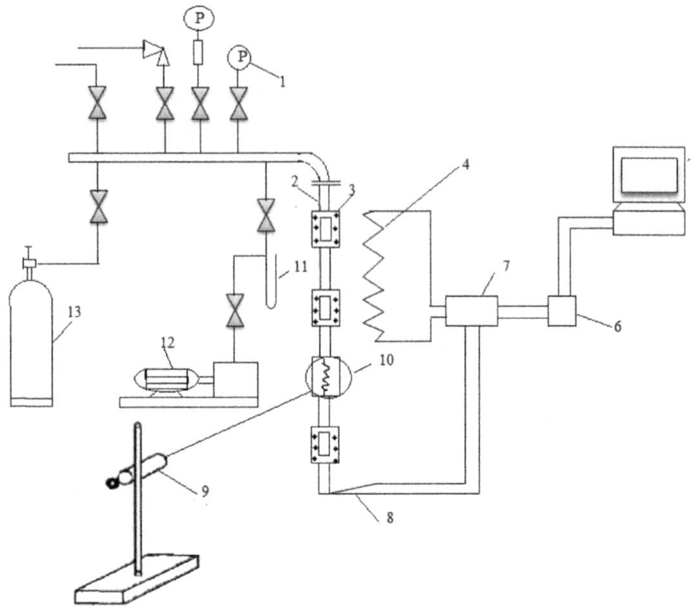

(1) pressure gauge; (2) measuring tube; (3) viewing glass; (4) electric heater;(5) computer; (6) electric source; (7) temperature controller; (8) thermocouple;(9) altimeter; (10) quartz spring, basket and adsorbent; (11) U shape pressure gauge; (12) vacuum pump; (13) liquid nitrogen cylinder.

Fig. 7.9 Test unit for the gravimetric method

7.8 Adsorption Cooling Systems

The basic adsorption cooling system consists of four main components, viz. a solid adsorbent bed, a condenser, an expansion valve and an evaporator.

The adsorbent bed is used to adsorb or desorb the refrigerant vapour according to the changes in the temperature of the adsorbent. It is heated to drive off the refrigerant vapour and is conveyed to the condenser. The refrigerant vapour is condensed into liquid in the condenser by rejecting latent heat of condensation. In the expansion valve, refrigerant is then expanded to lower pressure and passed to the evaporator. The refrigerant condensate is vapourised in the evaporator by utilising the latent heat taken from space to be conditioned. The adsorbent bed is subjected to cooling and the refrigerant vapour adsorbs in the bed to complete one cycle.

7.8.1 *Principle of Operation of Vapour Adsorption System and Cycle Analysis*

The operation of adsorption cooling system involves two phases. Phase I include pre-heating and desorption, while Phase II involves adsorption and pre-cooling. Figure 7.10a and b represents the schematic presentation of the working phases. The ideal adsorption cooling cycle is presented in Fig. 7.11.

Phase I

This phase includes two processes, namely, pre-heating (isosteric heating) and desorption (isobaric desorption).

(i) Pre-heating (isosteric heating, process 1-2)
At state point 1, the adsorbent is cold and saturated with the refrigerant, heat is applied to the adsorbent bed. Due to heating at a constant concentration, a certain amount of the refrigerant is desorbed from the adsorbent bed. During this process, the system pressure increases until the point 2 is reached. At point 2, the system pressure becomes equal to saturation pressure corresponding to the condensation temperature.

(ii) Desorption (isobaric desorption, process 2-3)
Desorption process starts at point 2 and continues until the adsorbent temperature reaches the maximum available temperature. The condensation of refrigerant takes place ideally at constant pressure (P_{con}). During this process, the refrigerant vapour driven off from the adsorbent bed is condensed and collected in the evaporator space.

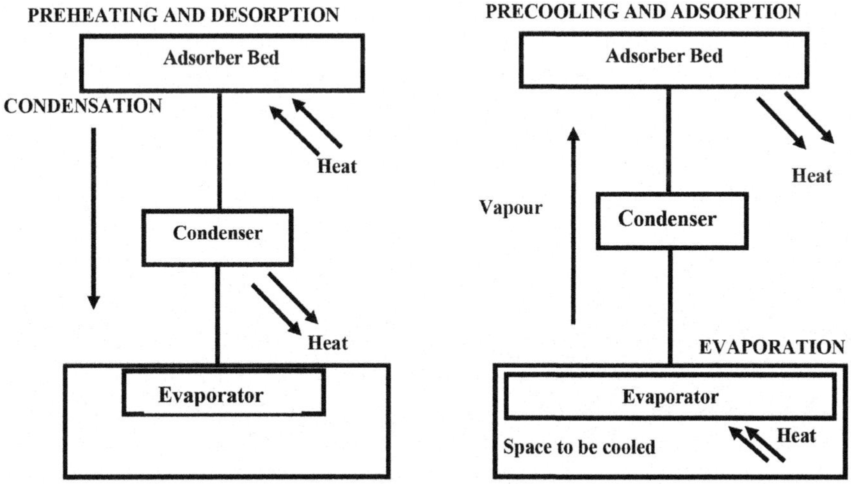

Fig. 7.10 Working of adsorption cooling system (Dawoud 2007)

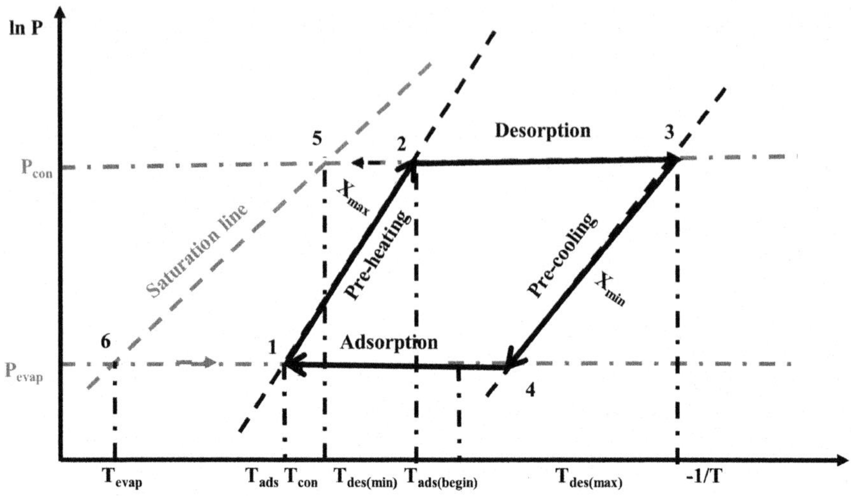

Fig. 7.11 Clausius Clapeyron diagram (Dawoud 2007)

Phase II

Phase II includes two processes, viz. pre-cooling (isosteric cooling) and adsorption (isobaric adsorption).

(i) Pre-cooling (isosteric cooling, process 3-4)
At the beginning of the pre-cooling process, the adsorbent bed is cooled by rejecting heat to ambient air. The adsorbent is pre-cooled and is able to adsorb the refrigerant vapour coming from evaporator. Since the adsorbent bed is further pre-cooled a portion of previously desorbed and the condensed refrigerant is adsorbed and the latent heat of vapourisation is drawn from the remaining liquid refrigerant in the evaporator. This leads to decreasing the refrigerant temperature at the end of the pre-cooling process (3-4).

(ii) Adsorption (isobaric adsorption, process 4-1)
The adsorption process starts at point 4 and proceeds by further cooling of adsorbent bed until the whole refrigerant is evaporated upon removing the cooling load from surrounding space and adsorbed in the adsorbent bed.

7.8.2 *Performance Parameters of Adsorption System*

The performance of an adsorption cooling system is normally expressed in terms of cycle coefficient of performance (COP) and specific cooling power (SCP).

The cycle COP depends on the cooling effect produced and total heat energy required for the desired cooling effect. The total heat input required is the sum of heat required for isosteric heating and desorption. Cycle COP is defined as the ratio of cooling effect to the total energy required for a desired cooling effect, i.e.,

$$\mathrm{COP} = \frac{\text{Cooling effect}}{\text{Total Energy input}} \tag{7.3}$$

$$\mathrm{COP} = \frac{Q_e}{Q_t} \tag{7.4}$$

The total energy input to the system is given by

$$Q_T = Q_{\text{isosteric heating}} + Q_{\text{desorption}} \tag{7.5}$$

The Specific cooling power relates the cooling effect produced and the mass of adsorbent charged into the unit, thus, it measures the size of the system. Higher SCP values indicate the compactness of the system.

$$\mathrm{SCP} = \frac{\text{Cooling effect}}{\text{Cycle time } \times \text{ Adsorbent mass}} \tag{7.6}$$

$$\mathrm{SCP} = \frac{Q_e}{m_a \tau_{\text{cycle}}} \tag{7.7}$$

In a solar-powered system, the solar coefficient of performance has to be estimated which is defined as the ratio of cooling effect to the net solar energy input and it is given below

$$\text{Solar COP} \ = \frac{Q_e}{Q_s} \tag{7.8}$$

On the other hand, the thermal uptake efficiency of an adsorption system is analogous to the volumetric efficiency of a mechanical reciprocating compressor and is defined as the ratio of actual uptake to the theoretically expected uptake (Fig. 7.12).

Uptake efficiency is given by

$$\eta = 1 + \left(\frac{\rho_b - \rho_d}{C_b - C_d}\right)\left(\frac{1}{\rho_{\text{eff}}} - \frac{1}{\rho_s}\right) \tag{7.9}$$

where $\Delta\rho = \rho_{\text{b}} - \rho_{\text{d}}$ is the density difference (points a and b) (Fig. 7.13). The value of change in density is negative since the specific volume decreases after compression $(\rho_{\text{b}} < \rho_{\text{d}}) \cdot \rho_{\text{eff}}$ is the effective packing density of the adsorbent and ρ_s the solid density. The effective packing density can be defined as the ratio of the mass of adsorbent that can be packed to the volume of the adsorbent bed.

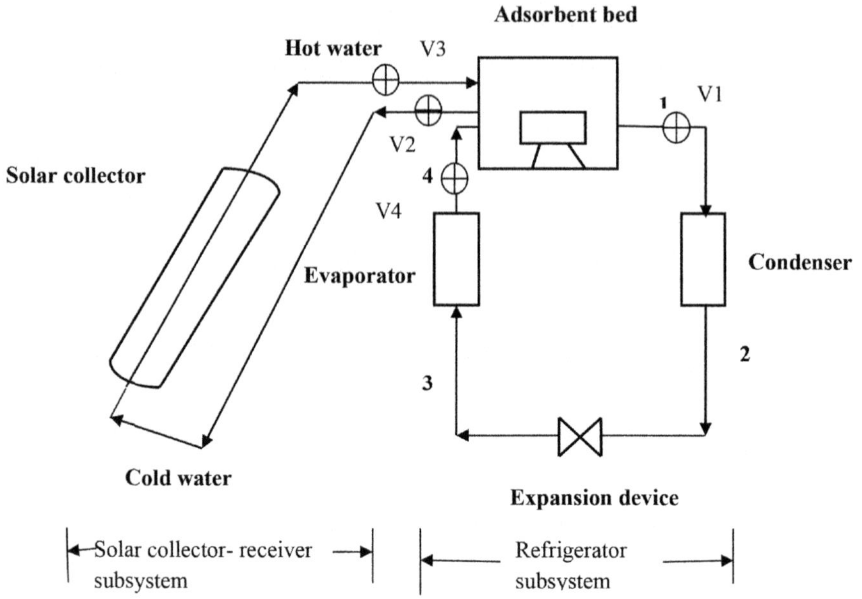

Fig. 7.12 A single bed solar adsorption cooling system (Baiju and Muraleedharan 2015)

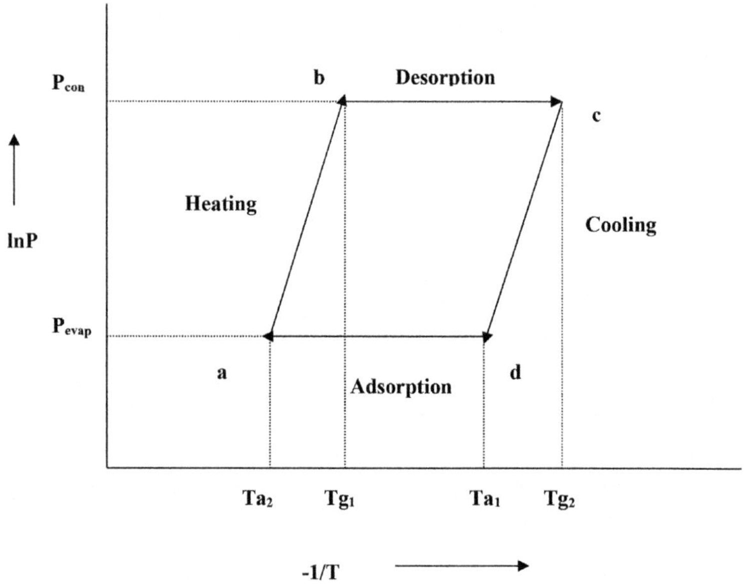

Fig. 7.13 Clapeyron diagram of ideal adsorption cooling cycle

7.8.3 Single Bed Adsorption Cooling System

The single bed solar adsorption cooling system is shown in Fig. 7.12. The system consists of two subsystems, namely, solar collector subsystem and refrigerator subsystem. The system mainly consists of an adsorbent bed which acts as the thermal compressor, water-cooled condenser, expansion device, evaporator and a solar collector. The assembly of the system starts with the preparation of adsorbent bed. The adsorbent bed is filled with designed quantity of adsorbent. The adsorbent bed along with the activated carbon is placed inside the water tank by using suitable stands for support. The charging line along with the pressure gauges and valves are connected to the top of the thermal compressor. The condenser inlet is connected to the refrigerant outlet of the adsorbent bed. The other end of condenser is connected to the expansion device. Inlet of the evaporator is connected to the outlet of the expansion device.

In the beginning, (Fig. 7.12) valves V1 and V4 are closed. The water tank is filled with the sufficient quantity of water. As the solar radiation falls on the solar thermal collector, water gets heated up due to the convective current formed in the flow path. The temperature of adsorbent bed increases ($T_{a2} - T_{g1}$). This causes the vapour pressure of the adsorbed refrigerant to reach up to the condensing pressure. The exit valve V1 is opened when the refrigerant temperature reaches to desorption temperature. The refrigerant is desorbed from the adsorbent bed matrix. The desorbed refrigerant enters the condenser where it condenses to liquid refrigerant and flows into the expansion device. After expansion through the expansion device, the low-pressure liquid refrigerant then enters the evaporator. The refrigerant absorbs heat from the space to be cooled as latent heat. In the evening, the hot water from the tank is drained off and is refilled with cold water.

The temperature of the bed reduces due to the isosteric cooling process ($T_{g2} - T_{a1}$) and the pressure in the adsorber drops below the evaporator pressure. The refrigerant coming from the evaporator will be adsorbed inside the adsorbent bed and completes one cycle of operation.

7.8.4 Two Bed Adsorption Cooling System

Two significant processes in the adsorption cooling cycle are adsorption and desorption. The single bed solar adsorption cooling system has the limitation that its working is intermittent in nature. To overcome the intermittency of an adsorption cooling system, two adsorbent beds working out of phase is necessary. Along with two beds in the adsorption cooler, a sensible heat storage unit can also serve the desired purpose. The schematic diagram of two bed VAD system integrated with a sensible heat storage system is shown in Fig. 7.14.

The system utilises a common evaporator, a condenser, an expansion device (capillary tube), solar concentrator and two adsorbent beds placed in two different water

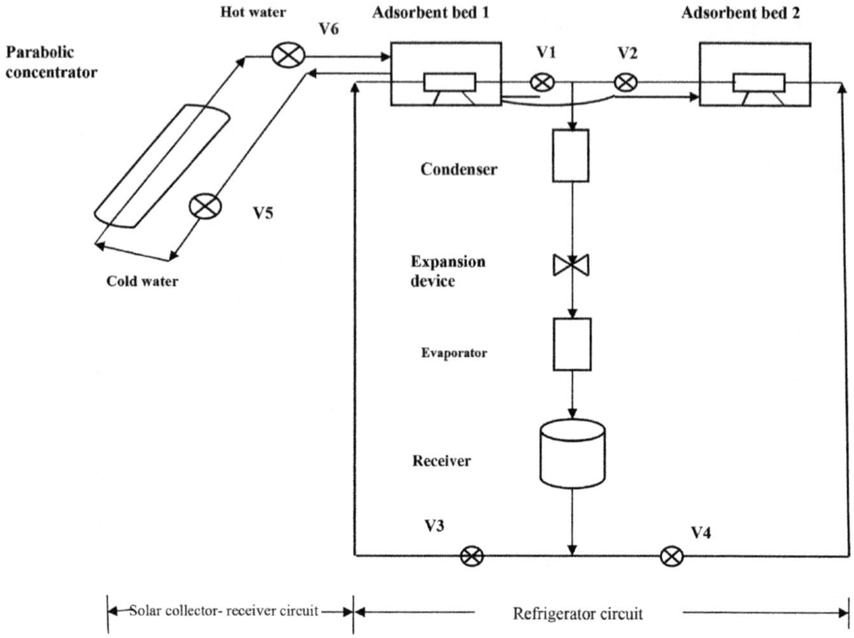

Fig. 7.14 Schematic of two-bed system

tanks. The heating of water starts in the morning through the solar concentrator by natural circulation. The temperature of water in the tank 1 increases, and hence the temperature of adsorbent bed 1. The heating causes the adsorbent bed temperature and hence the pressure of the adsorbate to increase. Valve V1 opens as the pressure of refrigerant equal to the condensing pressure of the refrigerant.

The refrigerant gets condensed in the condenser by rejecting its heat of condensation to the circulating fluid. The liquid refrigerant expanded through an expansion device to the evaporator pressure level. The low-pressure refrigerant then enters the evaporator, where the refrigerant takes the latent heat from water to be chilled and gets converted into the vapour. Hot water stored in the tank 1 is transferred to the tank 2. This stored water will act as a driving heat source for the adsorbent bed 2 in the night time. The function of two adsorbent beds are reversed in the night time. Hence, the adsorbent bed 1 is subjected to adsorption and adsorbent bed 2 is subjected to the desorption process. When the pressure in bed 2 is nearly equal to the evaporator pressure, valve V4 is opened by keeping valves V1, V2 and V3 closed. The low-pressure refrigerant vapour coming from the evaporator will be adsorbed in bed 2. In the evening, after the complete desorption of refrigerant vapour from bed 1, the hot water from tank 1 is transferred to tank 2, which acts as the heat source for the night time. At this time, tank 1 is filled with the cold water to cool bed 1 from the desorption temperature level to adsorption temperature. When the pressure in bed 2 reaches to the condensing pressure, valve V2 is opened by keeping valves

V1 and V4 closed. The refrigerant passes through the condenser, expansion device and evaporator. Finally, the refrigerant vapour is adsorbed in bed 1. During this time, bed 1 acts as the adsorber and the refrigerant will be adsorbed in adsorbent matrix 1. Now the system is ready for the next cycle of operation.

7.8.5 Four Bed Vapour Adsorption Cooling System

Basic four processes involved in an adsorption cooling system are isosteric heating, desorption process, isosteric cooling and adsorption process.

C_1–C_4: Discharge valves; E_1–E_4: Intake valves; HO_1–HO_4: Hot water outlet valves; CO_1–CO_4: Cold water outlet valves: HI_1–HI_4: Hot water inlet valves; CI_1–CI_4: Cold water inlet valves.

To allow all these four processes to occur in succession and to maintain a continuous flow of the refrigerant, it is necessary to use several adsorbent beds (typically 4). Figure 7.15 shows schematic of a four-bed adsorption cooling system driven by solar energy. Table 7.5 shows the sequence of operations in four thermal compressors of a continuous system.

Water contained in the water tank is heated by solar energy. Working of the system is similar to that of a two-bed system explained in Sect. 7.8.4.

7.8.6 Basic Thermal Wave Cycle

The system consists of two adsorbent beds, a cooler, heater, condenser, expansion device and an evaporator. Thermal wave cycle has the advantage of utilising high heat released during the adsorption process (adsorption is an exothermic reaction). The temperature of refrigerant or adsorbate rises or drops significantly during the process forming a large temperature difference in the form of a wave, and called as the thermal wave cycle. A single heating/cooling fluid is used to connect two adsorbent beds, viz. cooler and a heater. The circuit of fluid could transfer heat from the adsorbent bed to the desorption bed and to recover adsorption heat for the improvement of energy efficiency of the system. Study reveals that about 80% of the adsorption heat can be recovered by this process.

Let the bed A be at saturated adsorption state (T_e, P_e) and bed B at saturated desorption state (T_c, P_c). Hence, first the bed A to be subjected to isosteric heating and desorption, whereas the bed B switched over to isosteric cooling and adsorption mode.

Process 1

Heat transfer fluid heated in the heater enters bed A and transfers heat to the bed. The thermal wave will be formed within the bed. The fluid then enters the cooler, where the fluid will cool to a temperature T_L, and releases heat to the environment.

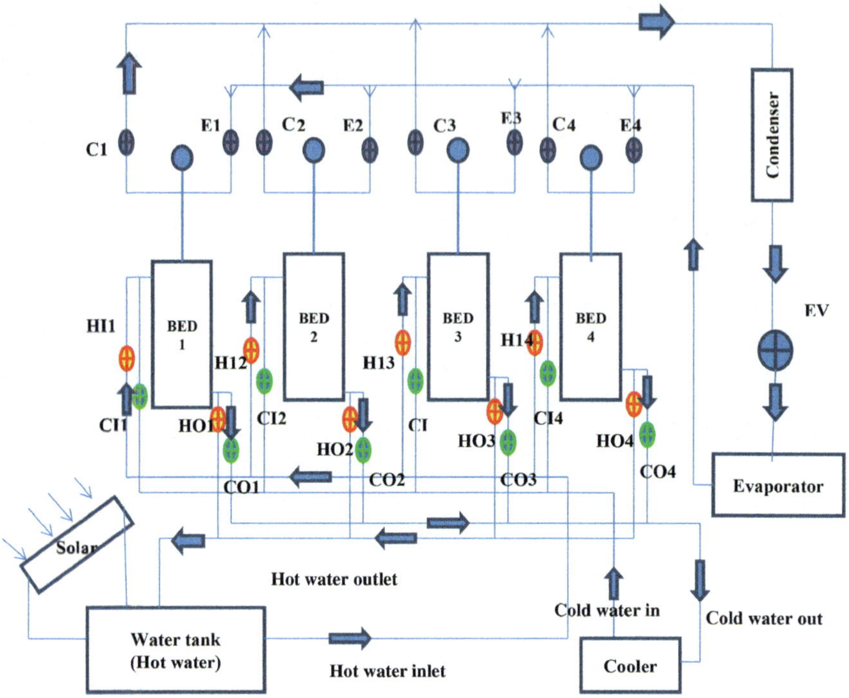

C1-C4: Discharge valves; E1-E4: Intake valves; HO1-HO4: Hot water outlet valves; CO1-CO4: Cold water outlet valves: HI1-HI4: Hot water inlet valves; CI1-CI4: Cold water inlet valves

Fig. 7.15 Schematic diagram of a four-bed adsorption cooling system

Table 7.5 Thermal compressor operating phases

	Phase 1	Phase 2	Phase 3	Phase 4
Compressor 1	Adsorbing	Heating	Desorbing	Cooling
Compressor 2	Heating	Desorbing	Cooling	Adsorbing
Compressor 3	Desorbing	Cooling	Adsorbing	Heating
Compressor 4	Cooling	Adsorbing	Heating	Desorbing

The fluid then enters bed B. The adsorbent bed B is subjected to isosteric cooling to a temperature of T_L, fluid flows back to the heater again for the next cycle. In this process, when the temperature of bed A increases a part of refrigerant begins to desorb, the pressure in bed A increases to the condenser pressure. Similarly, for bed B pressure decreases to P_e, temperature to T_e. When the pressure of bed A reaches P_c and B reaches P_e, open the valve between bed A and condenser and the valve between bed B and evaporator. Then bed A is ready for desorption and bed B for adsorption of refrigerant. As the thermal wave moves forward, refrigerant completes desorption from bed A and is being adsorbed in bed B. When this process completes

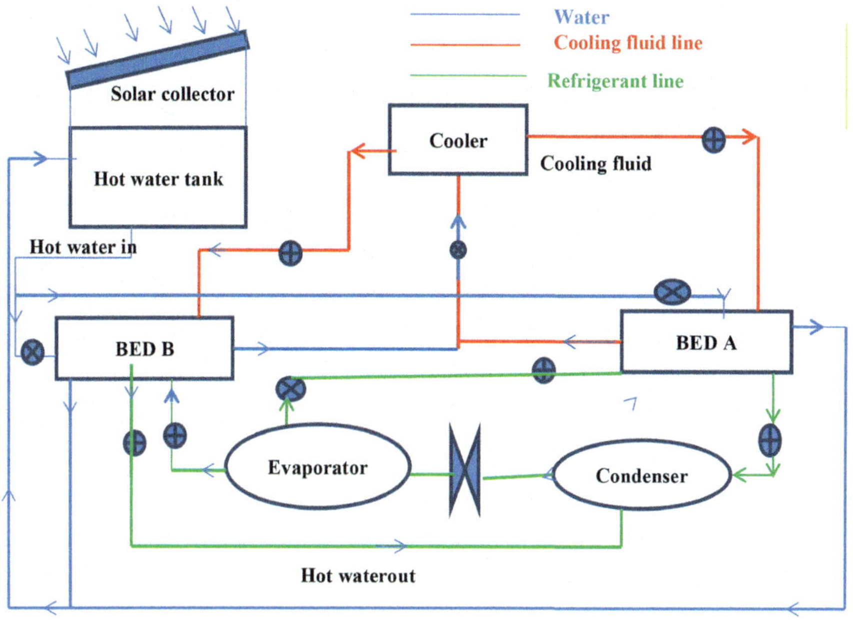

Fig. 7.16 Schematic of solar thermal wave cycle (Critoph 1998)

bed A reaches the starting point of B and bed B reaches the starting point of bed A (Fig. 7.16).

Process 2

In this process, bed A is subjected to isosteric cooling and adsorption, whereas bed B subjected to isosteric heating and desorption of the refrigerant. The direction of fluid flow will be switched to reverse direction. Thus the proceedings of two beds will be interchanged.

7.8.7 Convective Thermal Stream Cycle

Convective thermal stream cycle is the modification of the thermal wave adsorption cycle. The forced convection effect is used along with thermal wave to enhance the heat transfer in the adsorption refrigeration system. Critoph has presented a convective thermal stream cycle which is the modification of a thermal wave cycle. The basic concept of thermal wave cycle and convective thermal stream cycle is the same. In the convective thermal wave cycle, the thermal fluid for heating and cooling of bed is initiated by the refrigerant itself. Thus, a direct contact heat and mass transfer between the thermal fluid and adsorption bed are achieved.

The two active beds are packed with adsorbent and two inert beds are packed with non-reactive particles such as steel balls. Figure 7.17 shows the first half of the cycle

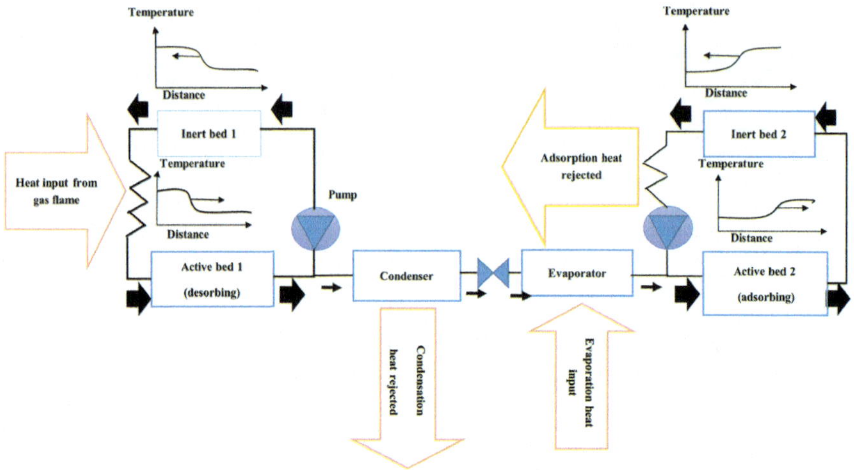

Fig. 7.17 Convective thermal stream cycle (Critoph 1999)

during which the active bed 1 is heated and desorbs the refrigerant while active bed 2 is cooled, adsorbing refrigerant. In fluid recirculation loop shown on the left, a low power pump circulates refrigerant stream through inert bed 1 which is initially hot. The gas stream is heated by the bed and a cold wave is passed through the bed from right to left. The refrigerant stream is being heated to a maximum cycle temperature in inert bed. The refrigerant gas is then passed to the active bed 1, where it heats the adsorbent. A hot thermal wave passes through the bed. The refrigerant gets desorbed from the adsorbent bed, which increases the pressure in the left loop. The desorbed refrigerant condenses in the condenser rejecting heat to the environment.

7.8.8 Solar VAD Cooling System Integrated with Heat Pipe

Figure 7.18 shows schematic of solar adsorption cooling system integrated with a heat pipe. The system mainly consists of a cylindrical adsorbent bed, evaporator, condenser, expansion device and a parabolic solar concentrating collector. In all the solar adsorption systems described so far, the adsorbent bed is heated by using sensible heat of water.

This system utilises the latent heat of condensation of water (working fluid inside heat pipe) rather than the sensible heat so that the desorption becomes more faster. This leads to decreased cycle time and hence increase in the COP and SCP values. The adsorbent is filled into the cylindrical adsorbent bed. The heat pipe is a closed evaporator-condenser system consisting of a sealed hollow tube whose inside walls are lined with a capillary wick structure. In this case, condenser section of heat pipe

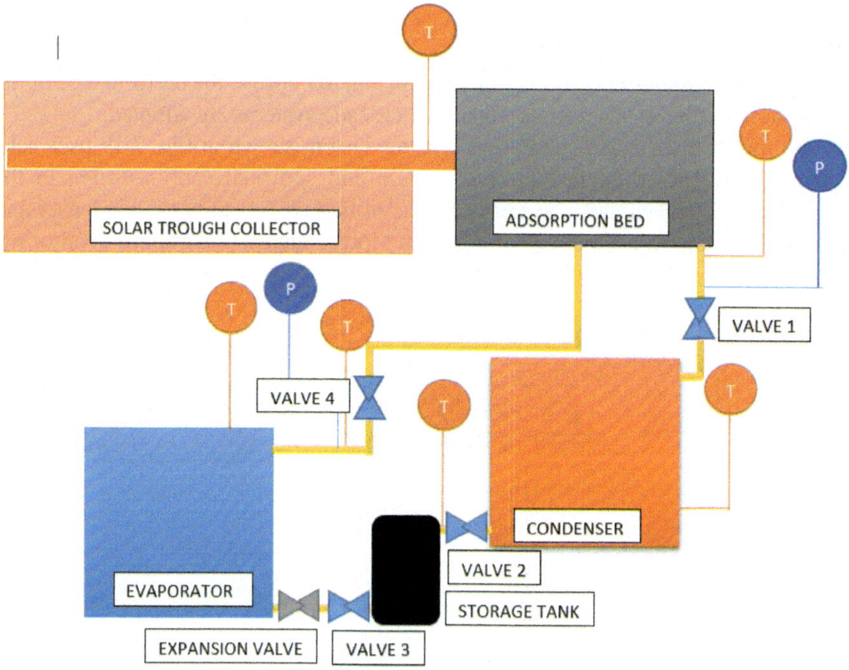

Fig. 7.18 A solar adsorption cooling system integrated with a heat pipe

is integrated with the adsorbent bed, whereas the receiver tube of parabolic collector acts as the evaporator section of the heat pipe. The adsorbent bed is connected to the condenser through another tank made of stainless steel. The water inside the evaporator section of the heat pipe is heated by solar energy. Then the water vapour forced to the condenser side of the heat pipe (inside the adsorbent bed) where it is subjected to condensation by rejecting its latent heat of condensation to the adsorbent bed matrix. Due to this, the bed temperature increases from T_e to T_C and pressure to P_C. The refrigerant vapour desorbed then enters to the condenser. In order to ensure condensation of refrigerant, heat accompanying the phase change has to be removed from the system; therefore the cooling water system of condenser is applied, while the inlet temperature of cooling water must be lower than the refrigerant condensation temperature at a given pressure.

7.8.9 Four Bed Cascading Adsorption Cooling System

Figure 7.19 shows the schematic diagram of four-bed cascading adsorption cooling cycle (Saha et al. 2003). Usually, the single-stage adsorption cooling system has lower COP and SCP. The single-stage adsorption cooling system cannot effectively utilise the high-temperature heat sources.

For the continuous operation of the system and to improve the performance of the adsorption cooling system, multi-stage and cascade systems are adopted.

The four-bed cascaded adsorption cooling system shown in Fig. 7.19 uses activated carbon-R507A at the bottoming cycle and activated carbon-R134a in the topping cycle. The cascaded system consists of two adsorbent beds, condenser and an evaporator at each stage, i.e., at upper and lower stages. The cooling load of the

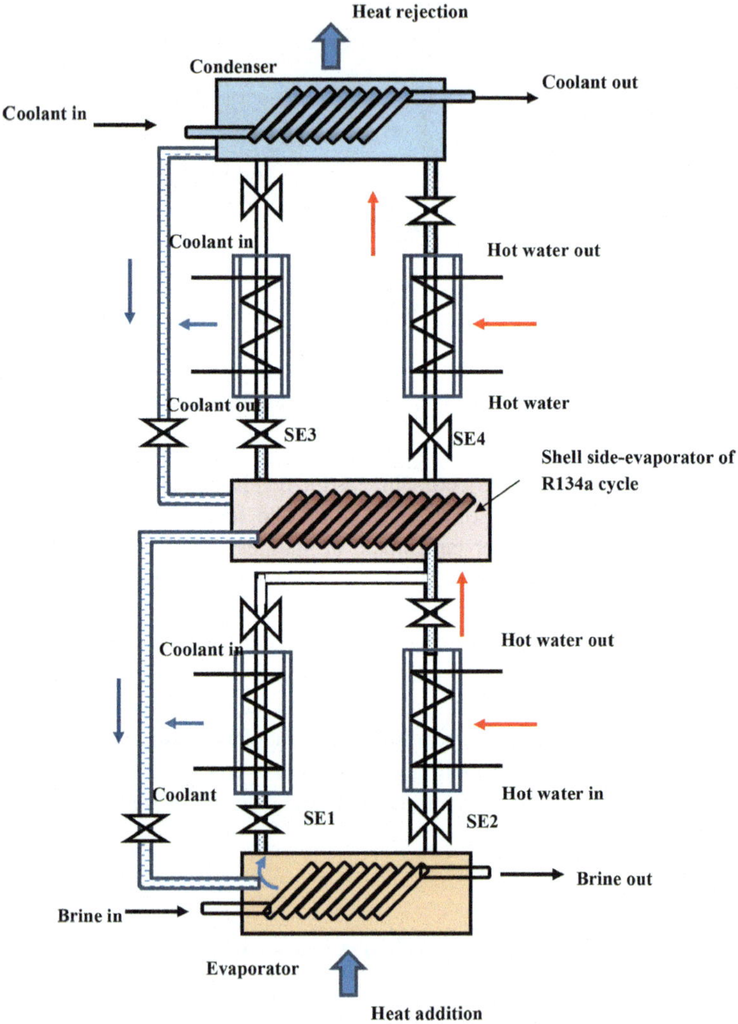

Fig. 7.19 System configuration of four-bed cascading adsorption cooling cycle (Saha et al. 2003; Shabir et al. 2020)

combined cycle is met by the lower cycle (activated carbon-R507A). The refrigerant vapour evaporated in the R507A cycle is adsorbed in the adsorbent bed (cold bed).

The hot bed is surrounded by a heat source to exhibit the desorption process in the adsorbent bed. The desorbed high-pressure refrigerant is subjected to condensation in the integrated condenser. The liquid refrigerant after expansion is refluxed back to the evaporator unit.

7.8.10 Adsorption Desalination System

Desalination process is defined as the separation of excess salts and other minerals from saline or brackish water. Figure 7.20 represents various processes involved in any desalination method.

Figure 7.21 shows various types of water desalination systems. Among these, adsorption desalination technology is emerging to produce both cooling and fresh water. The system utilizes a combination of flash evaporation and the thermal compression principle to achieve the specific task. Limitations of other solar desalination methods are:

(i) Solar still

- Requires large area
- Low overall performance
- Low distillate yield

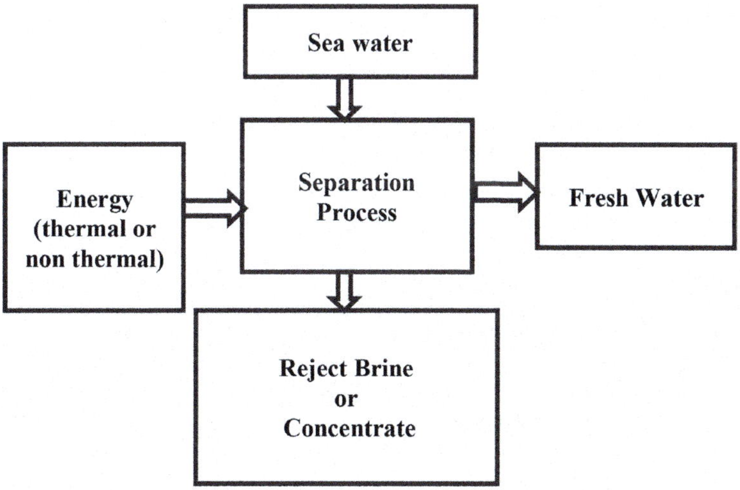

Fig. 7.20 Desalination process

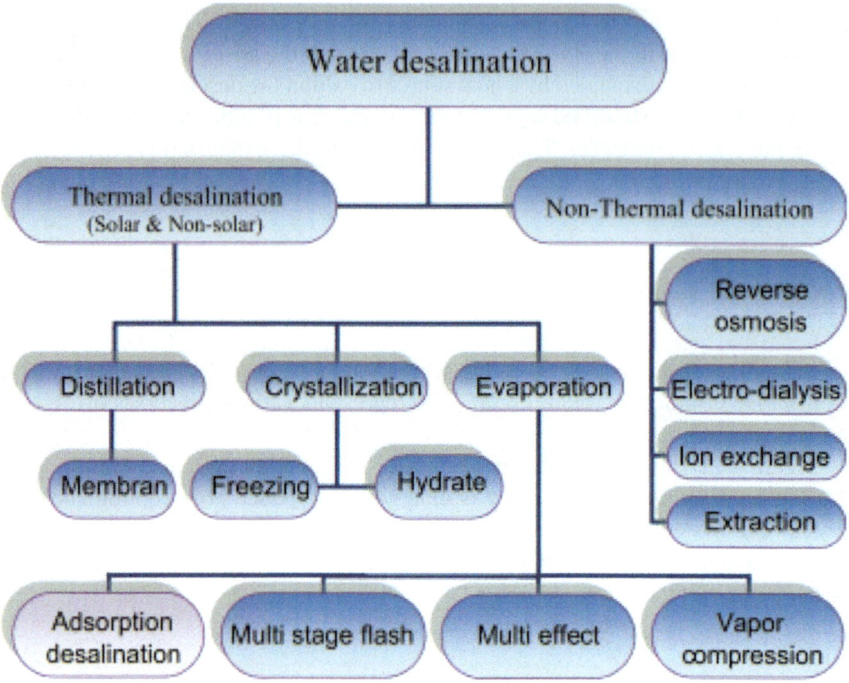

Fig. 7.21 Desalination systems

(ii) Solar humidification-dehumidification desalination

- Water production cost is higher
- Low energy efficiency

(iii) Solar multi-stage flash/multi-effect distillation

- Plant is heavy and costly
- Corrosion takes place due to high-temperature operation
- High energy consumption

(iv) Solar-powered reverse osmosis

- Pretreatment of feed water required
- Biological fouling of membrane

(v) Solar-powered electrodialysis

- Non-ionic substances and microorganism remain in the product

- Production cost increases with increase in salinity of feed water

(vi) Solar-powered vapour compression desalination

- Compressor may get corroded
- High water production cost

Principle of Adsorption Desalination

A typical adsorption plant comprises three major components, namely, evaporator, single to multi-reactor beds where the adsorbent is placed and the condenser. The following section discusses the working principle, components and the thermodynamic cycle of adsorption desalination. The adsorption desalination system is shown in Fig. 7.22. The system consists of three major parts, namely, (a) evaporator, (b) adsorbent bed and (c) condenser. First, the saline or brackish water is charged into the evaporator. In the evaporator, the water is flash evaporated at the low-pressure. The chilled water is circulated in the evaporator to maintain the evaporation process and it can be used for cooling purposes. Concentrated brine is discharged periodically from the evaporator to regulate the seawater level and salt concentration. Water evaporates and is then adsorbed in the silicagel matrix during the adsorption process. The adsorption continues until the silicagel gets saturated with water vapour. Once it gets saturated close the valve between the adsorbent bed and evaporator. The saturated silicagel is regenerated by using solar heated hot water. The water vapour desorbed then enters the condenser. The water vapour gets condensed in the condenser by rejecting heat of condensation and the condensate is collected in the condenser. A collecting tank is provided to collect the desalinated water.

The adsorbent material must be a hydrophilic porous material having a higher surface area with transient bonding to water molecules via hydrogen bonds. Silica gel is a popular adsorbent because it is able to adsorb significant levels of water (up to 40% by mass) without significant structural or volume change and readily release it under mild heating. Silica gel can be selected as adsorbent material because it is able to perform regeneration at the lowest temperature since the adsorption desalination cycle is designed to recover solar heat and the main advantages are as given below:

Advantages

- Low production cost.
- Low maintenance cost.
- Environmental-friendly (as almost no chemicals are used for cleaning of adsorbent).
- Hybrid system is able to produce a cooling effect as well as potable drinking water.

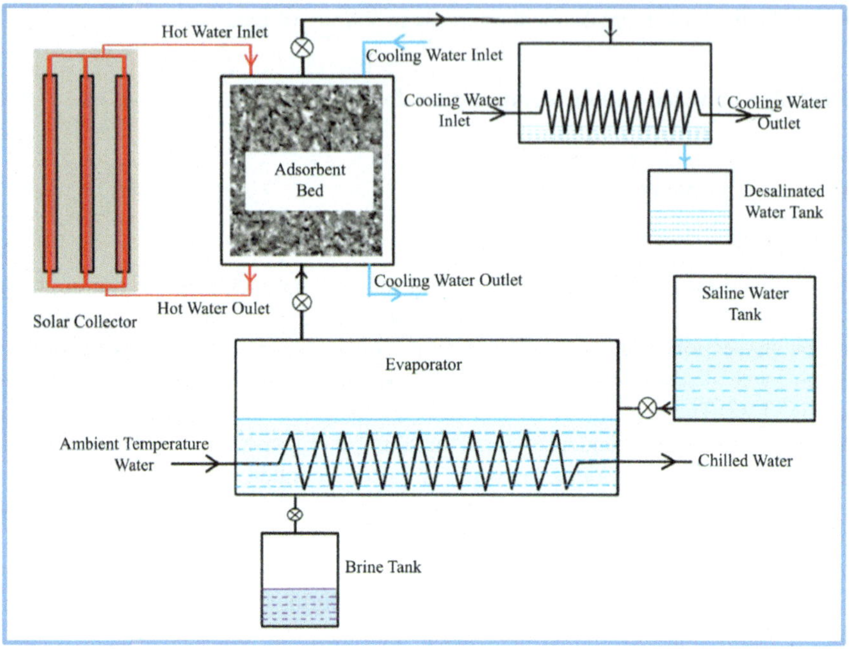

Fig. 7.22 Adsorption desalination system

7.8.11 Metal Hydride Systems

The metal hydride systems are similar to conventional adsorption cooling systems. The major difference between the two systems is that the metal hydride systems use hydrogen as working fluid unlike the refrigerants in the adsorption cooling systems. The thermodynamic cycle of the metal hydride system is similar to that of the adsorption cooling systems. The metal hydride systems utilise two metal reactors of different characteristics in order to replace the condenser/evaporator and the generator/absorber of the adsorption cooling systems. Based on the operating pressure levels, the metal hydride systems are classified into three groups, namely, (a) single-stage, (b) double-stage and (c) multi-stage (Muthukumar and Groll 2010).

Single-Stage Single Effect Metal Hydride System

Cooling effects produced in such systems are in single stage only. Figure 7.23 shows schematic of a single stage metal hydride system. Its Vant's Hoff plot is given in the Fig. 7.24. A single stage metal hydride system consists of two reactors at different temperature levels—one at very high temperature (metal hydride M) and the other at low temperature (hydride N).

The working of the system consists of two-half cycles. During the first half cycle, hydride M_1 desorbs hydrogen at high-pressure P_2 by taking up driving heat Q_{des} at T_{des}. At this time, coupled hydride N_1 adsorbs hydrogen from reactor M_1 and the heat

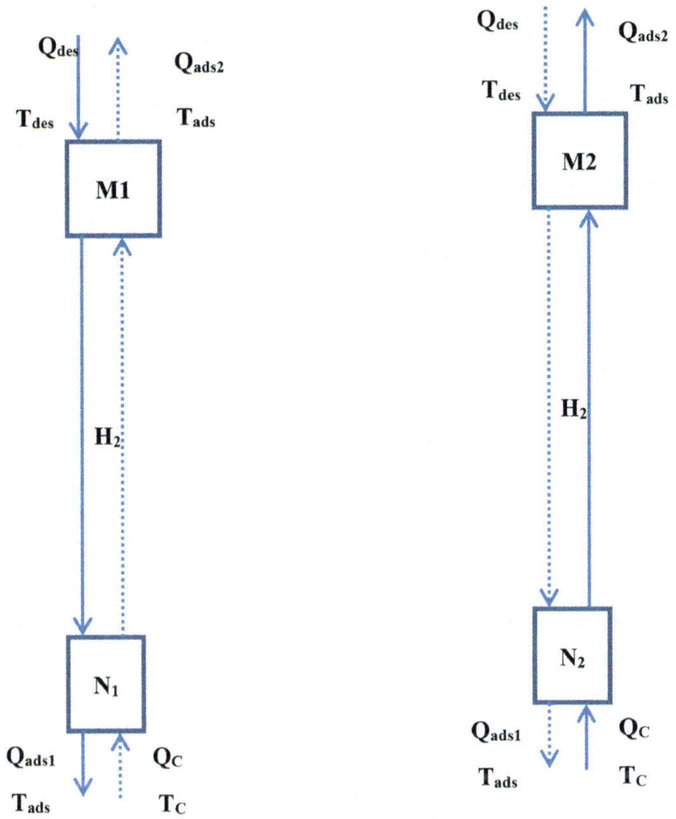

Fig. 7.23 A single-stage, single effect metal hydride system operating with two pairs of reactors

of absorption Q_{ads1} is released to the sink at T_{abs}. The hydride N_2 desorbs hydrogen at very low-pressure P_1 by taking up the heat Q_c from space to be refrigerated to produce useful cooling. The hydride M_2, absorbs this desorbed hydrogen by releasing the absorption heat Q_{abs2} at T_{abs2}. Then the metal reactors M_2 and N_2 are subjected to sensible heating and M_1 and N_1 are sensibly cooled to T_c and T_{abs}, respectively, to complete the first half of the cycle.

During the second half cycle, the hydrogen adsorbed by the reactor M_2 to be desorbed at high-pressure P_2. This can be achieved by supplying the driving heat Q_{des} at T_{des} to reactor M_2. The desorbed hydrogen is being absorbed by the reactor N_2, releasing heat Q_{ads1} at T_{ads}. As one, hydride N_1 desorbs hydrogen at low pressure P_1 by taking up useful cold Q_c at T_c, which is absorbed by M_1 releasing heat Q_{ads2} at T_{ads}. The reactors M_2 and N_2 are then subjected to the sensible cooling and reactors M_1 and N_1 to sensible heating. Thus one cycle of operation is completed. For metal hydride heat pumps, usually, the performance is expressed in terms of Coefficient of Amplification (COA) like COP. In case of a single stage single effect system, these coefficients are defined as (without consideration of sensible heat losses):

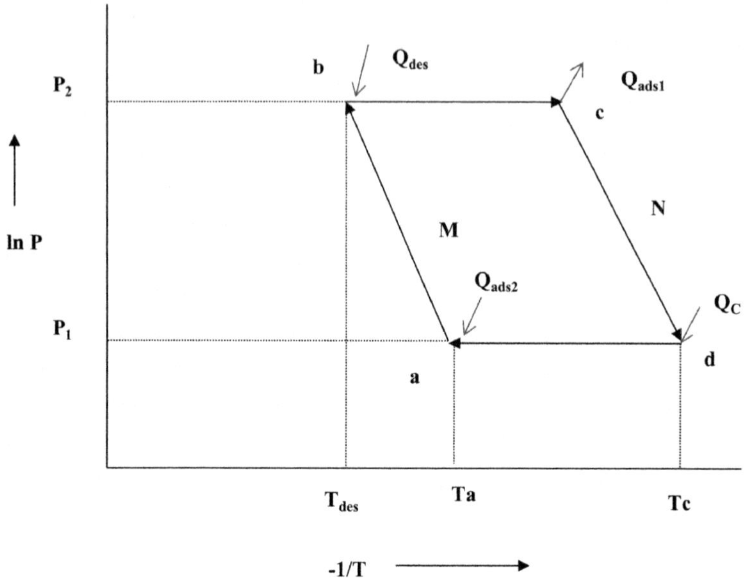

Fig. 7.24 Thermodynamic cycle of a single-stage, single effect metal hydride system on Van't Hoff plot

$$\mathrm{COA} = \frac{Q_{\mathrm{ads1}} + Q_{\mathrm{ads2}}}{Q_{\mathrm{des}}}$$

$$\mathrm{COP} = \frac{Q_c}{Q_{\mathrm{des}}} \tag{7.10}$$

If the thermal properties such as the enthalpy of metal hydrides (ΔH) are considered, the expression becomes

$$\mathrm{COA} = \frac{\Delta H_{\mathrm{M,ads}} + \Delta H_{\mathrm{N,des}}}{\Delta H_{\mathrm{M,des}}} \tag{7.11}$$

$$\mathrm{COP} = \frac{\Delta H_{\mathrm{N,des}}}{\Delta H_{\mathrm{M,des}}} \tag{7.12}$$

The subscripts *M* and *N* represents high and low-temperature hydrides, respectively.

Double Stage Double Effect Metal Hydride System

Figure 7.25 shows the schematic of the system with corresponding Van't Hoff plot (Fig. 7.26). As compared to the single effect system, a double effect system has three reactors. As seen from Fig. 7.25 the metal hydride reactors are connected in star scheme. This method in the reaction bed is introduced by Suda et al. (1991). In this configuration, each hydride reactor is connected by the other two hydride reactors.

As in the case of a single effect system, this star connected scheme also operates in two halves and permits continuous cold generation.

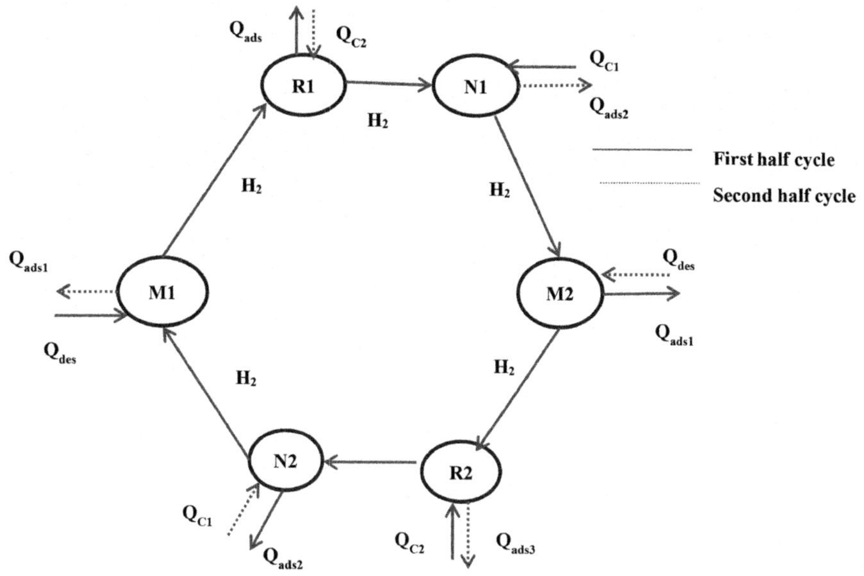

Fig. 7.25 A double stage/double effect metal hydride system

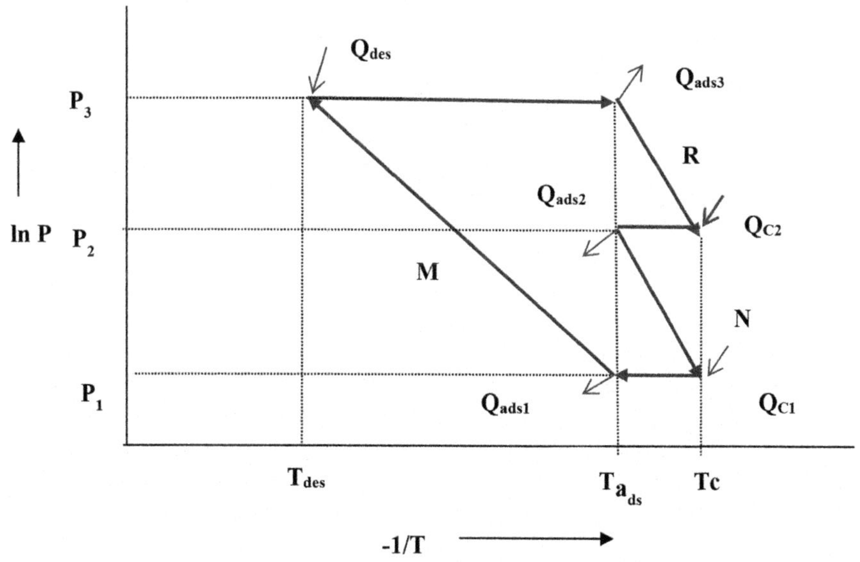

Fig. 7.26 Thermodynamic cycle on Van't Hoff plot (Suda et al. 1991)

The system operates at three pressure levels as indicated in the Van't Hoffs plot. The reactor M_1 is connected by R_1 and N_2. During the first half, the reactor M_1 desorbs hydrogen by taking the heat of desorption Q_{des} at T_{des} and this hydrogen is being adsorbed by reactor R_1 by releasing the heat of absorption Q_{ads} at T_{ads}. The hydrogen from reactor N_2 is desorbed by using the heat from the space to be refrigerated (Q_{c1}), useful cold at the temperature T_c. The reactor N_2 is connected by M_2 and R_1. This desorbed hydrogen from N_2 is adsorbed in reactor M_2 at T_{ads} and the hydride R_2 desorbs hydrogen at T_c by producing cold Q_{c2}. The R_2 is connected with N_1, hence the desorbed hydrogen from R_2 is being adsorbed by N_1 at T_{ads}. The performance parameters of the double stage or double effect system are given by

$$\mathrm{COA} = \frac{Q_{\mathrm{ads1}} + Q_{\mathrm{ads2}} + Q_{\mathrm{ads3}}}{Q_{\mathrm{des}}} \tag{7.13}$$

$$\mathrm{COP} = \frac{Q_{c1} + Q_{c2}}{Q_{\mathrm{des}}} \tag{7.14}$$

and in terms of hydride properties

$$\mathrm{COA} = \frac{\Delta H_{M,\mathrm{ads}} + \Delta H_{N,\mathrm{des}} + \Delta H_{R,\mathrm{ads}}}{\Delta H_{M,\mathrm{des}}}$$

$$\mathrm{COP} = \frac{\Delta H_{N,\mathrm{des}} + \Delta H_{R,\mathrm{des}}}{\Delta H_{M,\mathrm{des}}} \tag{7.15}$$

7.9 Applications of Solar Adsorption Cooling Systems

Nowadays the interest in adsorption systems have increased for the applications in refrigeration and air conditioning purposes which can easily make use of solar energy. The important applications include ice making, airconditioning, medical applications, thermal management of defence electronic equipment etc. In thermal management of defense electronic equipment, suppression of infrared signature becomes one of the design criteria. On one hand, there are numerous thermally sensitive components, which need localised cooling and on the other, considerable on-board heat generated from other electronic equipment such as travelling wave tubes (TWT) needs to be dissipated. It is logical to use the heat from the hot source to cool the other. A similar situation exists in medical appliances, such as scanners, where considerable heat is generated and also there are some thermally sensitive components. In particular, in the area of cooling of electronics, where electromagnetic interference effects have to be considered, and adsorption cooling systems could have an edge over other active cooling methods such as thermoelectric refrigeration. With the ever shrinking size of electronic systems, a phenomenal increase in volumetric and real densities of heat dissipation has occurred. This needs the development of application specific cooling technologies. Adsorption cooling can also meet the cooling requirements across a range of temperatures, ranging from cryogenic to near ambient.

7.10 Current Status and Developments in VAD Systems

Currently, adsorption type systems have very little commercial success, and most of the systems are in the stage of demonstration and prototyping. A few of reported commercial systems are:

1. The first commercial adsorption refrigeration system was introduced by Nishiyodo Kuchouki Co. Ltd. in 1986. A temperature of 3 °C was reported for the silica gel-water system developed by the company. The chiller was driven by a heat source temperature fluctuating between 50 and 90°C.
2. Qianchuan Co. in Canada developed a silica gel-water system capable of producing chilled water at a temperature of 14 °C. The chiller developed by the company was driven by a heat source temperature of 60–80 °C.
3. Chen et al. (2010) developed an adsorption chiller of capacity 10–200 kW for Maltresar Hospital in Germany. A Japanese company Mayekawa developed an adsorption chiller of capacity 105 kW and installed at St Johannes hospital. The system was driven by the hot water produced from 116 m^2 solar panels installed. The system was installed in 2000 and it consists of a fuel cell, heat storage vessel, ice storage tank, solar collector and cooled ceilings.
4. Bry Air (Asia) Pvt. (Ltd Company) India, introduced a silica gel-water adsorption chiller utilising low-grade heat source of 50–100 °C of capacity 40 kW.
5. New Leaf Dynamic Technologies, India, supplying adsorption bulk milk coolers.
6. Sortech's adsorption chillers are available in two nominal capacities. The system works with silica gel–water and zeolite–water with capacities 1–7.5 and 7.5–15 kW.

In adsorption technology, combining the adsorption technology with nano science is a new research trend. This will help to increase the adsorption characteristics and hence its coefficient of performance. The research in the area of combined systems of solar adsorption cooling and desalination (cogenerate type of systems) is a new area to be explored. Certainly, these research options will help to make the solar adsorption systems much more attractive and cost-effective. The next few years of research in the area of adsorption will be decisive for the success of these adsorption systems. The following areas need to be focussed: (1) Solar adsorption system with 24 h of operation is not yet proved. (2) Develop an off-shelf low cost/compact 24 h cooling low temperature VAD systems for Indian dairy farms for milk cooling, and rural/village agricultural for the preservation of agricultural products.

Present R&D in solar cooling options concentrates on VAD systems due to the following:

1. VAD systems can be easily powered by the low-temperature heat sources.

2. In comparison with the vapour compression refrigeration systems, adsorption refrigeration systems have the benefits of energy savings if powered by waste heat or solar energy
3. The working fluids satisfy the Montreal protocol on ozone layer depletion and the Kyoto protocol on global warming.
4. Comparing with an absorption system, an adsorption cooling system has less parasitic energy consumption since it doesn't require a solution pump for its operation. The range of heat source temperature is high (50–500 °C). It can be easily powered by the low heat source temperature; hence the initial cost of the solar collector systems can be reduced.
5. The adsorption cooling systems are suitable for low, medium and high capacity applications.
6. The main drawback of solid vapour adsorption cooling and heat pump systems is their poor performance in terms of cooling capacity and COP. So, R and D should be focused on the aspects of improving adsorption characteristics of adsorbent-refrigerant working pair by integrating the current technology with nano technology.

References

Alam, K. C. A., Khan, M. Z. I., Uyun, A. S., Hamamoto, Y., Akisawa, A. & Kashiwagi, T. (2007). Experimental study of a low temperature heat driven re-heat two-stage adsorption chiller. *Applied Thermal Engineering, 27*(10), 1686–1692.

Anyanwu, E. E. (2000). Environmental pollution: Restructuring the refrigeration industry as a way out. *Environment protection engineering, 26*(4), 17–27.

Anyanwu, E .E. (2003). Review of solid adsorption refrigerator I: An over view of the refrigeration cycle, *Energy Conversion and Management, 44,* 301–312.

Baiju, V. & Muraleedharan, C. (2015). Experimental Analysis on Adsorption Characteristics of Methanol and R134A by Activated Carbon in Adsorption Refrigeration System. *Journal of Thermal Science and Engineering Applications, 7*(1).

Chen, C. J., Wang, R. Z., Xia, Z. Z., Kiplagat, J. K. & Lu, Z. S. (2010). Study on a compact silica gel–water adsorption chiller without vacuum valves: Design and experimental study. *Applied Energy, 87*(8), 2673–2681.

Critoph, R. E. (1998). Forced convection adsorption cycles. Gold Applied Thermal Engineering, *18*(9–10), 799–807.

Critoph, R. E. (1999). Forced convection adsorption cycle with packed bed heat regeneration: Cycle à adsorption à convection forcée avec régénérationthermique du lit fixe. *International Journal of Refrigeration, 22(*1), 38–46.

Dawoud, B. (2007). A hybrid solar-assisted adsorption cooling unit for vaccine storage. *Renewable Energy, 32*(6), 947–964.

Gurdeep Raj. (2002). *Advanced Physical Chemistry*, GOEL Publishing House, Meerut.

Mahesh, A. & Kaushik, S. C. (2012). Solar adsorption cooling system: An overview. *Journal of Renewable and Sustainable Energy, 4*(2), 022701.

Muthukumar, P. & Groll, M. (2010). Erratum to "Metal hydride based heating and cooling systems: a review" [International Journal of Hydrogen Energy (2010) *35,* 3817–3831]. *International journal of hydrogen energy, 35*(16), 8816–8829.

Saha, B. B., Koyama, S., Lee, J. E. A., Kuwahara, K., Alam, K. C. A., Hamamoto, Y., Akisawa, A. & Kashiwagi, T. (2003). Performance evaluation of a low-temperature waste heat driven multi-bed adsorption chiller. *International Journal of Multiphase Flow, 29*(8), 1249–1263.

Sha, A. A. & Baiju, V. (2019). Simulation of single bed adsorption refrigeration system driven by solar energy. In Proceedings of the 25th National and 3rd International ISHMT-ASTFE Heat and Mass Transfer Conference (IHMTC-2019). Begel House Inc.

Shabir, F., Sultan, M., Miyazaki, T., Saha, B. B., Askalany, A., Ali, I., ... & Shamshiri, R. R. (2020). Recent updates on the adsorption capacities of adsorbent-adsorbate pairs for heat transformation applications. *Renewable and Sustainable Energy Reviews, 119*, 109630.

Srinivasan, K. & Dutta, P. (2019). Activated carbon–carbon dioxide based two stage adsorption compression Brayton cycle power generation. *Adsorption, 25*(8), 1663–1672.

Suda, S., Komazaki, Y., Narasaki, H. & Uchida, M. (1991). Development of a double-stage heat pump: experimental and analytical surveys. *Journal of the Less Common Metals, 172*, 1092–1110.

Chapter 8
Desiccant Cooling Systems

Desiccant cooling systems are being explored as an alternative to conventional cooling systems. The desiccant system is environmentally friendly and can be used to condition the indoor environment of buildings. The system can be effectively driven by low-grade energy sources such as solar energy, biomass heating systems, industrial waste heat etc. Basic principles, different types of desiccant cooling systems and its applications are presented in this chapter.

8.1 Introduction

The desiccant cycle partially converts latent heat of moist air to sensible load to reduce the moisture content and then meet this load with evaporative coolers/VCR systems. The desiccant dehumidification removes humidity from the ventilation air. The humidity control (reduction during the cooling season) in the air handling units and building spaces improves the air quality by preventing condensation in equipment and reducing the growth and propagation of micro-organisms. For such situation, air conditioning is to be based on the use of adsorbents, which can adsorb the moisture content of the air. For places near the seashore where temperature and humidity are high during summer and it rains during winter, an air conditioning system based on dehumidification followed by adiabatic evaporative cooling is a viable proposition. Thus, in general, for thermal comfort, air conditioning comprises of proper control of temperature, humidity, air movement and purity of air.

When the dew point temperature of the entering air is low the refrigeration systems are preferred. After the water content has been removed from the moist air, it may be necessary to cool it further, possibly using evaporative cooling, which is essentially an adiabatic process, with no heat flow. During the adiabatic saturation process, some of the water may be evaporated by increasing the latent heat content of air while lowering the sensible heat in the air.

S. C. Kaushik et al., *Solar Cooling*, https://doi.org/10.1007/978-3-031-42410-6_8

The outlet air leaving from the evaporative cooler has low dry bulb temperature as indicated in the psychrometric chart. The relative humidity of the air is seems to be on higher side. The process can be approximated by a constant wet bulb temperature line on a psychrometric chart.

8.2 Dehumidification

Whenever the warm and humid air contacts any surface including water that has been cooled below the dew point temperature of the air/water mixture, dehumidification takes place. Water will condense on the cooled surface, thus drying the air/water vapour mixture. Dehumidification can be done by (a) conventional airconditioning systems and (b) use of desiccant materials.

The conventional airconditioning systems also reduce the humidity of the air before being supplied into the targeted space. In such a system, the expanded refrigerant from the expansion valve flows through a finned evaporator coil, where the humid air is allowed to pass over the cold evaporator coils below its dew point temperature. The saturated air leaves the wet evaporator coil at a lower temperature. Therefore, the processed air stream is comparatively drier due to condensation of excess moisture in this process. Another, but most commonly used method for dehumidification is the use of a material called desiccants, which has a high affinity for water and they maybe in the form of solid or liquid. There are different types of desiccant materials and methods of dehumidification, are discussed later in this chapter.

8.3 Evaporative Cooling Options

The evaporative cooling systems are based on the principle of water evaporation, which reduces the temperature of the air. Thus, water acts as a refrigerant in these systems and the cooling is achieved through several ways as given below:

(a) Direct evaporative cooling
(b) Indirect evaporative cooling
(c) Combined evaporative coolers
(d) Regenerative cooling.

8.3.1 Direct Evaporative Cooling

The system consists of a water-saturated medium. In this case, the outside air enters space through a water-saturated medium and cooled by direct evaporation of water.

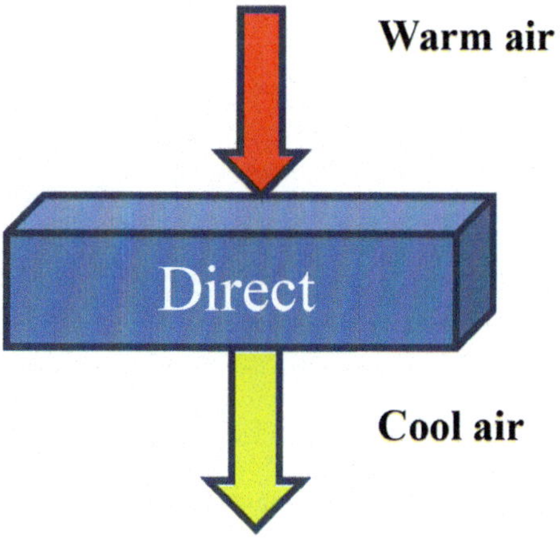

Fig. 8.1 A direct evaporative cooling system

This cool air is supplied to the space to be cooled using a fan or a blower as shown in Fig. 8.1. Since the incoming air is unsaturated, water vapour is added to it until it is saturated and the process takes place adiabatically. Many commercial systems available now are using the Earth's unsaturated atmosphere indirectly by discharging heat to a stream of water, which in turn is cooled in an open cycle by direct contact with the atmosphere in a wet cooling tower. During direct evaporative cooling, the dry bulb temperature is reduced, while the wet bulb temperature stays the same. The high moisture content of the cool air results in conditions that are only marginally improved because it can only lower the temperature and not the moisture content, and therefore its use is limited to arid or semi-arid regions only.

8.3.2 Indirect Evaporative Cooling

In this system, water is used to cool the secondary stream of air, while a heat exchanger is installed within the system to cool down the primary air stream. The primary air stream is cooled by the secondary air stream, which passes through the heat exchanger. The cooled primary air stream is circulated by a blower into the space to be cooled. Indirect evaporative cooling can change the dry bulb and wet bulb temperature of the air. The exhaust air is used as the secondary air stream in the indirect evaporative cooler heat exchanger. During the heating season, this secondary air stream can preheat the outside air in the heat exchanger of the indirect evaporative cooler, as can be seen clearly from Fig. 8.2.

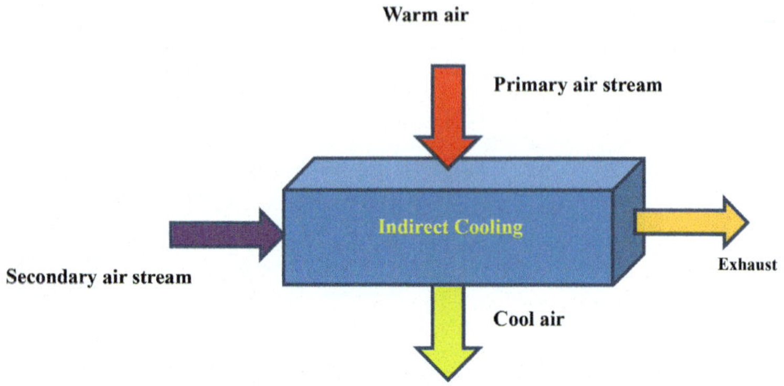

Fig. 8.2 Indirect evaporative cooling (Jia et al. 2006a, b)

8.3.3 Combined Evaporative Coolers

The combined evaporative cooler is actually a combination of direct and indirect evaporative cooling systems. In this system, the air is first cooled down indirectly, and then the dry air is further cooled directly before being supplied to the space to be cooled as shown in Fig. 8.3.

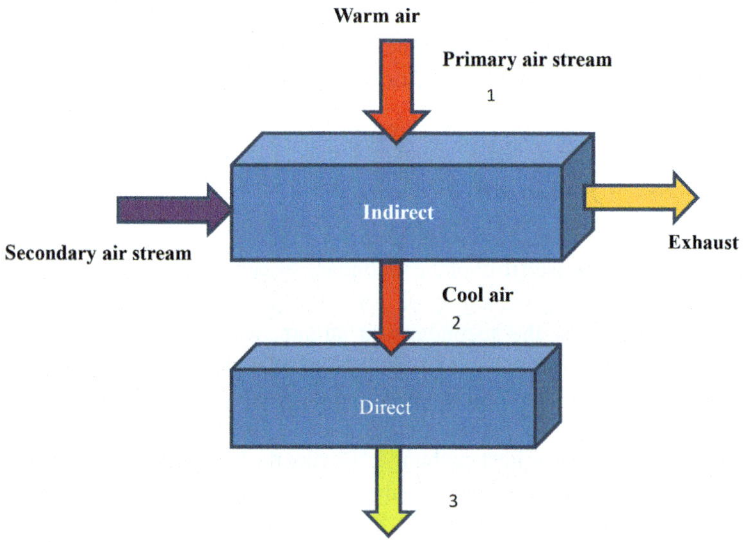

Fig. 8.3 Combined evaporative cooling

8.3.4 *Regenerative Cooling*

The regenerative cooling can be used in slightly humid, over heated or high temperature regions. The direct evaporative coolers could not obtain the temperature below the wet bulb temperature of the process air streams as discussed earlier. However, the regenerative cooling systems could achieve temperature close to the dew point temperature, thus resulting in lowering of the enthalpy of the process stream without additional expenditure in the thermal energy. The schematic of a regenerative cooling system is as shown in Fig. 8.4.

8.4 Desiccant Materials and Their Characteristics

The selection of desiccant materials plays a key role in the performance of the systems. Commonly used desiccant materials are solid, liquid and consolidated/composite desiccant materials. Some of the commonly used solid desiccants are silica gel, activated alumina, natural and synthetic zeolite, titanium silicates, synthetic polymers etc. and have been already dicussed in Sect. 7.5. The commonly used liquid desiccants are lithium chloride methyl glycol, triethylene glycol, lithium bromide, potassium acetate, brine solution, potassium formate etc.

Lithium bromide, lithium chloride and triethylene glycol are the most commonly used liquid desiccants. The surface vapour pressure of the desiccant is one of the most important parameters that needs to be investigated extensively. For the desiccants like lithium chloride, lithium bromide etc., the surface vapour pressure at low temperature and high concentration are lower than that of the humid process air.

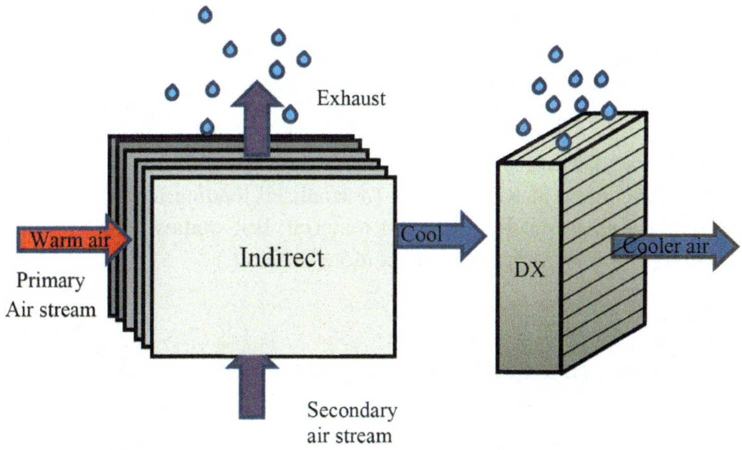

Fig. 8.4 Regenerative cooling

Among all the liquid desiccants available, $CaCl_2$ is the cheapest and most readily available. The wide use of $CaCl_2$ is limited by the drawbacks of relatively high vapour pressure at a given temperature. The other liquid desiccant is lithium chloride, which is the most stable desiccant with low vapour pressure. Triethylene glycol is the earliest used desiccant in liquid desiccant dehumidification systems, but it has high viscosity. Besides, triethylene glycol has a very low surface vapour pressure, which causes it to evaporate into the process air.

Researchers are working for the development of consolidated or composite desiccant materials. The consolidated desiccant materials are prepared by impregnating host porous materials like vermiculite or silica gel with hygroscopic salts such as lithium chloride or calcium chloride. The research shows that the developed consolidated desiccants may have three times the sorption capacity as that of pure host materials (Liu et al. 2004). Some of the developed composite materials are:

1. Impregnating silica gel with calcium chloride. Liu and Wang (2003) introduced a composite adsorbent $SiO_2.xH_2O.yCaCl_2$ that was fabricated with macroporous silica gel and calcium chloride. Researches on composite materials with silica gel such as impregnating with $CaCl_2$, LiCl etc. are in progress.
2. The other composite material available is SiO_2 and synthesized zeolite based composite desiccant.
3. The composite adsorbents based on inorganic salt such as $SrCl_2$, $CaCl_2$, LiBr, $NaSO_4$ and silica gel has been proposed by Aristov (2007).
4. Sepiolite-based composite desiccants, namely, sepiolite-activated carbon and sepiolite-calcium chloride are developed by González et al. (2001).
5. The low-cost adsorbent is proposed by Tokarev et al. (2002) is about clay–$CaCl_2$.
6. Mathiowitz et al. (2001) have introduced a novel concept for the preparation of desiccants based on polymeric blends.

   The factors to be considered for the selection of desiccant materials are:

   (1) Ideal desiccant material should have very high adsorption property.
   (2) The study reveals that Type 1M material represents the optimum shape of desiccant material suitable for the air-conditioning application.

Figure 8.5 gives a comparison of different types of desiccant materials and clearly shows that the actual desiccant water content at corresponding RH or maximum desiccant water content at RH = 100% (normalised loading fraction) is higher for Type 1E material than that for Type 1M material. But studies revealed that due to low RH value of material regeneration is more difficult.

8.5 Principle of Desiccant Cooling

The basic principle applied in desiccant cooling is the attraction of moisture content in the air towards the desiccant materials due to the moisture vapour pressure difference that exists between air and the desiccant surface. The system utilises the ability

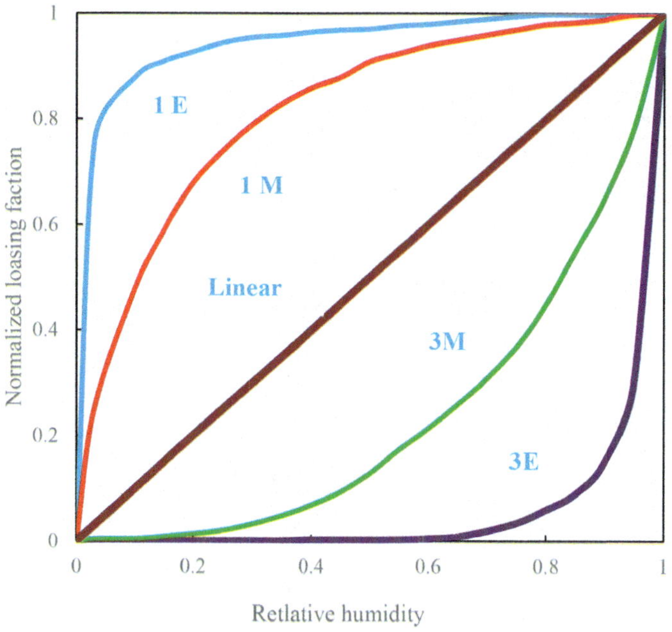

Fig. 8.5 Comparison between adsorption isotherms of Type 1 extreme (1E), Type 1 moderate (1M), linear, Type 3 moderate (3M) and Type 3 extreme (3E) (Jia et al. 2006a, b)

of desiccant materials to absorb air moisture content by the sorption process. The sorption process, which is a combined effect of absorption and desorption process is an interaction between the adsorbent and adsorbate molecules. As the desiccant materials are at very low concentration of water vapour content, the air moisture content is attracted to desiccant materials due to the moisture vapour pressure difference between air and the desiccant surface. The desiccant material needs to be activated with the thermal energy source for regeneration. The basic working of desiccant cooling technologies is shown in Fig. 8.6.

The temperature of process air from the desiccant dehumidifier rises due to rejection of the heat of sorption. This energy is recovered by providing suitable energy recovery devices for regenerating the desiccant in conjunction with other sources of thermal energy. The air warm and dry air thus produced is typically above the thermal comfort temperature, so that evaporative cooling process is done by either indirect/direct process.

Figure 8.7 shows the block diagram of thermally driven desiccant cooling technologies.

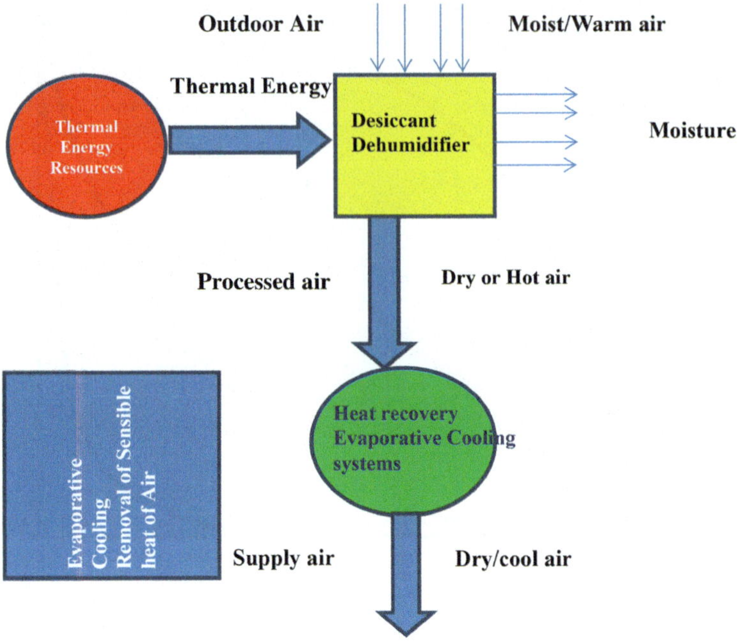

Fig. 8.6 Schematic of desiccant cooling technology

8.6 Types of Desiccant Cooling Systems

The system utilise the desiccant materials, which can attract and remove water vapour from an air stream. Based on the type of desiccant material used and working cycle, the desiccant cooling systems are classified into three types viz., solid, liquid and hybrid systems, as shown in Fig. 8.8.

The solid desiccant systems are available in single wheel and multiwheel. The single wheel consists of a desiccant wheel, a cooling coil and a heating coil that can be driven by electric energy, hot source or steam. In this system, the outside air is dehumidified by the desiccant wheel provided. The temperature of the air reduces to the desired level by a cooling coil integrated to the unit. The dry bulb temperature of the air is increased by using a reheat coil provided. The basic principle involved in a liquid desiccant-based system is at the system uses a liquid desiccant for removing the moisture from the air. The hybrid desiccants systems are combinations of desiccant wheel and VCR or VAB or VAD systems. The most commonly used hybrid cooling systems are based on the vapour compression systems.

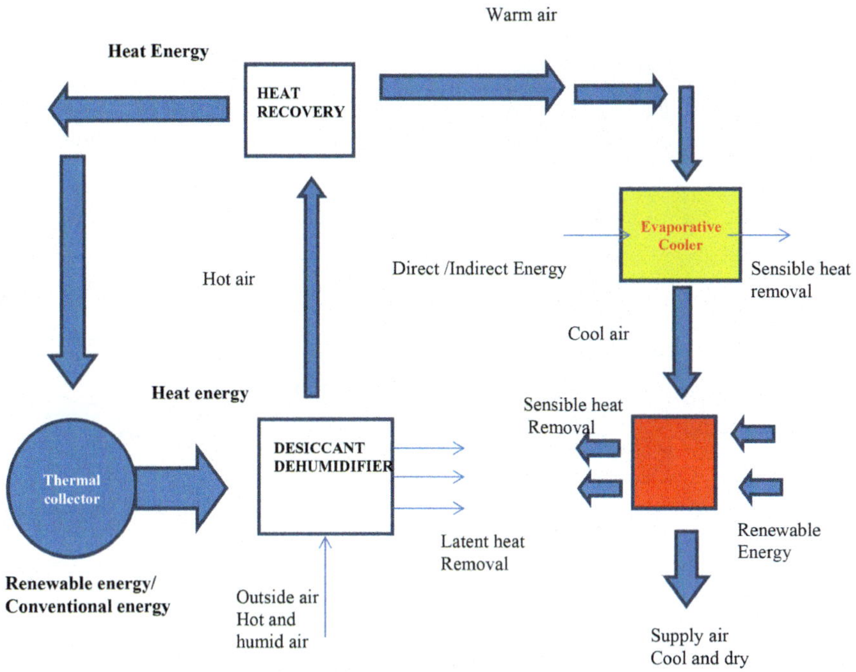

Fig. 8.7 Block diagram showing the general concept of the thermally activated desiccant cooling technologies

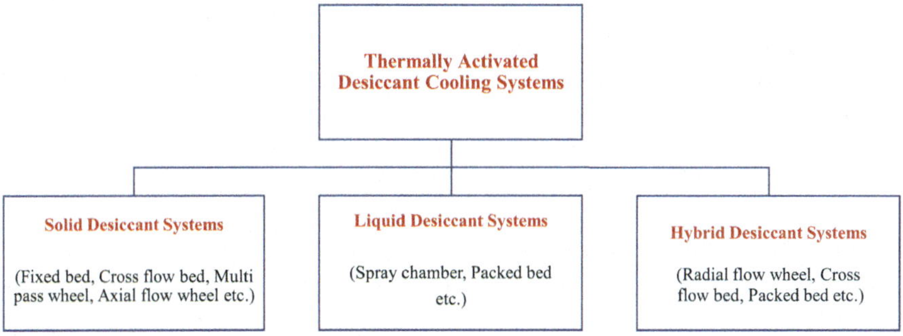

Fig. 8.8 Classification of thermally activated desiccant systems

8.6.1 Solid Desiccant Cooling Systems

The early work on a solid desiccant cooling system was undertaken by Pennington in 1955, Dunkle (1965) at CSIRO in Australia, and also by Lunde (1976) in USA. A solid desiccant system shown in Fig. 8.9 is a solid desiccant cooling system consisting of desiccant material, heat exchangers and an evaporative cooler.

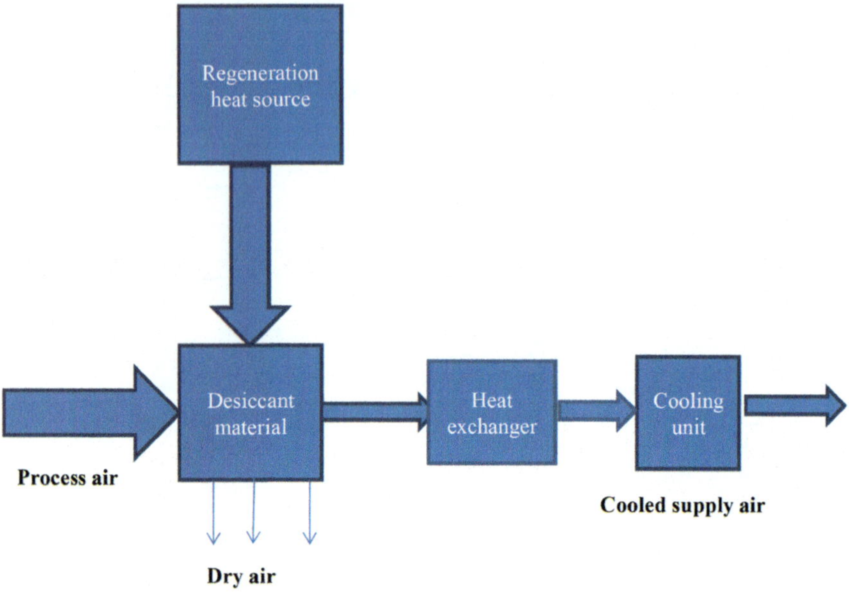

Fig. 8.9 Solid desiccant cooling system

The most important component in the solid desiccant system is the desiccant dehumidifying bed, which removes moisture by adsorption from the air. Air from the space to be conditioned or from outside is passed through the dehumidifier, so that the emerging air stream is hot and dry. This air is then cooled in a heat exchanger (recuperator) followed by evaporative cooling. This cooled air is sent in the area to be air conditioned. To regenerate the desiccant bed, ambient air is passed through the heat exchanger where it is heated by the dehumidified dry and hot process air. Heat energy input is also made available at the second heat exchanger, bringing the temperature to a level, at which the desiccant bed is well regenerated.

Dehumidifier (Desiccant Material)

The main component in a solid desiccant cooling system is the desiccant material. In this system, a solid desiccant is employed to adsorb the moisture content of the air. The slow rotating wheel also known as a desiccant dehumidifier is used as adsorbent bed.

The hydrophilic materials are considered to be an adsorbent to absorb the moisture from the air. It should be porous and strong hydrophilic material.

Regeneration Heat Source

The moisture content in the desiccant wheel is to be removed for its further usage. Energy in the form of heat is necessary to remove the moisture content that is being adsorbed in the desiccant wheel. The thermal energy source can be solar energy, exhaust gases, waste heat, natural gas etc. For industrial applications, where the condition of moderate to low temperature with low humidity is needed, the solar desiccant dehumidification system can be coupled with compression, absorption or adsorption refrigeration system. Such a combination is more efficient and less expensive. Since the process of adsorption is exothermic, the temperature of the adsorbent bed increases, which reduces the adsorption capability of the bed. It is desirable to cool the adsorption bed during the adsorption process. The loss of adsorption capacity due to heating of the bed (in the adiabatic process) can be reduced by using a material which retains its capacity at a higher temperature, such as molecular sieve. However, because molecular sieve beds retain moisture holding capacity at elevated temperature, they must be regenerated at even more elevated temperature. Another possible solution to avoid the decrease in bed adsorption and desorption capacity due to excessive temperature swing is to use interstage cooling and interstage heat exchangers.

Roy and Gidaspaw (1972) proposed cross cooling and cross flow regenerator operating with non-linearly coupled heat and mass exchange and Worek and Lavan (1982) later developed and tested the unit of a cross flow dehumidifier as shown in Fig. 8.10. The system consists of two fixed bed cross-cooled silica gel dehumidifiers (one for providing cooling and the other for regeneration) and rectangular channels in cross flow arrangement. The process stream channels are lined with silica gel sheets (Fig. 8.10).

In the dehumidification cycle, air from the room flows through the process channels and humidified ambient air is passed through the cooling channels. This lowers the desiccant temperature and in turn proves the sorption process.

In the regeneration cycle, hot air from the solar collectors is passed through the process channels. The advantage of the cross-cooled concept is the ability to provide adequate cooling with relatively low regeneration temperatures.

Another common configuration of a desiccant bed is a rotary type, and the working of a rotary desiccant dehumidifier can be explained with the help of Fig. 8.11. The rotary wheel is divided into the two segments, viz. process air side and regeneration side. The process air is dried by the adsorption. The regeneration air gets humidified on the other side of the wheel, after being heated and desorbing the water from the

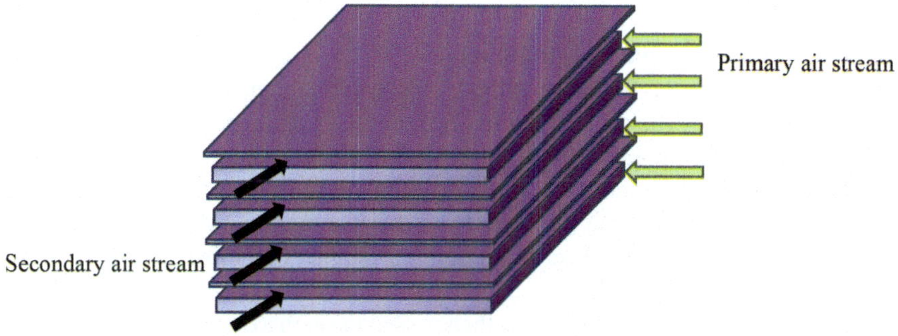

Fig. 8.10 Cross-flow dehumidifier (Dai et al. 2001)

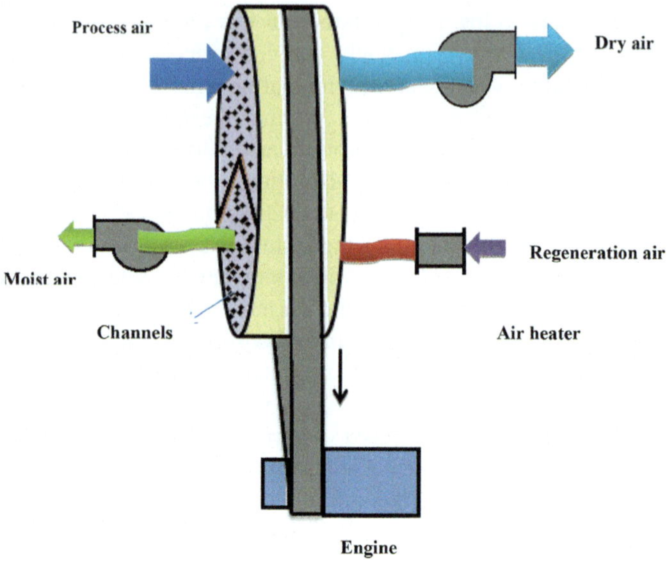

Fig. 8.11 Rotary desiccant dehumidifier (Yashpalbajaj and Swaty 2014)

wheel. The rotary wheel desiccant dehumidification process only converts latent heat of air to sensible heat; there is no useful cooling effect produced. Thus, a cooling unit must be included along with the desiccant dehumidifier to achieve the cooling effect.

The various modes of operation of solid desiccant systems are as follows, which are briefly explained further.

(i) Ventilation cycle (Pennington cycle)
(ii) Recirculation cycle
(iii) Dunkle cycle
(iv) Modified Dunkle cycle (Dhar and Singh 2001)
(v) Multi-pass desiccant wheel system.

(i) Ventilation Cycle (Pennington Cycle)

Figure 8.12 shows the Pennington cycle and its psychometric representations of various processes. The system basically comprises of a desiccant wheel, heat exchanger, direct evaporative cooler and heat source.

The ambient air at state point 1 passes through a desiccant wheel (DW). The desiccant wheel absorbs the moisture from the incoming atmospheric air. Due to this adsorption, the temperature of process air (ambient air) increases. During the process 2 to 3 in a HE, the hot dry air coming from the desiccant wheel is subjected to sensible cooling. The process air is then passed through a direct evaporative cooler (DEC), which cools the process air to supply air state. The return air at the state point 5 is cooled and humidified in another DEC, on the regeneration air side. This air then

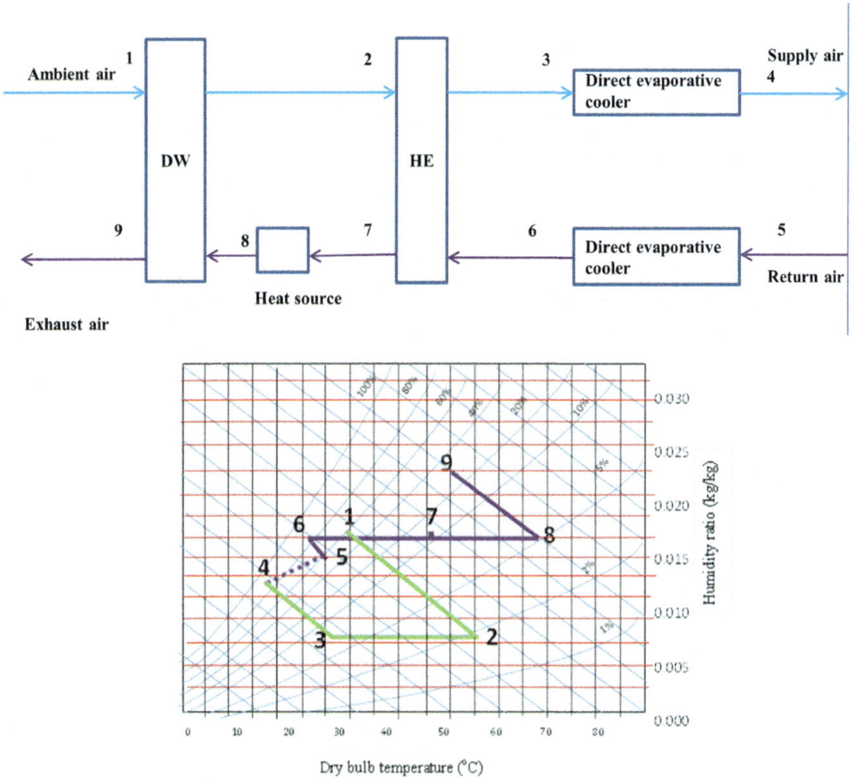

Fig. 8.12 Ventilation cycle (Pennington cycle) (Dai et al. 2001)

exchanges heat with incoming ambient air to precool the process air and preheat itself. The warm air leaves then heated from the state points 7 to 8 by the heat source. After regenerating the desiccant wheel, the air is exhausted at state point 9.

(ii) Recirculation Mode

In this process, the return air from the conditioned space is used as the inlet air, whereas, the regeneration air is used as the outdoor air. The recirculation cycle is also termed as recirculation cycle.

The recirculation cycle is a modified form of Pennington cycle. In recirculation mode of operation, the system restates return air as process air. Ambient air is used for regeneration in this cycle (Fig. 8.13).

(iii) Dunkle Cycle

Dunkle cycle combines the merits of ventilation cycle and recirculation cycle. The schematic of the system is shown in Fig. 8.14. Ventilation cycle provides cold air with relatively low temperature for the heat exchanger, and recirculation cycle, which can provide the conditioned space with relatively large amount of cooling capacity.

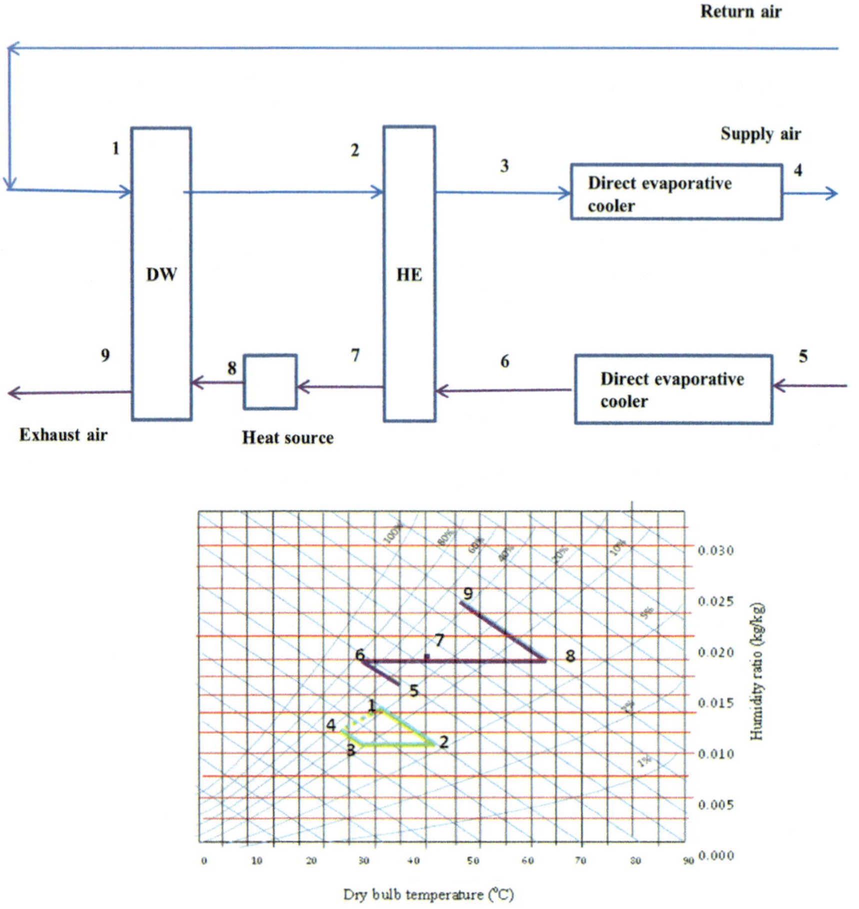

Fig. 8.13 Recirculation mode

(iv) Modified Dunkle Cycle

The schematic of a modified Dunkle cycle is shown in Fig. 8.15. A modified Dunkle cycle uses ambient air for regeneration. Since the humidity ratio and temperature of ambient air is higher than that of return air, modified cycle has better coefficient of performance (COP) than the standard ventilation cycle.

(v) Multi-pass Desiccant Wheel System

The advanced desiccant cooling system proposed by Kodama et al. (2005) is shown in Fig. 8.16. This system basically consists of a rotary dehumidifier, passage for hot and cold streams. The unit has two sides, namely, adsorption and desorption side. The area ratio of the two sides is approximately one. The adsorber unit has a sandwich structure of the alternative arrangement of honeycomb zeolite block and aluminium made passages.

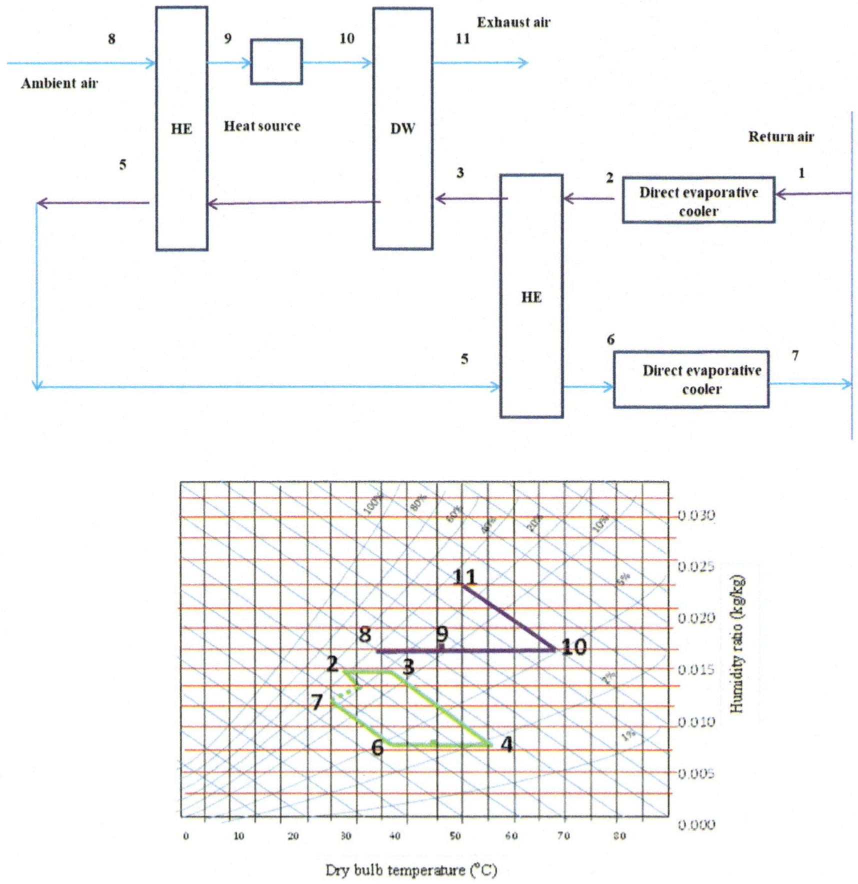

Fig. 8.14 Dunkle cycle

8.6.2 Liquid Desiccant Cooling Systems

The liquid desiccant cooling system employs a liquid desiccant as the absorbent to remove the moisture content of the atmospheric air. The basic principle of liquid desiccant control is the process of absorption.

As early as 1985, Lof and Tybont (1985) developed a liquid desiccant system using triethylene glycol (TEG) solution. Lof and Tybont (1985) studied the solar operation of a dehumidification system in which the drying agent liquid triethylene glycol is sprayed into an absorber where it picks up moisture from the building air. It is then pumped through the sensible heat exchanger to a stripping column where it is sprayed into a stream of solar-heated air (in the solar operation).

Sheridan (1961) proposed a refrigeration/air conditioning system for industrial cooling application (Fig. 8.17). The system consists of two spray towers. One spray

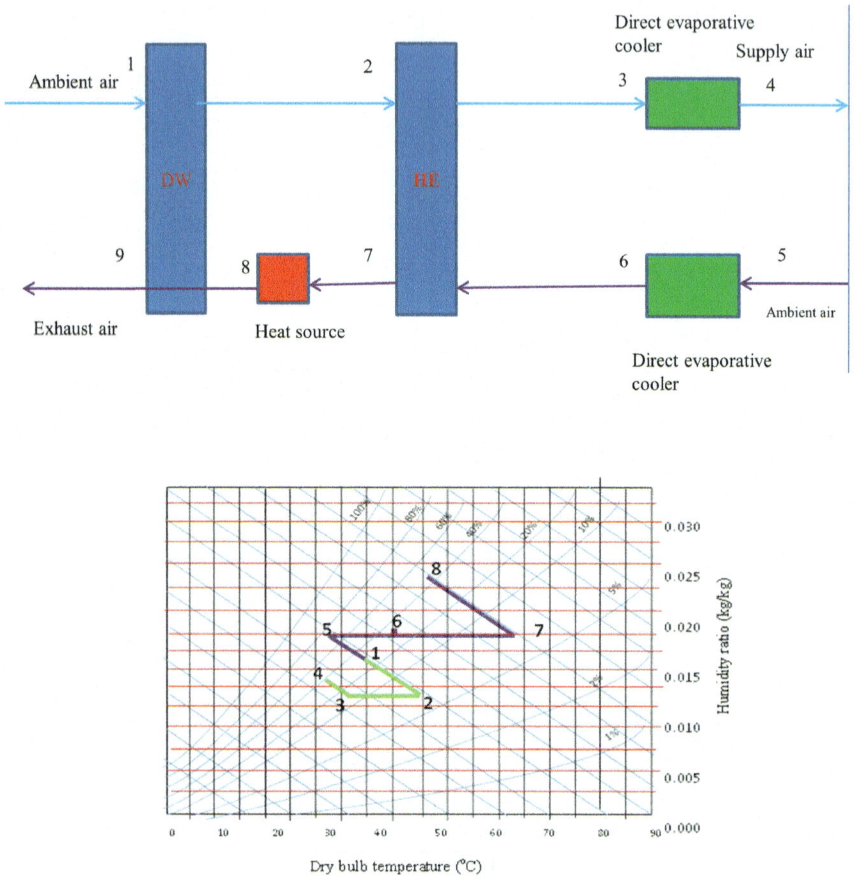

Fig. 8.15 Modified Dunkle cycle

tower is used for dehumidification of room and the other for regeneration of the absorbent solution. The lithium chloride is used as a desiccant in this system. The weak lithium chloride solution pumped through a heat exchanger absorbs heat from the strong solution returning from the regenerator. The weak solution is then further heated by the hot air from ~~the~~ regenerator before being sprayed into the regeneration tower.

Here, it contacts atmospheric air, which has been heated by hot water from the solar collector. The weak solution is then concentrated and returned to the humidifier. On being sprayed into the dehumidifying tower, the solution extracts moisture from the room air because of its low vapour pressure. The room air after being dehumidified is passed to an evaporative cooler. Air in the regenerator gains moisture and then gives up heat to the incoming fluid.

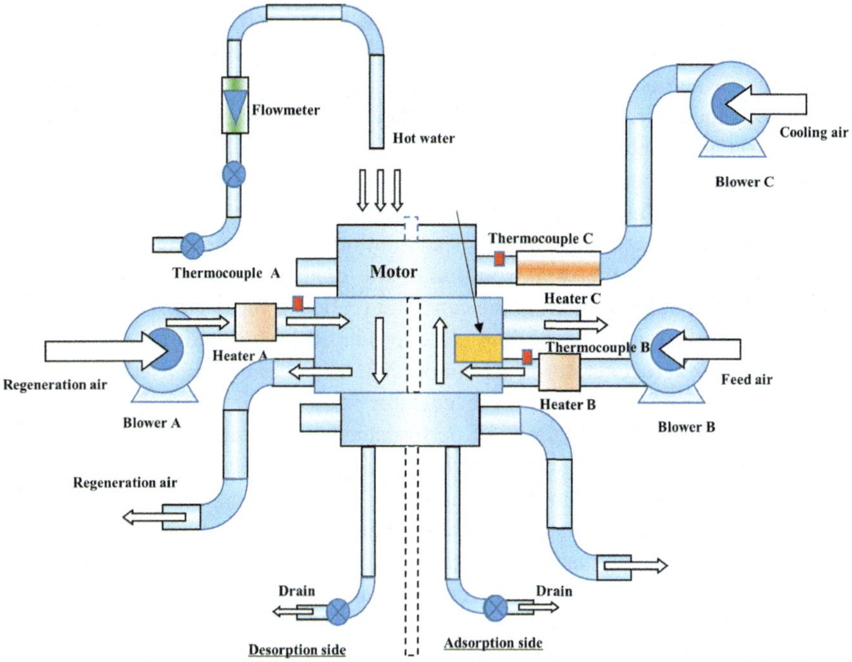

Fig. 8.16 Multi-pass desiccant wheel

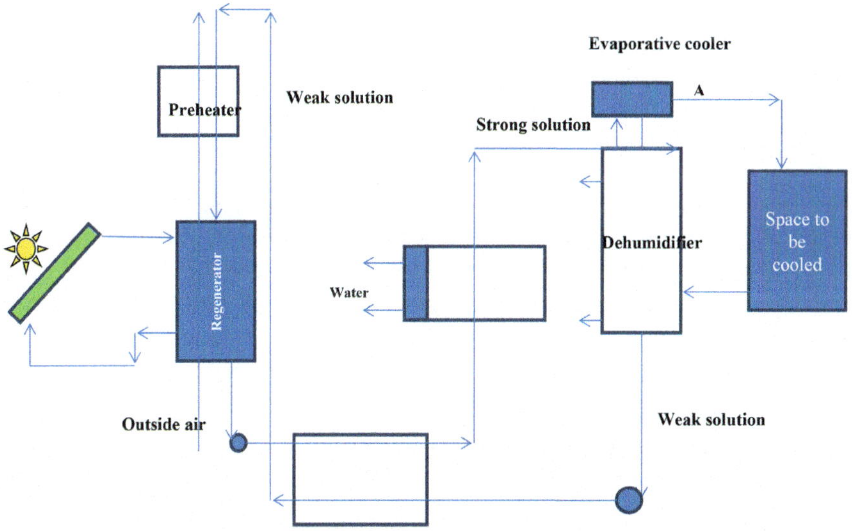

Fig. 8.17 Absorption-dehumidification solar air conditioning system (Sheridan 1961)

Lof and Tybont (1985) designed and developed advanced liquid desiccant cooling systems using EG as a desiccant solution and operated the systems in different modes. The cooling water is generated internally by evaporative cooling by processing the dry air over the recirculation water. The other components in both the flow modes of operation are identical. Absorber is an ordinary finned tube heat exchanger with a falling film distributor on the top. The desiccant solution flows down the finned tube surface as a falling film while the air flows in the counter flow direction through the flow channels between the fins. The solution regenerator involves desorption of water vapour from the weak solution due to contact with atmospheric air. The amount of water evaporated from the generator (called as the degree of regeneration) is the decisive factor of the cooling effect produced by a liquid desiccant system. The selection of a regenerator is also important. Most commonly used regenerators are either open surface falling film regenerator or closed air circulation regenerator as explained in Chap. 6. In terms of performance, air circulation type regenerator has much higher rate of regeneration and needs smaller regenerator length for the same process conditions. This regenerator involves higher capital cost and increased operating cost. Open surface regenerator on the other hand, is of low cost and simple design but requires high maintenance due to dirt and rains. Nevertheless, these type of systems are promising and needs further development in this direction.

Figure 8.18 shows a schematic diagram of a basic liquid-desiccant dehumidification and air conditioning system. The system consists of four main components, namely, a dehumidifcation unit, regeneration unit liquid desiccant storage container and a sensible heat recovery system.

The strong solution is stored in liquid desiccant storage unit. The purpose of a sensible heat removing system is to extract the sensible heat from the conditioned air leaving from the dehumidification part as per the indoor condition requirement standards. The low-grade energy sources can be integrated to the unit to preheat the diluted solution flowing from the dehumidification part. The driving potential for the mass transfer is the difference in vapour pressure between the liquid desiccant film and air. This is achieved by preheating the diluted solution. Also the inlet liquid desiccant required to be precooled to reduce its vapour pressure. The preheating and precooling are achieved by means of a heat exchanger, which is integrated to the system. The desiccant container is used for the purpose of storing desorbed strong solution.

8.6.3 Hybrid Desiccant Cooling Systems

A hybrid desiccant cooling system refers to a desiccant system which combines various technologies to improve the cooling performance of the base system. Some of the hybrid concepts in desiccant cooling are discussed as below.

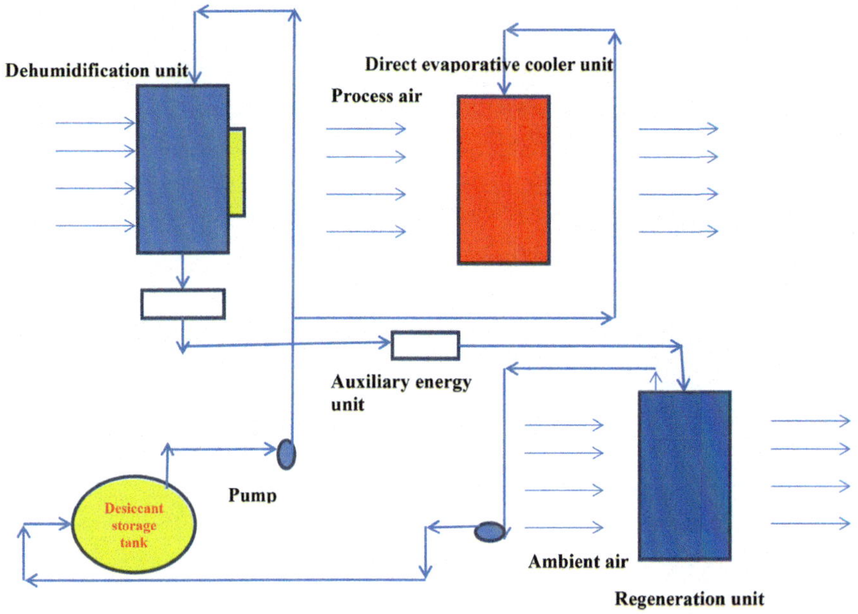

Fig. 8.18 Schematic diagram of liquid-desiccant dehumidification airconditioning system

A trigeneration plant with liquid desiccant is depicted in Fig. 8.19. The system basically consisting of (a) an internal combustion cogenerator (Natural gas NG) (b) a liquid desiccant system operating on fresh air (c) a cooling tower to dissipate heat from the desiccant conditioner (d) Two heat exchangers connected for the recovery of heat from the cooling water and from the flue gases of the engine (e) an additional hot water circuit that allows the desiccant unit regenerators to be fed with the heat recovered from the engine (f) Humidifiers. One humidifier is used on the supply air and the other on the return air and a rotary heat exchanger between the supply and return airstreams.

The desiccant system consists of two main sections, viz. conditioner and regenerator (Fig. 8.19). In conditioner-the outside air is dehumidified by passing through the LiCl–water solution in a porous matrix. The LiCl solution in the conditioner is subjected to cooling by means of water supplied from the cooling tower. The cooled LiCl matrix then absorbs the moisture. The processed air from the unit is then subjected to evaporative cooling process. This air is passed through a rotary air-to-air heat exchanger, where heat is transferred from the supply airstream, to the return airstream extracted from the building. The airstream to the building is then humidified.

Figure 8.20 shows an another approach with a cogeneration unit. The most crucial part of the system is liquid desiccant dehumidification system integrated with a heat energy storage unit and a desiccant storage unit, that improves the operating hours of the system, reduced the heat requirement in the boiler, reducing power for the

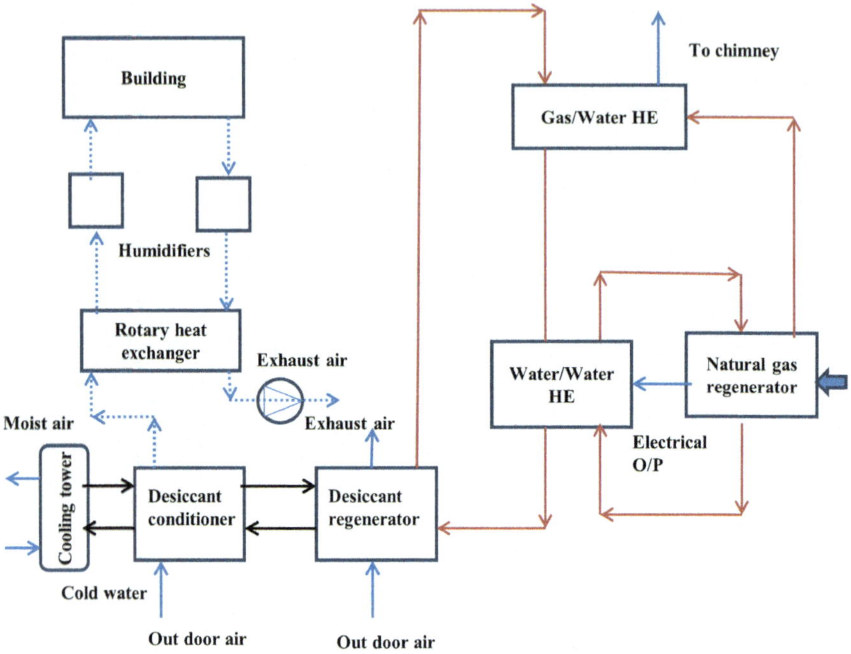

Fig. 8.19 A trigeneration plant with liquid desiccant (Saman and Alizadeh 2002)

compressor in the chiller unit and for the desiccant cooling system. The system basically consists of a gas engine, regenerator, desiccant storage tank, exhaust gas heat exchanger, thermal storage, a high-pressure gas generator and low-pressure gas unit. The exhaust gases from the gas engine are directly coupled to the high-pressure (HP) generator. A part of the exhaust gases leaving the gas engine is used to produce hot water. Hot water leaving the gas engine enters the regenerator side of the unit. The outlet of the heat exchanger is also coupled to the regenerator section. The chilled water coming out from the system can be used in the buildings. The concentrated solution leaving the regenerator is stored in the desiccant storage tank integrated to the system. The power produced from the system can be used to meet building energy demand. Exhaust heat source can be used to heat the building in winter or to drive desiccant system in summer.

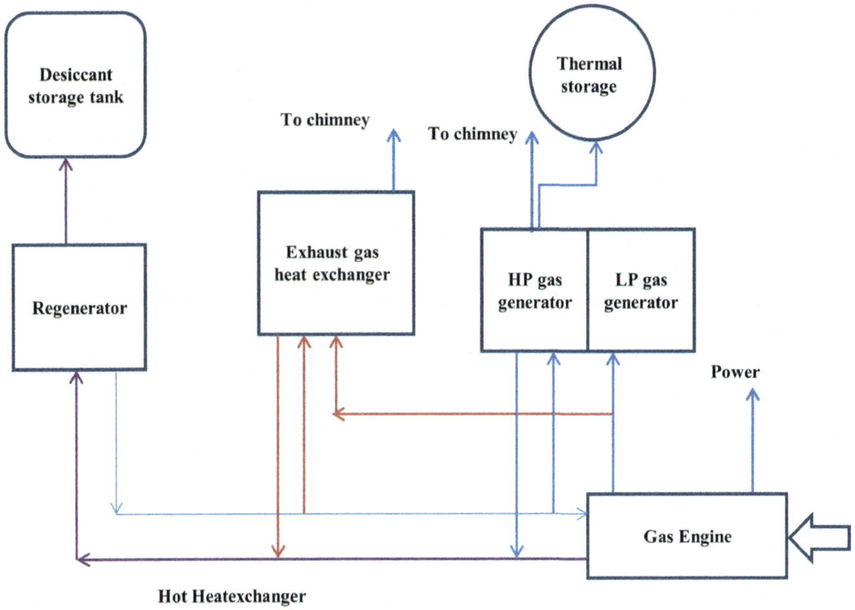

Fig. 8.20 Cogeneration with the liquid desiccant system (Liu et al. 2004)

8.7 Advantages and Disadvantages of Desiccant Cooling Systems

Advantages

1. The regeneration cost is low in comparison with the cost of energy to dehumidify the air by cooling it to below the dew point temperature.
2. Desiccant cooling systems are cost effective than conventional vapour compression refrigeration systems when the space to be cooled requires the continuous delivery of air.
3. There is no condensation on different sections such as cooling coils and in drip pans.
4. Energy consumed by the desiccant system is very less.

Disadvantages

1. High initial and maintenance cost of the system.
2. Cost of the desiccant material used in the systems.
3. The heat source for regeneration of the desiccant materials. Suggested to use solar energy as the driving heat source for the system.
4. Necessity of piping system for cool water to remove the heat released and precools the hot air.

8.8 Applications of Desiccant Cooling Systems

The desiccant systems are suitable in situations where low humidity is important. Important applications are given below:

1. In the industrial applications where the required humidity is very low.
2. The storage of food items which can be spoiled due to the moisture content in it.
3. Drying of air to enhance the drying of heat-sensitive products.
4. Desiccant systems are used to create very low humidity environments for pharmaceuticals and other manufacturing hygroscopic products requiring low humidity operating environment.
5. The desiccant cooling systems are also employed in clean rooms, hospital operating rooms, laboratories, and some restaurants and supermarkets.

References

Aristov, Y. I. (2007). New family of solid sorbents for adsorptive cooling: Material scientist approach. *Journal of Engineering Thermophysics, 16*, 63–72.

Dai, Y. J., Wang, R. Z., Zhang, H. F. & Yu, J. D. (2001). Use of liquid desiccant cooling to improve the performance of vapor compression air conditioning. *Applied Thermal Engineering, 21*(12), 1185–1202.

Dhar, P. L. & Singh, S. K. (2001). Studies on solid desiccant based hybrid air-conditioning systems. *Applied Thermal Engineering, 21*(2), 119–134.

Dunkle, R. B. (1965). A method of solar air conditioning, *Transactions of Mechanical and Chemical Engineering, 1*: 73–78.

Gonzalez, J.C., Molina-Sabio, M. & Rodrıguez-Reinoso, F. (2001). Sepiolite-based adsorbents as humidity controller. *Applied Clay Science, 20*(3), 111–118.

Jia, C. X., Dai, Y. J., Wu, J. Y. & Wang, R. Z. (2006a). Analysis on a hybrid desiccant air-conditioning system. *Applied Thermal Engineering, 26*(17–18), 2393–2400.

Jia, C. X., Dai, Y. J., Wu, J. Y. & Wang, R. Z. (2006b). Experimental comparison of two honeycombed desiccant wheels fabricated with silica gel and composite desiccant material. *Energy Conversion and Management, 47*(15–16), 2523–2534.

Kodama, A., Watanabe, N., Hirose, T., Goto, M. & Okano, H. (2005). Performance of a multipass honeycomb adsorber regenerated by a direct hot water heating. *Adsorption, 11*, 603–608.

Liu, X. H., Geng, K. C., Lin, B. R. & Jiang, Y. (2004). Combined cogeneration and liquid-desiccant system applied in a demonstration building. *Energy and buildings, 36*(9), 945–953.

Liu, Y. & Wang, R. (2003). Pore structure of new composite adsorbent $SiO_2\cdot$ x $H_2O\cdot$ y $CaCl_2$ with high uptake of water from air. *Science in China Series E: Technological Sciences, 46*, 551–559.

Lof, G. O. G. & Tybont, R. A. (1985). House heating and cooling with the solar energy, Solar Energy Notes, University of Wisconsin, pp. 33–46.

Lunde (1976). Solar desiccant air conditioning system using silica gel. Proceedings of second workshop on use of solar energy for cooling of buildings, California, USA, pp. 280.

Mathiowitz, E., Jacob, J. S., Jong, Y. S., Hekal, T. M., Spano, W., Guemonprez, R., ... & Langer, R. (2001). Novel desiccants based on designed polymeric blends. *Journal of Applied Polymer Science, 80*(3), 317–327.

Roy, D. & Gidaspow, D. (1972). A cross flow regenerator—a Green's matrix representation. *Chemical Engineering Science, 27*(4), 779–793.

Saman, W. Y. & Alizadeh, S. (2002). An experimental study of a cross-flow type plate heat exchanger for dehumidification/cooling. *Solar Energy, 73*(1), 59–71.

Sheridan, N. R. (1961). Industrial evaporative cooling solar air conditioning, *The International Journal of Engineering*, 47–52.

Tokarev, M., Gordeeva, L., Romannikov, V., Glaznev, I. & Aristov, Y. (2002). New composite sorbent $CaCl_2$ in mesopores for sorption cooling/heating. *International Journal of Thermal Sciences, 41*(5), 470–474.

Worek, W. M. & Lavan, Z. (1982). Performance of a cross-cooled desiccant dehumidifier prototype.

Yashpalbajaj & Swaty (2014). Desiccant cooling system—a novel technique for air conditioning, *International Journal of Engineering Research and Applications (IJERA),* ISSN: 2248-9622, National Conference on Advances in Engineering and Technology (AET-29th March 2014).

Chapter 9
Thermoelectric Cooling Systems

9.1 Introduction

Thermoelectric cooling is another potential option for refrigeration that uses electrical energy directly without converting it to mechanical form to obtain the refrigerating effect. The process of thermoelectric cooling is based on the principles of reversible thermoelectric effects, i.e., a temperature difference and heat absorption/rejection may be established between two junctions of dissimilar conductor materials charged by a direct electric current. Since there is no mechanical transfer, no moving parts are present, and hence, the system is free from wear and tear problem as well as noise etc. The major attractions of thermoelectric coolers are simplicity in design, lightweight, and static components as compared to other cooling devices. Historically, Peltier in 1834 observed that if a current is passed through a junction of two dissimilar materials, absorption or rejection of thermal energy occurs at the junctions. In 1857, Thomson related these thermoelectric effects and discovered that for a single material with a uniform temperature gradient along its length, an amount of heat is absorbed or rejected if a current is passing through the conductor. Seebeck in 1821 observed that an electric current is produced in a closed circuit made of dissimilar conductors at different temperatures.

Thermoelectric cooling is one of the techniques in which no refrigerant and mechanical compressor is required to produce cooling. The cooling in the thermoelectric cooling system can be achieved by electrons transporting the heat from the space to be cooled to the environment (heat sink reservoir). This system occupies less space (compact), no maintenance because of no moving parts, can be oriented in any direction, and it can be used for low cooling power applications (electronic chip

S. C. Kaushik et al., *Solar Cooling*, https://doi.org/10.1007/978-3-031-42410-6_9

cooling). Some of the key aspects of thermoelectric phenomena are briefly discussed below.

Seebeck Effect

Seebeck effect was discovered by Thomas Seebeck in 1821, when he observed a magnetic needle being deflected when brought near a circuit made of two different conductors.

Consider a circuit with two different materials (A and B) as shown in Fig. 9.1. The junction J_3 is maintained at higher temperature T_1, junctions J_2 and J_4 at lower temperature T_2 and J_1 and J_5 at atmospheric temperature T_o, which can be measured in absolute temperature scale (Kelvin). At steady state, when there is no current flowing through the conductors, and a finite temperature difference present between the junctions, the potential difference across J_1 and J_5 is called the Seebeck voltage defined as

$$|\Delta V|_{I=0} = \alpha \Delta T \tag{9.1}$$

where α is called the Seebeck coefficient with the unit of V K^{-1}. If α is a function of temperature, then integrate Eq. (9.1) to get the Seebeck voltage.

Applying Eq. (9.1) in each segment of the circuit shown in Fig. 9.1 and assuming α is constant, we get

$$V_{J2} - V_{J1} = \alpha_{Cu}(T_2 - T_o) \tag{9.2}$$

$$V_{J3} - V_{J2} = \alpha_A(T_1 - T_2) \tag{9.3}$$

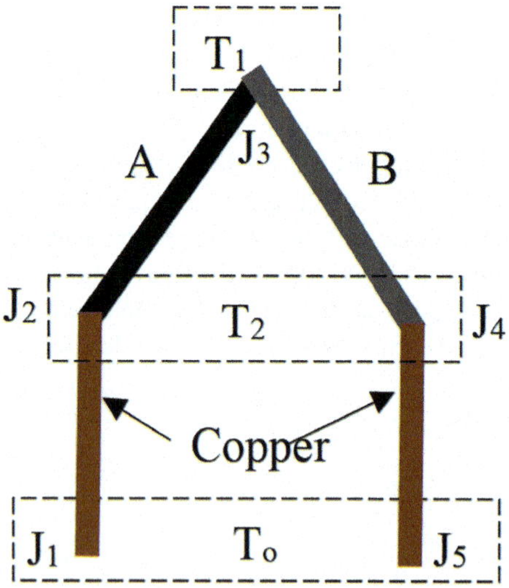

Fig. 9.1 Basic thermocouple circuit

$$V_{J4} - V_{J3} = \alpha_B(T_2 - T_1) \tag{9.4}$$

$$V_{J5} - V_{J4} = \alpha_{Cu}(T_o - T_2) \tag{9.5}$$

Adding the above equations, the potential difference between J_5 and J_1 can be given as

$$V_{J5} - V_{J1} = (\alpha_B - \alpha_A)(T_1 - T_2) \tag{9.6}$$

where $(\alpha_B - \alpha_A)$ is called the Seebeck coefficient of a thermocouple. If the temperature difference between junctions $(T_1 - T_2)$ is negative, i.e., $T_2 > T_1$, the potential difference between J_5 and J_1 will also be negative. Therefore, the Seebeck effect can be termed as the reversible thermoelectric phenomenon.

It can be seen from Eq. (9.6) that, to get higher Seebeck voltage, we must use materials with different Seebeck coefficient. Seebeck coefficient of *n*-type material is negative and the Seebeck coefficient of *p*-type material is positive. Therefore, in commercial thermoelectric generator and coolers, semiconductor materials are used.

Peltier Effect

Jean Charles Athanase Peltier in 1834 discovered that when an electric current passes through a junction formed by two different conductors, heat is liberated and/or absorbed by it. It is also proved that the heat liberated and/or absorbed at the junctions is proportional to the electric current (I) flowing through it, i.e.,

$$Q_p = \pi_{AB} I \tag{9.7}$$

where π_{AB} is called the Peltier coefficient and has the unit of W A^{-1}.

The basic circuit to demonstrate the Peltier effect is shown in Fig. 9.2. When electric current flows from conductor B to A (I is positive), heat (Q_1) is absorbed in the junction, thereby cooling the junction J_1. When an electric current flows from conductor B to A (I is negative) at junction J_2, heat is liberated (Q_2) at this junction, thereby heating the junction. The heat absorbed at junction J_1 and heat liberated at junction J_2 can be given as

$$Q_1 = \pi_{AB} I \tag{9.8}$$

$$Q_2 = \pi_{AB}(-I) \tag{9.9}$$

Peltier effect is the principle behind the thermoelectric cooling. It is evident from Eqs. (9.8) and (9.9) that, if the direction of current flow is reversed, the same amount of heat can be absorbed/liberated. Therefore, the Peltier effect is also termed as the reversible thermoelectric phenomenon.

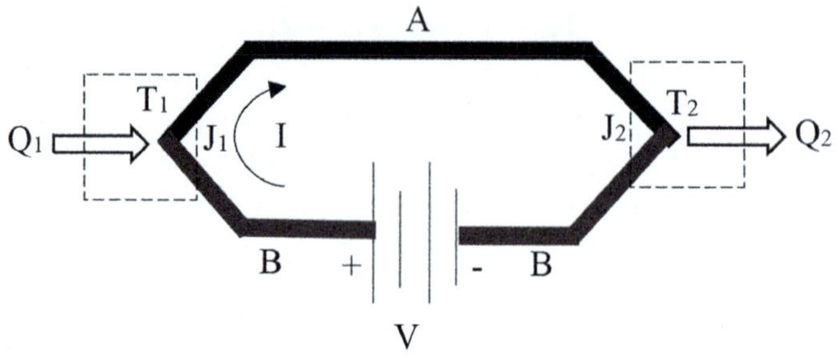

Fig. 9.2 Basic Peltier circuit

Thomson Effect

Seebeck effect and Peltier effect are junction phenomenon, whereas Thomson effect is a volume phenomenon and was discovered by William Thomson (Lord Kelvin) in 1854.

When an electric current flows through a conducting material with a temperature gradient along its length, heat absorbed (or) released is proportional to the electric current flow and the temperature gradient of the material, i.e.,

$$Q_t = \tau I \frac{dT}{dl} dl \tag{9.10}$$

where τ is called the Thomson coefficient, having unit V K^{-1}.

The current I is positive in the increasing direction of length l and Q is positive if the heat is absorbed from the surroundings and vice-versa. Thomson effect is also a reversible thermoelectric phenomenon.

Joule Heating

Joule heating is an irreversible phenomenon caused due to inherent resistivity of the conductor material. It is always dissipative[1] and the joule heat is proportional to the square of the current flowing through the conductor, i.e.,

$$Q_j = I^2 \frac{\rho l}{A} \tag{9.11}$$

where ρ is the electrical resistivity of the material, A the cross-sectional area of the conductor and l the length of the conductor. The thermoelectric element has its inherent electrical resistivity, therefore, when electric current flows through the thermoelectric element Joule heat is dissipated.

Fourier Heat Conduction

Energy transfer within a homogeneous medium due to the temperature gradient in the medium can be calculated by Fourier law. For one-dimensional heat transfer (along

[1] Q_j is always positive even if the current direction is changed, so Joule heat is dissipative.

the length (l) of path of heat flow), the heat transfer rate is given by the Fourier law as:

$$Q_f = -kA\frac{dT}{dl} \tag{9.12}$$

where k is the thermal conductivity of the material and A the area of heat transfer (usually cross-section area). The Fourier heat conduction is a spontaneous process and is dissipative in nature.

In any thermoelectric devices (generator, cooler or heat pump), all the above five effects namely Seebeck effect, Peltier effect, Thomson effect, Joule heating and Fourier heat conduction may be present. If the temperature difference between the hot and cold junctions ($T_1 - T_2$) is lower than T_2, Thomson heat is not prominent, hence, can be neglected for the analysis. The presences of irreversible effects in the thermoelectric circuit make the energy conversion efficiency of the thermoelectric devices lower than the Carnot efficiency.

Kelvin's Relations

Lord Kelvin derived the relation between the three reversible thermoelectric effects, known as second Kelvin's relation between the Seebeck, Peltier and Thomson coefficients. The relation is derived by using the first and second law of thermodynamics by considering only the reversible thermoelectric effects and hence ignoring the irreversible effects occurring in the thermocouple.

Figure 9.3 shows the basic thermoelectric engine, in which temperature T_1 is higher than T_2, therefore, the current will flow in the circuit because of the Seebeck effect. The electric current is supplied to the battery (load) for charging. Consider the energy balance of this circuit for an operating period of one second.

Considering the temperature-dependent thermoelectric properties, at junction J_1 and J_2, the heat input is given by

$$Q_1 = \pi_{AB}(T_1)I \tag{9.13}$$

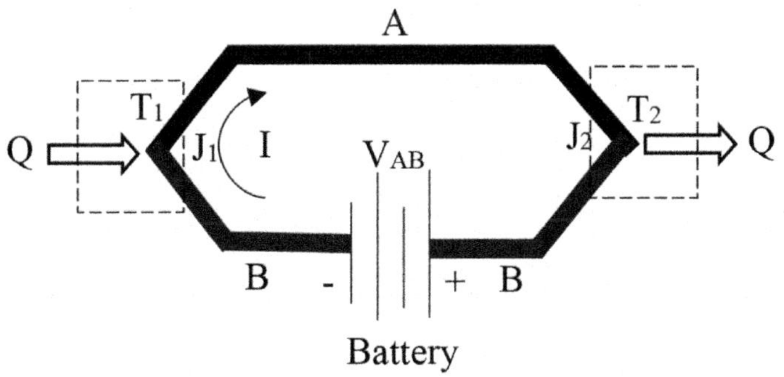

Fig. 9.3 Basic thermoelectric engine

$$Q_2 = \pi_{AB}(T_2)(-I) \tag{9.14}$$

Let us assume the direction of increasing length L is from junction J_2 to J_1. Therefore, in thermo-element, A the direction of current flow is negative. Then, the Thomson effect in the thermoelement A and B can be given by

$$\begin{aligned} Q_A &= -I\int_2^1 \tau_A(T)\frac{dT}{dl}dl = -I\int_{T_2}^{T_1} \tau_A(T)dT \\ Q_B &= I\int_2^1 \tau_B(T)\frac{dT}{dl}dl = I\int_{T_2}^{T_1} \tau_B(T)dT \end{aligned} \tag{9.15}$$

and the total Thomson heat input in the thermocouple by

$$Q_A + Q_B = I\int_{T_2}^{T_1} [\tau_B(T) - \tau_A(T)]dT \tag{9.16}$$

Then, by the first law of thermodynamics, the electrical work done by the thermocouple is the sum of heat input (Peltier and Thomson heat) to the thermocouple circuit

$$V_{AB}I = \pi_{AB}(T_1)I - \pi_{AB}(T_2)I + I\int_{T_2}^{T_1} [\tau_B(T) - \tau_A(T)]dT \tag{9.17}$$

Applying the second law of thermodynamics in the thermocouple circuit (Fig. 9.3), another relation can be obtained. Note that we have considered only the reversible effects so that the entropy generation in the thermocouple circuit is zero (i.e., no change in entropy).

$$\Delta S = -\frac{Q_1}{T_1} - \frac{Q_2}{T_2} - \int_2^1 \frac{Q_A}{T} - \int_2^1 \frac{Q_B}{T} = 0 \tag{9.18}$$

substituting Eqs. (9.13)–(9.17) in Eq. (9.7), yields

$$\frac{\pi_{AB}(T_1)}{T_1} - \frac{\pi_{AB}(T_2)}{T_2} + I\int_{T_2}^{T_1} [\tau_B(T) - \tau_A(T)]\frac{dT}{T} = 0 \tag{9.19}$$

Equations (9.18) and (9.19) are the two basic equations of the thermoelectric system, derived on the basis of the first and second laws of thermodynamics. Differentiating Eq. (9.17) with respect to temperature T_1, gives:

$$\frac{V_{AB}}{dT_1} = \frac{d\pi_{AB}(T_1)}{dT_1} - \tau_A(T_1) + \tau_B(T_1) \tag{9.20}$$

Equation (9.20) is only the function of temperature T_1 and it is valid for all range of temperatures. Rewriting Eq. (9.20) gives:

$$\frac{V_{AB}}{dT} = \frac{d\pi_{AB}(T)}{dT} - \tau_A(T) + \tau_B(T) \tag{9.21}$$

Let us consider right-hand side of Eq. (9.21) as $S_{AB}(T)$, then

$$S_{AB}(T) = \frac{d\pi_{AB}(T)}{dT} - \tau_A(T) + \tau_B(T) \tag{9.22}$$

$$\frac{V_{AB}}{dT} = S_{AB}(T) \tag{9.23}$$

By integrating Eq. (9.23) between the limits T_1 and T_2, we get

$$V_{AB}(T_1, T_2) = \int_{T_2}^{T_1} S_{AB}(T)dT \tag{9.24}$$

$$V_{AB}(T_1, T_2) = S_{AB}(T_1) - S_{AB}(T_2) \tag{9.25}$$

If $T_1 - T_2$ is sufficiently small, then

$$V_{AB} = (T_1 - T_2)S_{AB}(T_m) \tag{9.26}$$

where T_m is the mean temperature between T_1 and T_2.

Differentiating Eq. (9.19) with respect to T_1, we get

$$\frac{d}{dT_1}\left(\frac{\pi_{AB}(T_1)}{T_1}\right) + \frac{[\tau_B(T_1) - \tau_A(T_1)]}{T_1} = 0 \tag{9.27}$$

Equation (9.27) is also a function of temperature T_1, hence valid for all range of temperature and can rewrite as a function of temperature T:

$$\frac{d}{dT}\left(\frac{\pi_{AB}(T)}{T}\right) = \frac{[\tau_A(T) - \tau_B(T)]}{T} \tag{9.28}$$

Left hand side of Eq. (9.28) can be written as

$$\frac{d}{dT}\left(\frac{\pi_{AB}(T)}{T}\right) = \frac{1}{T}\left(\frac{d\pi_{AB}(T)}{dT}\right) - \frac{\pi_{AB}(T)}{T^2} \tag{9.29}$$

Substituting Eq. (9.29) in (9.28), we get

$$\left(\frac{d\pi_{AB}(T)}{dT}\right) + \tau_B(T) - \tau_A(T) = \frac{\pi_{AB}(T)}{T} \tag{9.30}$$

Equation (9.30) is called as the first Kelvin's relation.

Comparing Eqs. (9.22) and (9.30), we get

$$\frac{\pi_{AB}(T)}{T} = S_{AB}(T) \tag{9.31}$$

Substituting Eq. (9.31) in Eq. (9.28), we get

$$T\frac{dS_{AB}(T)}{dT} = \tau_A(T) - \tau_B(T) \tag{9.32}$$

Equation (9.32) is called the second Kelvin's relation. Kelvins relations are the basic equations used in the thermodynamic design of thermoelectric devices.

It can be seen from Kelvin's second relation Eq. (9.32), that if the Seebeck coefficient is assumed to be constant (not a function of temperature), then the Thomson coefficient will become zero.

9.2 Thermodynamic Analysis of Thermoelectric Cooling Devices

In this section, the thermodynamic design of the thermoelectric cooling system is described based on the first and second law of thermodynamics. The thermo-elements are connected electrically in series and thermally in parallel to form a thermocouple. Electrically in series because all the electrical resistance are connected electrically in series, and in the thermal resistance network, the thermal resistance of the thermo-elements are connected parallel between the hot and cold temperature reservoir as shown in Fig. 9.4.

In Fig. 9.4a, R_c, R_n and R_p represent the electrical resistance of the contact wires, *n*-type thermocouple and *p*-type thermo-element respectively. V_n and V_p are the Seebeck voltage of *n*- and *p*-type thermo-element. In Fig. 9.4b, $R_{th,n}$ and $R_{th,p}$ represent the thermal resistance of the thermoelement in a thermocouple.

Certain assumptions made for the thermodynamic design and analysis of the thermoelectric cooler are as follows:

- The thermoelectric cooler is operating between the two temperatures, T_H and T_C, which are the hot and cold junction temperatures.
- Heat losses from the lateral surface of the thermoelements are assumed negligible.
- Electric and thermal contact resistances between the metal conductor and thermoelements do not affect the performance. However, the performance of electrical and thermal contact resistance reduces the performance of actual thermoelectric cooling systems.
- One dimensional steady-state heat transfer along the length (L) of the thermoelement is considered for the modelling and analysis.

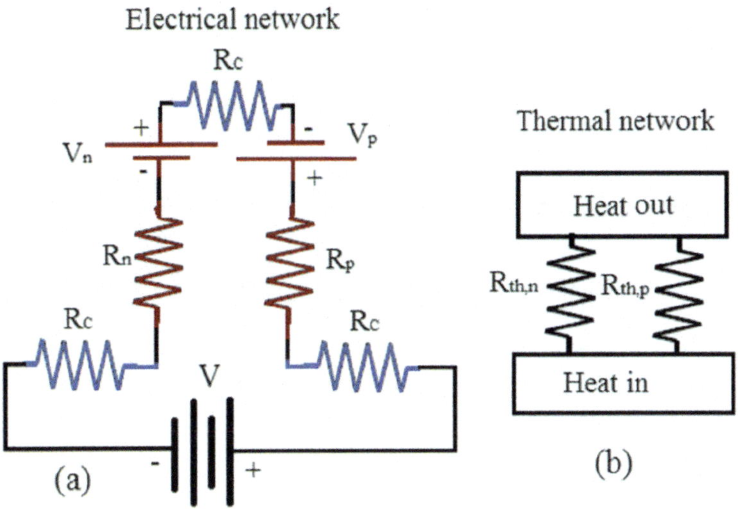

Fig. 9.4 Electrical and thermal analogy

- Electrical resistivity, thermal conductivity and Seebeck coefficient are assumed to be independent of temperature although these properties are temperature-dependent.
- The cross-section area of the thermo-element is constant.

9.2.1 Thermodynamic Analysis of Thermoelectric Cooler

The basic schematic of a thermoelectric cooler with a pair of thermocouples is shown in Fig. 9.5. A battery supplies the input voltage V and input current I to the thermocouple. The thermocouple elements are connected with copper conductors. The ceramic layer is placed at the top and bottom of the copper conductors at the hot and cold side respectively to protect the user from the open electric connection. The ceramic layer provides the low resistance path for heat flow and a high resistance path for electric current flow.

Cross-section view of an n-type thermo-element is shown in Fig. 9.6. The length of the thermo-element is L and the cross-section area of the thermoelement is A.

Applying first law of thermodynamics at the hot side (sink) of the thermoelectric cooler, the energy balance in the p- and n-type thermo-element can be written as

$$Q_p = \alpha_p I T_H - k_p A_p \frac{dT}{dl} \tag{9.33}$$

$$Q_n = -\alpha_n I T_H - k_n A_n \frac{dT}{dl} \tag{9.34}$$

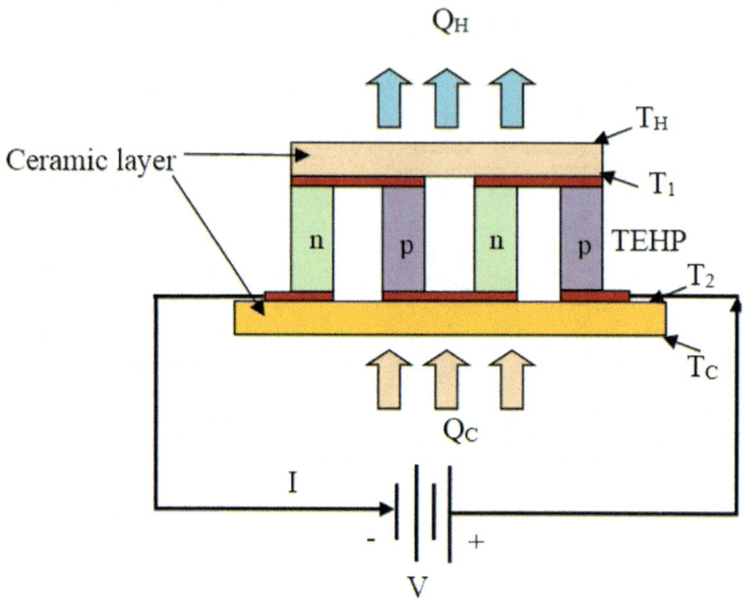

Fig. 9.5 Thermoelectric cooler

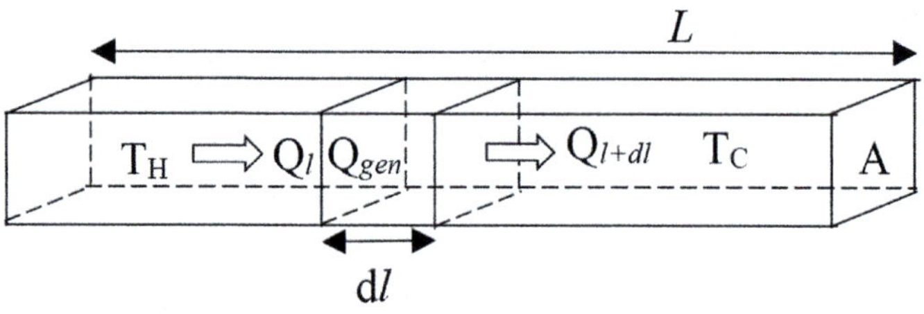

Fig. 9.6 Cross-section view of a thermo-element

Note that the Seebeck coefficient of *n*-type thermo-element is negative, but the heat is generated by the Peltier effect at the hot side. Therefore, the generated heat is opposed by the heat conduction by Fourier's heat transfer phenomenon.

Q_{gen} is the heat generated in the thermo-element because of Joule heating (volume phenomenon). Then by doing energy balance in the infinitesimal length (dl) of the thermo-element (Fig. 9.6), we get

$$Q_l + Q_{\text{gen}} = Q_{l+dl} \tag{9.35}$$

$$\left| k_n A_n \frac{dT}{dl} \right|_l = -\frac{\rho_n dl}{A_n} I^2 + \left| k_n A_n \frac{dT}{dl} \right|_{l+dl} \tag{9.36}$$

Simplifying Eq. (9.36), we get

$$\frac{d^2T}{dl^2} = -\frac{\rho_n}{k_n A_n^2} I^2 \tag{9.37}$$

Integrate Eq. (9.37) to find the temperature distribution in the thermo-element

$$\int_0^l \frac{d^2T}{dl^2} = -\int \frac{\rho_n}{k_n A_n^2} I^2 \tag{9.38}$$

$$\frac{\rho_n l^2}{2k_n A_n^2} I^2 + C_1 l + C_2 \tag{9.39}$$

To evaluate the integration constants C_1 and C_2, we apply the fixed temperature boundary conditions as

$$T_{(l)} - T_C \text{ at } l = 0 \tag{9.40}$$

$$T_{(l)} - T_H \text{ at } l = L \tag{9.41}$$

Then the integration constants C_1 and C_2 are evaluated as

$$C_1 = \frac{T_H - T_C}{L} + \frac{\rho_n L}{2k_n A_n^2} I^2 \tag{9.42}$$

$$C_2 = T_C \tag{9.43}$$

Substituting C_1 and C_2 in Eq. (9.29) and differentiating it with respect to length, we get

$$\frac{dT}{dl} = -\frac{l\rho_n}{k_n A_n^2} I^2 + \frac{T_H - T_C}{L} + \frac{\rho_n L}{2k_n A_n^2} I^2 \tag{9.44}$$

The length and area of cross-section of the p-type thermo-element are same as n-type thermo-element. Therefore, Eq. (9.44) with necessary changes in the material properties can also be applied for p-type thermo-element.

Therefore, substituting Eq. (9.44) in Eqs. (9.33) and (9.34), we get

$$Q_p = \alpha_p I T_H - k_p A_p \left(\frac{T_H - T_C}{L} \right) - \frac{\rho_p}{A_p} I^2 \left(\frac{L}{2} - l \right) \tag{9.45}$$

$$Q_n = -\alpha_n I T_H - k_n A_n \left(\frac{T_H - T_C}{L} \right) - \frac{\rho_n}{A_n} I^2 \left(\frac{L}{2} - l \right) \tag{9.46}$$

At $l = L$, i.e., at the hot side, the total heat is the sum of the contributions of n- and p-type thermo-element, and therefore, we get

$$Q_H = (\alpha_p - \alpha_n)IT_H - \left(\frac{k_n A_n}{L} + \frac{k_p A_p}{L}\right)(T_H - T_C) + \left(\frac{\rho_n L}{A_n} + \frac{\rho_p L}{A_p}\right)\frac{I^2}{2} \quad (9.47)$$

Similarly, when $l = 0$ (cold side), the total heat input is the sum of contributions of n- and p- type thermo-element. Hence, we get

$$Q_H = (\alpha_p - \alpha_n)IT_H - \left(\frac{k_n A_n}{L} + \frac{k_p A_p}{L}\right)(T_H - T_C) - \left(\frac{\rho_n L}{A_n} + \frac{\rho_p L}{A_p}\right)\frac{I^2}{2} \quad (9.48)$$

Equations (9.47) and (9.48) are the energy balance equations of the thermoelectric cooler at the hot and cold junction, respectively, and can be written as

$$Q_H = \alpha IT_H - K(T_H - T_C) + \frac{I^2 R}{2} \quad (9.49)$$

$$Q_C = \alpha IT_C - K(T_H - T_C) - \frac{I^2 R}{2} \quad (9.50)$$

where R is the total electrical resistance of the thermocouple, K the total thermal conductance of the thermocouple and α the combined Seebeck coefficient of the thermocouple.

In the thermoelectric cooler, the electrical power input required to absorb Q_C (W) of energy from the cold side can be given by

$$P_{\text{in}} = (Q_{\text{H}} - Q_{\text{C}}) = \alpha I(T_{\text{H}} - T_{\text{C}}) + I^2 R \quad (9.51)$$

The energy efficiency or COP of the thermoelectric cooler can be written as

$$COP = \frac{Q_C}{P_{\text{in}}} = \frac{(\alpha IT_C - K(T_H - T_C) - 0.5I^2 R)}{(\alpha I(T_H - T_C) + I^2 R)} \quad (9.52)$$

It can be seen from Eq. (9.52) that the COP of the thermoelectric cooler is a strong function of current I but it is not clear that how the COP changes with current, is there any optimum current for maximum COP. To make it clear, we differentiate Eq. (9.52) with respect to current and equate it to zero, to find the optimum current for maximum COP,[2] to get

[2] To find the maximum of a function, the derivative of the function with respect to its differential variable must be negative.

$$I_{\mathrm{COP}} = \frac{\alpha(T_H - T_C)\left[\sqrt[2]{1 + ZT_m} + 1\right]}{RZT_m} \tag{9.53}$$

where ZT_m can be given by

$$ZT_m = \frac{\alpha^2}{KR}\frac{(T_H + T_C)}{2} \tag{9.54}$$

We will study more about the Z parameter, which is called a figure of merit.

Similarly, there is an optimum current for maximum cooling power, which can be obtained by differentiating Eq. (9.52) with respect to current. Thus, the obtained optimum current for maximum cooling power is given by

$$I_{Q_{c,\max}} = \frac{\alpha T_C}{R} \tag{9.55}$$

Substituting Eq. (9.53) in Eq. (9.52), we get the maximum COP of the thermoelectric cooler, which can be given as

$$COP_{\max} = \left(\frac{T_C}{T_H - T_C}\right)\left(\frac{\sqrt[2]{1 + ZT_m} - \frac{T_H}{T_C}}{\sqrt[2]{1 + ZT_m} + 1}\right) \tag{9.56}$$

Substituting Eq. (9.55) in (9.52), we get the COP of the thermoelectric cooler at maximum cooling power, which can be given as

$$COP_{Q_{c,\max}} = \left(\frac{0.5T_C^2 Z - (T_H - T_C)}{ZT_H T_C}\right) \tag{9.57}$$

It should be noted that the COP at maximum cooling power is lower than the maximum COP of the thermoelectric cooler.

Here, we have not derived the energy balance at hot and cold junctions, and the energy efficiency of thermoelectric generator (direct heat to electricity conversion), but for the reference of the readers, it is given below by following similar modelling techniques.

$$Q_H = \alpha I T_H + K(T_H - T_C) - \frac{I^2 R}{2} \tag{9.58}$$

$$Q_C = \alpha I T_C + K(T_H - T_C) + \frac{I^2 R}{2} \tag{9.59}$$

$$\eta_{\max} = \left(\frac{T_H - T_C}{T_H}\right)\left(\frac{\sqrt[2]{1 + ZT_m} - 1}{\sqrt[2]{1 + ZT_m} + \frac{T_C}{T_H}}\right) \tag{9.60}$$

In the thermoelectric cooler, there is a limit for minimum cold side temperature, up to which it could operate with the fixed hot side temperature. The maximum temperature difference in which the thermoelectric cooler can operate can be found out by equating Eq. (9.40) to zero when the current is $IQ_{c,\max}$, since it is known that the cooling power will be minimum if the temperature difference (temperature lift) is maximum. Thus, the maximum temperature difference in a single stage (only one stage) thermoelectric cooler can be given as

$$(T_H - T_C)_{\max} = (\Delta T)_{\max} = \left(\frac{ZT_C^2}{2}\right) \tag{9.61}$$

It should be noted that the ΔT is maximum when T_C is the minimum for a fixed hot side temperature T_H. When Z is 2.5×10^{-3} K^{-1}, and with the hot side temperature of 300 K, the obtained minimum cold side temperature T_C, min is nearly 237 K. This cold side temperature limit can be further reduced by using cascaded (multi-stage) thermoelectric coolers.

9.2.2 Figure of Merit (FOM) of Thermoelectric Devices

We have seen an important parameter Z (FOM) in the expressions for maximum COP of a thermoelectric cooler and maximum efficiency of a thermoelectric generator. It can be shown that COP of a thermoelectric cooler and energy efficiency of the thermoelectric generator can be made maximum (Carnot limit) if FOM reaches positive infinity or maximum. FOM can be made dimensionless by multiplying it with the average temperature T_m as shown by Eq. (9.54). It is clear from the expression for the dimensionless figure of merit, for any thermoelectric devices, the thermo-elements should be selected with higher Seebeck coefficient and with the product of electrical resistance and thermal conductance should be minimum.

The thermal conductance and electrical resistance of the thermocouple is given as

$$K = \left(\frac{k_n A_n}{L_n} + \frac{k_p A_p}{L_p}\right) \tag{9.62}$$

$$R = \left(\frac{\rho_n L_n}{A_n} + \frac{\rho_p L_p}{A_p}\right) \tag{9.63}$$

The Seebeck coefficient, thermal conductivity and the electrical resistivity are the material property, therefore, only by changing the material, we can alter the figure of merit of the thermoelectric device. It can also be seen from the expressions for electrical resistance and thermal conductance that, there is certain geometry (length and area) of the thermo-element, the FOM can be made maximum.

Assume $\frac{A_n L_p}{L_n A_p}$ of the thermo-elements γ, and then the denominator term of the FOM can be written as

$$KR = \left(\rho_n k_n + \gamma \rho_p k_n + \frac{\rho_n k_p}{\gamma} + \rho_p k_p \right) \tag{9.64}$$

It should be noted that the minimum of Eq. (9.64) also can lead to maximum FOM. Therefore, by mathematical induction, we get

$$\gamma = \sqrt[2]{\frac{\rho_n k_p}{\rho_p k_n}} \tag{9.65}$$

Substituting in Eq. (9.65), we get the minimum of *KR* product as

$$KR_{\min} = \left(\sqrt{\rho_n k_n} + \sqrt{\rho_p k_p} \right)^2 \tag{9.66}$$

Then the figure of merit of the thermoelectric device for optimized geometry can be given as

$$ZT_m = \frac{\alpha^2 T_m}{\left(\sqrt{\rho_n k_n} + \sqrt{\rho_p k_p} \right)^2} \tag{9.67}$$

If the properties of *n*- and *p*-type thermo-element are same, that is, $\alpha_p = -\alpha_n$, $k_n = k_p$, $\rho_n = \rho_p$, then ZT_m can be written as

$$ZT_m = \frac{\alpha^2 T_m}{k\rho} \tag{9.68}$$

Equation (9.68) represents the FOM of the thermoelectric material.

9.2.3 Cascaded Thermoelectric Cooler Systems

The two-stage thermoelectric cooler with electrical series configuration is shown in Fig. 9.7. There is a total *M* number of thermocouples in the two-stage thermoelectric cooler. The number of thermocouples in the first stage is *n*- (not with *n*-type thermo-element) and in second-stage is *m*, and therefore, $M = n + m$.

The cold side of stage 2 will act as the heat sink for stage 1. Therefore, the temperature at the cold side of stage 1 will be further lower than the cold side temperature of stage 2. This similar technique can be put forth to reduce the cold side temperature further by adding more number of stages, but the thermal resistances between the junctions make the infinite stage of thermoelectric cooler impractical.

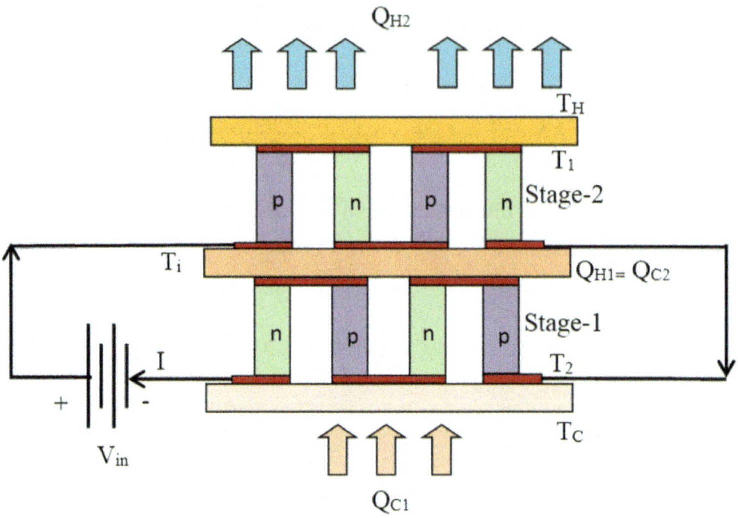

Fig. 9.7 Two-stage thermoelectric cooler

Considering Fig. 9.7, by the first law of thermodynamics, the energy balance at the hot and cold junctions of the two stage thermoelectric cooler, connected electrically in series is given as follows:

$$Q_{C1} = \alpha_1 T_C I - \frac{I^2 R_1}{2} - K_1(T_i - T_C) \tag{9.69}$$

$$Q_{H1} = \alpha_1 T_i I + \frac{I^2 R_1}{2} - K_1(T_i - T_C) \tag{9.70}$$

$$Q_{C2} = \alpha_2 T_i I - \frac{I^2 R_2}{2} - K_2(T_H - T_i) \tag{9.71}$$

$$Q_{H2} = \alpha_2 T_H I + \frac{I^2 R_2}{2} - K_2(T_H - T_i) \tag{9.72}$$

The inter-stage temperature T_i can be calculated by equating $Q_{H1} = Q_{C2}$, since the heat rejected in the first stage is the heat absorbed by the second stage, and the inter-stage temperature can be derived as follows:

$$T_i = \frac{\left[\frac{I^2(R_1+R_2)}{2} + K_2 T_H + K_1 T_C\right]}{[I(\alpha_2 - \alpha_1) + K_1 + K_2]} \tag{9.73}$$

Since all the thermocouples are connected electrically in series, the current flowing through the two stages of the thermoelectric cooler will be equal and can be given as

$$I = \left[\frac{V_{\text{in}}}{R_1 + R_2}\right] - \left[\frac{\alpha_1 \Delta T_1 + \alpha_2 \Delta T_2}{R_1 + R_2}\right] \tag{9.74}$$

where ΔT_1, ΔT_2 are the temperature difference between the first and second stages of the thermoelectric cooler, respectively, and V_{in} the input voltage to the TEC. The $\alpha_1 \Delta T_1$ and $\alpha_2 \Delta T_2$ are the total Seebeck voltage generated in the first and second stages, respectively, and it is opposing to the applied input voltage V_{in} as seen from Fig. 9.4a.

$$V_{\text{in}} = V_1 + V_2 = \alpha_1 \Delta T_1 + \alpha_2 \Delta T_2 + I(R_1 + R_2) \tag{9.75}$$

The power input to the two-stage thermoelectric cooler can be given as

$$W_{in} = I(\alpha_1 \Delta T_1 + \alpha_2 \Delta T_2) + I^2(R_1 + R_2) \tag{9.76}$$

The COP of two-stage electrically series thermoelectric cooler can be obtained as the ratio of heat absorbed at the cold side of stage 1 to the work input to the two-stage thermoelectric cooler.

$$\text{COP} = \frac{Q_{C1}}{W_{\text{in}}} = \frac{\alpha_1 T_C I - \frac{I^2 R_1}{2} - K_1(T_i - T_C)}{I(\alpha_1 \Delta T_1 + \alpha_2 \Delta T_2) + I^2(R_1 + R_2)} \tag{9.77}$$

9.2.4 *COP of Cascaded Thermoelectric Cooler*

The cascaded thermoelectric cooler can be realised as many single-stage thermoelectric cooler operating in cascaded configuration, in which the hot side of stage 1 is acting as the cold side of stage 2.

Considering the overall energy balance of stage 1 of the two-stage thermoelectric cooler, we get

$$W_1 = Q_2 - Q_1 \tag{9.78}$$

$$\text{COP}_1 = \frac{Q_1}{W_1} \tag{9.79}$$

The subscripts correspond to the stages of the multistage thermoelectric cooler. Therefore, heat rejected by the first stage or heat absorbed by the second stage Q_2 can be given as

$$Q_2 = Q_1\left(1 + \frac{1}{\text{COP}_1}\right) \tag{9.80}$$

Similarly, for the second stage, the heat rejected by the second stage or heat absorbed by the third stage Q_3 is given as

$$Q_3 = Q_2\left(1 + \frac{1}{COP_2}\right) \tag{9.81}$$

Substituting Eq. (9.80) in Eq. (9.81), we get,

$$Q_3 = Q_1\left(1 + \frac{1}{COP_1}\right)\left(1 + \frac{1}{COP_2}\right) \tag{9.82}$$

Similarly, for n stages, the heat rejected at the nth stage can be given as

$$Q_n = Q_1\left(1 + \frac{1}{COP_1}\right)\left(1 + \frac{1}{COP_2}\right)\cdots\left(1 + \frac{1}{COP_n}\right) \tag{9.83}$$

Then the total work input to the n stage thermoelectric cooler and its overall COP can be written as

$$W = Q_n - Q_1 \tag{9.84}$$

$$COP_{\text{overall}} = \frac{Q_1}{W} \tag{9.85}$$

Therefore, the heat rejected at the nth stage can also be given as

$$Q_n = Q_1\left(1 + \frac{1}{COP_{\text{overall}}}\right) \tag{9.86}$$

Then, from Eq. (9.83) and Eq. (9.86), we get

$$\left(1 + \frac{1}{COP_{\text{overall}}}\right) = \prod_{i=1}^{n}\left(1 + \frac{1}{COP_i}\right) \tag{9.87}$$

Then the overall COP of cascaded thermoelectric cooler can be written as by rearranging Eq. (9.87)

$$\text{COP}_{\text{overall}} = \frac{1}{\prod_{i=1}^{n}\left(1 + \frac{1}{\text{COP}_i}\right) - 1} \tag{9.88}$$

If the COP of all stages are same, then

$$\text{COP}_{\text{overall}} = \frac{1}{\left(1 + \frac{1}{\text{COP}_{\text{individual}}}\right)^n - 1} \tag{9.89}$$

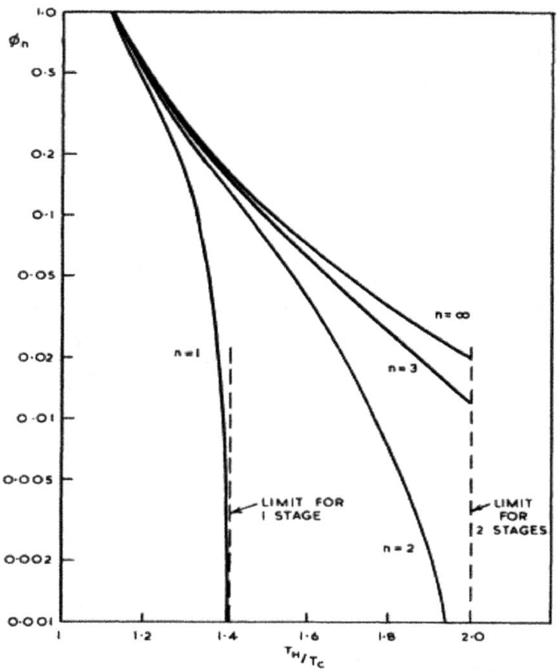

Fig. 9.8 COP of the cascaded thermoelectric cooler when for $ZT = 1$, n is a number of stages (Goldsmid 1964)

The cascaded thermoelectric cooler operates at higher temperature lift than the single-stage thermoelectric cooler. Properties of the thermoelectric materials may vary for this higher range of temperature variations in the hot and cold side of the cascaded thermoelectric cooler. Therefore, different thermoelectric materials can be used for different stages which possess a higher figure of merit for the optimum temperature stages. This is an advantage of the cascaded thermoelectric cooler. Figure 9.8 shows the variation in COP (f_n) of the cascaded thermoelectric cooler with the number of stages. It can be seen that the COP of the multistage thermoelectric cooler is higher than the single-stage and it is not economical and impractical to go for more number of stages and the rate of improvement in the COP is also low with an increase in a number of stages. The multistage thermoelectric coolers with 2–4 stages are commercially available with the thermoelectric cooler manufacturers like Marlow Industries and Liard Technologies etc.

9.2.5 Second Law Analysis of Thermoelectric Devices

We have seen the usefulness of the second law of thermodynamics in deriving Kelvin's relation for the thermoelectric devices. Moreover, the second law of thermodynamics in combined with the first law is very useful in analysing the performance of the thermal energy systems. In this section, the exergy analysis technique has been elaborated and its usefulness in analysing the thermoelectric cooler has been given.

Consider the thermoelectric cooler shown in Fig. 9.5, and then by applying the first law of thermodynamics, the energy balance can be given as

$$Q_H - Q_C = P_{in} \tag{9.90}$$

According to the second law of thermodynamics, the entropy generation is given by

$$S_{\text{gen}} = \left(\Delta S_{\text{system}} + \Delta S_{\text{surrounding}}\right) \geq 0 \tag{9.91}$$

$$S_{\text{gen}} = \frac{Q_H}{T_H} - \frac{Q_C}{T_C} \geq 0 \tag{9.92}$$

Combining the first and second law of thermodynamics, the exergy balance in any thermodynamic system at steady state can be given as

$$Ex_{\text{in}} = Ex_{\text{out}} + Ex_{\text{lost}} + \text{Irreversibilities} \tag{9.93}$$

where Ex_{in} is the exergy input to the system, Ex_{out} the exergy output of the system, Ex_{lost} the exergy lost in the output stream (if present) and irreversibilities are the exergy destroyed in the process. The lost exergy has two quantities, viz. recoverable exergy waste and non-recoverable exergy waste. In the thermoelectric systems, we can classify the irreversibilities into two categories, viz. internal and external.

The exergy input to the system is the electrical power input P_{in} and it is 100% exergy. Exergy output is the thermal exergy E_{Qc} deposited at the cold side of the thermoelectric cooler, which is defined as

$$E_{Qc} = Q_C\left(\frac{T_o}{T_C} - 1\right) \tag{9.94}$$

where T_o is the environment temperature. Then exergy balance becomes

$$P_{\text{in}} = Q_{\text{C}}\left(\frac{T_o}{T_{\text{C}}} - 1\right) + \text{Irr} + Q_{\text{H}}\left(1 - \frac{T_o}{T_{\text{H}}}\right) \tag{9.95}$$

where the first term in the right-hand side is the exergy deposited at the cold side of the thermoelectric cooler E_{Qc}. The second term is the exergy destruction and the third term is the exergy rejected at the hot side of the thermoelectric cooler (E_{Qh}). Rearranging Eq. (9.95) we get

$$Irr = P_{\text{in}} - Q_C\frac{T_o}{T_C} + Q_H\frac{T_o}{T_H} + Q_C - Q_H \tag{9.96}$$

Substituting Eq. (9.90) in Eq. (9.96) and solving for irreversibilities *Irr*, we get

$$Irr = T_o\left(\frac{Q_H}{T_H} - \frac{Q_C}{T_C}\right) \tag{9.97}$$

Comparing Eqs. (9.92) and (9.97), the irreversibilities (*Irr*) becomes

$$Irr = T_o S_{gen} \tag{9.98}$$

Hence, it is clear that the irreversibilities in the system are dependent on the environmental temperature. In a thermoelectric cooling system, hot side temperature T_H is the environment temperature $T_{o.}$ Then the exergy efficiency can be defined as

$$\Psi = \frac{Exergy\ Out}{Exergy\ in} = \frac{E_{Qc}}{P_{\text{in}}} = 1 - \frac{T_o S_{\text{gen}}}{P_{\text{in}}} \tag{9.99}$$

Thus, by knowing the irreversibilities in the system we can find the exergy efficiency of a thermoelectric system. Applying the concept of exergy analysis, the exergy efficiency of the thermoelectric cooler system can be given as

$$\Psi_{\text{TEC}} = \frac{\left(\alpha I T_C - \frac{I^2 R}{2} - K(T_H - T_C)\right)\left(\frac{T_H}{T_C} - 1\right)}{\alpha I (T_H - T_C) + I^2 R} \tag{9.100}$$

It can be seen from the above equation, there will be optimum current for maximum exergy efficiency (I_{exergy}) similar to the optimum current for the maximum COP. Hence, the similar technique involved in finding I_{COP} is also followed to find I_{exergy}. Thus, it is found that the optimum current for maximum COP and exergy efficiency conditions are the same. Then, by substituting I_{exergy} in Eq. (9.100) we get,

$$\Psi_{\text{TEC}} = \left(\frac{\sqrt{1 + ZT_m} - \frac{T_H}{T_C}}{1 + \sqrt{1 + ZT_m}}\right) \tag{9.101}$$

Comparing the COP of thermoelectric coolers as given by Eqs. (9.52) and (9.101) for exergy efficiency of the thermoelectric cooler, we can see that the exergy efficiency is lower than the COP by the Carnot factor $\left(\frac{T_C}{T_H - T_C}\right)$ for the thermoelectric cooler.

Similarly, the exergy efficiency of the thermoelectric generator can be written as

$$\Psi_{\text{TEG}} = \left(\frac{\sqrt{1 + ZT_m} - 1}{\sqrt{1 + ZT_m} + \frac{T_C}{T_H}}\right) \tag{9.102}$$

The exergy efficiency of the thermoelectric generator is higher than the energy efficiency by the Carnot factor.

9.2.6 Irreversibilities in Thermoelectric Cooler System

As mentioned in Sect. 9.2.5, the thermoelectric cooler has irreversibilities or exergy losses. Irreversibilities are the exergy losses associated with the entropy generation in the system. There are four thermodynamic systems based on the assumptions of reversibility, and are shown in Fig. 9.9.

Internal irreversibilities are the exergy losses happening inside the system boundary and the external irreversibilities are exergy losses happening outside the system boundary (very close to the system boundaries).

Reversible or ideal thermodynamic system has no internal and external irreversibilities. Endo reversible thermodynamic system is one which is internally reversible and externally irreversible. Exoreversible thermodynamic system is externally reversible and having internal irreversibilities. Here, we have assumed the thermoelectric cooler as an exoreversible thermodynamic system. Irreversible thermodynamic systems possess internal as well as external irreversibilities. All the real systems are the near irreversible thermodynamic systems.

The entropy generation *Sgen* in the thermoelectric cooler can be calculated as per the entropy balance obtained by the second law of thermodynamics given by Eq. (9.97)

$$\mathrm{Irr} = T_o\left(\left(\frac{\alpha I T_H}{T_H} + \frac{I^2 R}{2T_H} - \frac{K(T_H - T_C)}{T_H}\right) - \left(\frac{\alpha I T_C}{T_C} - \frac{I^2 R}{2T_C} - \frac{K(T_H - T_C)}{T_C}\right)\right) > 0 \tag{9.103}$$

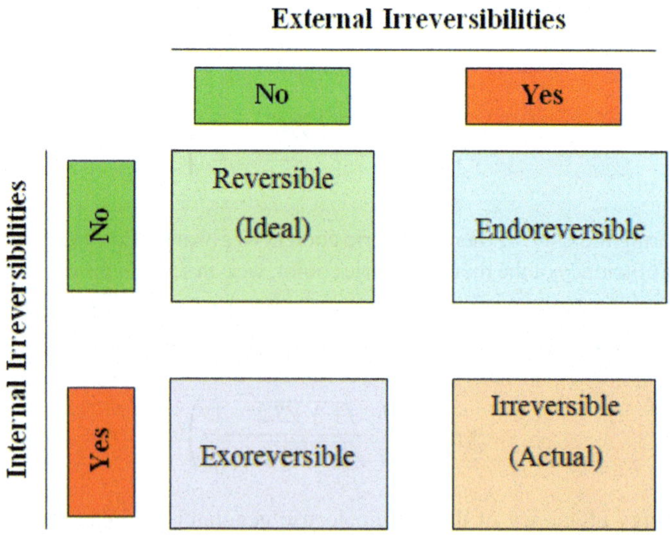

Fig. 9.9 Types of thermodynamic thermoelectric systems

Solving Eq. (9.103), we get

$$\text{Irr} = T_o\left(\frac{I^2RT_m}{T_HT_C} + \frac{K(T_H - T_C)^2}{T_HT_C}\right) > 0 \tag{9.104}$$

Note that the irreversibilities in thermoelectric cooler system is caused because of the electrical resistance R and the thermal conductance K of the thermoelectric material alone, and it is no wonder that the Peltier heat term got cancelled while calculating the irreversibilities in the thermoelectric cooler system, since the Peltier heat is a reversible phenomenon.

9.3 Solar Operation of Thermoelectric Cooler

The thermoelectric coolers have the advantages of solid-state device with no moving parts, precise temperature control, it can be operated in any orientation, it can be operated as a heater in the cold conditions by revising the direction of the current, it occupies less space and requires less or no maintenance. Having all these advantages but not making this device commercially viable is because of its low COP, i.e., more electrical energy input is required to get the desired cooling effect and the electrical input is not free of cost. Solar energy is available for free of cost and if we couple the solar energy conversion devices to operate thermoelectric cooler, the operating cost of the thermoelectric cooler can be nearly zero (nevertheless it increases capital cost!). The lifetime of the solar energy conversion devices (photovoltaic devices) is higher (approximately 20 years) and may get the payback within the lifetime of these devices. With the advancements in material science research, the figure of merit of a thermoelectric material can be made higher, so the COP of TEC can be improved, and thus making the solar thermoelectric cooler commercially viable in near future. The two routes for the solar operation of the thermoelectric cooler are solar thermal and solar photovoltaic (electric) conversion route. These two techniques are briefly explained in the following section.

9.3.1 Solar Thermoelectric Generator-Cooler System

In this technique, the thermoelectric generator is operated by the solar heat and the electrical energy output of the thermoelectric generator is given as the input to the thermoelectric cooler either directly or with DC/DC converter.

In early days, without the development of power electronics, the electrical power output from the solar thermoelectric generator was directly supplied as the input to thermoelectric cooler (imagine Fig. 9.10 without the DC/DC converter) and the heat

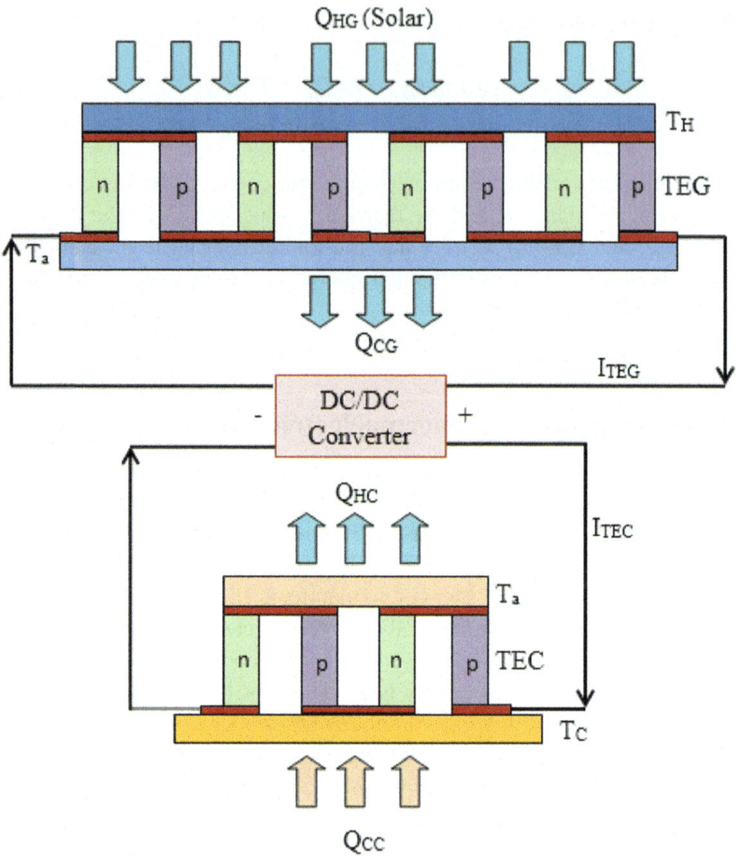

Fig. 9.10 Solar thermoelectric generator-cooler system

rejected at the hot side of a thermoelectric cooler and the cold side of the thermoelectric generator is same, and both contributions of these heat are rejected to the environment at temperature T_a (K). In this configuration (without DC/DC converter), we will not be able to control the current flowing through the thermoelectric cooler (i.e., current flowing through the thermoelectric cooler and thermoelectric generator are same), and we could not operate the thermoelectric generator at its maximum power output/maximum energy efficiency condition, unless it is designed for maximum efficiency conditions. Nevertheless, this method is simple and important to understand the solar operation of thermoelectric cooler system.

The overall efficiency of the solar thermoelectric generator-cooler system without DC/DC converter can be given as

$$\eta_{\text{overall}} = \frac{Q_{CC}}{Q_{HG}} = \left(\frac{Q_{CC}}{P_o}\right)\left(\frac{P_o}{Q_{HG}}\right) \quad (9.105)$$

where heat energy absorbed at the cold side of the thermoelectric cooler is Q_{CC}, the heat input to the thermoelectric generator is the solar energy input Q_{HG}, and the power output of the thermoelectric generator (power input to the thermoelectric cooler) is P_o. The maximum overall energy efficiency and exergy efficiency of the combined system can be given as

$$\eta_{\text{overall}} = \left(\frac{T_C}{T_H}\right)\left(\frac{T_H - T_a}{T_a - T_C}\right)\left(\frac{M_G - 1}{M_C + 1}\right)\left(\frac{M_C - \frac{T_a}{T_C}}{M_G + \frac{T_a}{T_H}}\right) \tag{9.106}$$

$$\Psi_{\text{overall}} = \left(\frac{M_G - 1}{M_C + 1}\right)\left(\frac{M_C - \frac{T_a}{T_C}}{M_G + \frac{T_a}{T_H}}\right) \tag{9.107}$$

where $M = \sqrt[2]{1 + ZT_m}$ and the subscripts G and C are the thermoelectric generator and thermoelectric cooler, respectively. The solar thermoelectric generator cooler system have shown very low overall energy conversion efficiency and this efficiency can be improved by using thermoelectric devices made of materials with a higher figure of merit and with solar concentrated collectors.

Now, consider the system as presented in Fig. 9.10 (with DC/DC converter). In this system, the electric power output of the thermoelectric generator is fed to the thermoelectric cooler via a DC/DC converter. In such a system the voltage input and current input to the thermoelectric cooler is different from the voltage output and current output of the thermoelectric generator keeping the power constant. This can be done by changing the duty cycle (D) of the DC/DC converter.

The basic configuration of DC/DC converter is shown in Fig. 9.11. Using the DC/DC converter the load impedance of the thermoelectric generator, i.e., internal resistance of the thermoelectric cooler can be altered by changing the duty cycle ratio and can be matched with the internal resistance of the thermoelectric generator (source), thus transferring the maximum power from the thermoelectric generator to the thermoelectric cooler (by maximum power transfer theorem). The relation between the load impedance and the input resistance seen by the different DC/DC converter configurations are given in Table 9.1.

Then the overall energy efficiency and the exergy efficiency of the solar thermoelectric generator-cooler system with DC/DC converter can be given as

$$\eta_{\text{overall}} = \frac{Q_{CC}}{Q_{HG}} = (COP_{TEC})(\eta_{TEG})\left(\eta_{\text{DC/DC}}\right)(\eta_{\text{solar}}) \tag{9.108}$$

$$\eta_{\text{overall}} = \frac{\alpha T_C I - \frac{I^2 R}{2} - K(T_a - T_C)}{G_{\text{solar}} A_{\text{collector}}} \tag{9.109}$$

$$\Psi_{\text{overall}} = \frac{\left(\alpha T_C I - \frac{I^2 R}{2} - K(T_a - T_C)\right)\left(\frac{T_H}{T_C} - 1\right)}{G_{\text{solar}} A_{\text{collector}}\left(1 - \frac{T_O}{T_{\text{SUN}}}\right)} \tag{9.110}$$

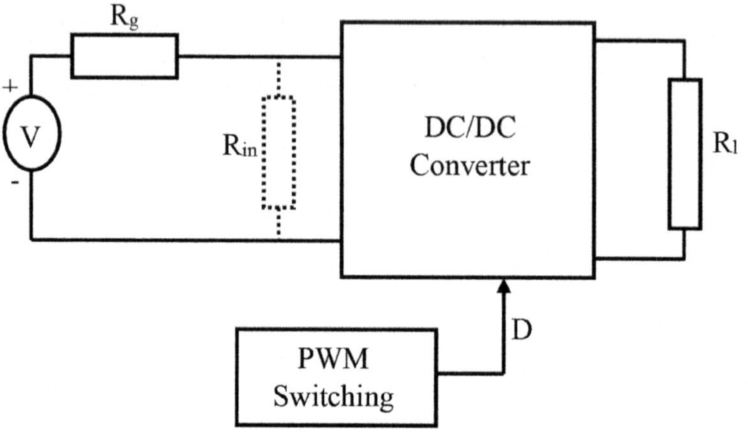

Fig. 9.11 Configuration of DC/DC converter.

Table 9.1 Impedance relations of DC/DC converter

DC/DC converter	Impedance relation
Buck converter	$R_{\text{in}} = R_l/D^2$
Boost converter	$R_{\text{in}} = R_l(1 - D)^2$
Buck boost converter	$R_{\text{in}} = R_l(1 - D)^2/D^2$

where $G_{\text{solar}}\ A_{\text{collector}}$ is the solar energy input to the system and the numerator corresponds to the amount of heat energy absorbed by the thermoelectric cooler.

With this technique, the COP of thermoelectric cooler can be greatly improved, because the cooling power output of the thermoelectric cooler is higher. The power output of the thermoelectric generator is higher because of the use of maximum power point tracking techniques in the DC/DC converter. Thus, the energy efficiency of the thermoelectric generator also can be improved. The conversion efficiency of the DC/DC converter is higher than 90% (for a good converter!). The solar evacuated tube collector can operate at a temperature range of 90–180 °C, which can be used for solar thermoelectric generator-cooler combined system. Temperature difference between the hot and cold side of the thermoelectric generator increases its power output, therefore, concentrating solar collectors can also play an important role in solar thermoelectric energy conversion.

Though these systems have lower conversion efficiency, these systems have an important role to play in sustainable development and having practical relevance, if the thermoelectric materials with a higher figure of merit are found.

Table 9.2 Energy efficiency of the solar thermoelectric cooler for $ZT_m = 1$ (Zhao et al. 2017)

S. no	Solar collector	Temperature range (Approx.)	Energy efficiency of TEG (%)	Combined system efficiency (TEG-TEC) (%)
1	Flat plate collector	70–90 °C	3.2	3.6
2	Evacuated tube collector	80–180 °C	6.7	7.6
3	Linear concentrator	90–200 °C	7.3	8.3
4	Dish collector	180–250 °C	8.8	10.0

9.3.2 Solar Collector Options for Thermoelectric Energy Conversion

Solar collector is an integral part of the solar thermoelectric energy conversion system. The energy efficiency of the thermoelectric generator and the combined system efficiency of the thermoelectric generator-cooler system presented in Table 9.2 for the ZT_{m} of 1, when the temperature of the hot and cold side of the thermoelectric generator is at the maximum temperature of solar collector and 300 K, respectively, and the hot and cold side temperature of the thermoelectric cooler is at 300 and 270 K for different solar collectors.

9.3.3 Solar Photovoltaic Thermoelectric Cooler

Another promising route of solar thermoelectric energy conversion is by cascading the solar photovoltaic module with the thermoelectric cooler. The basic schematic of the solar PV thermoelectric cooler is shown in Fig. 9.12.

The PV thermoelectric cooler (Fig. 9.13) can be operated in two modes, viz. with battery and without battery. Configuration with battery can provide cooling during the night also. In this, we will discuss the design techniques involved for a small box type solar *PV* thermoelectric cooler.

The power input to the thermoelectric cooler can be taken care of by the solar PV panel. The power output of the solar PV array can be given as

$$P_{\mathrm{in}} = P_{\mathrm{solar}} = V_{\alpha} I_{\mathrm{SC}}(\mathrm{FF}) \tag{9.111}$$

where FF is the fill factor of the solar PV panel and can be given as

$$\frac{V_m I_m}{V_{OC} I_{SC}} \tag{9.112}$$

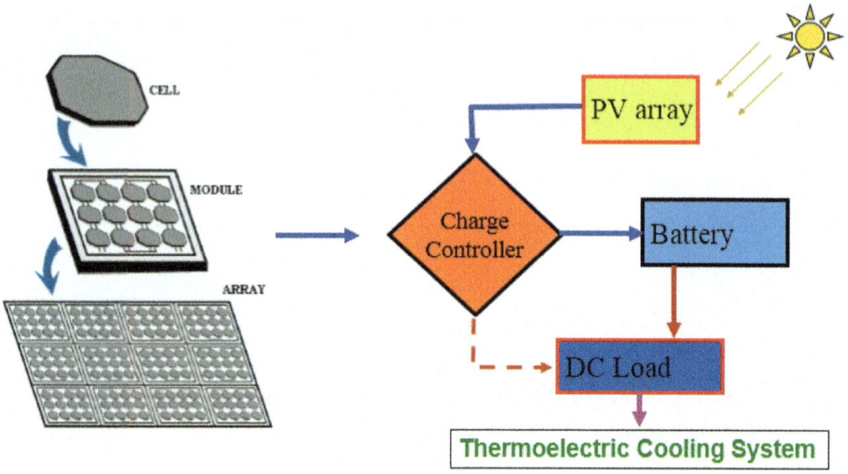

Fig. 9.12 A solar photovoltaic thermoelectric cooler

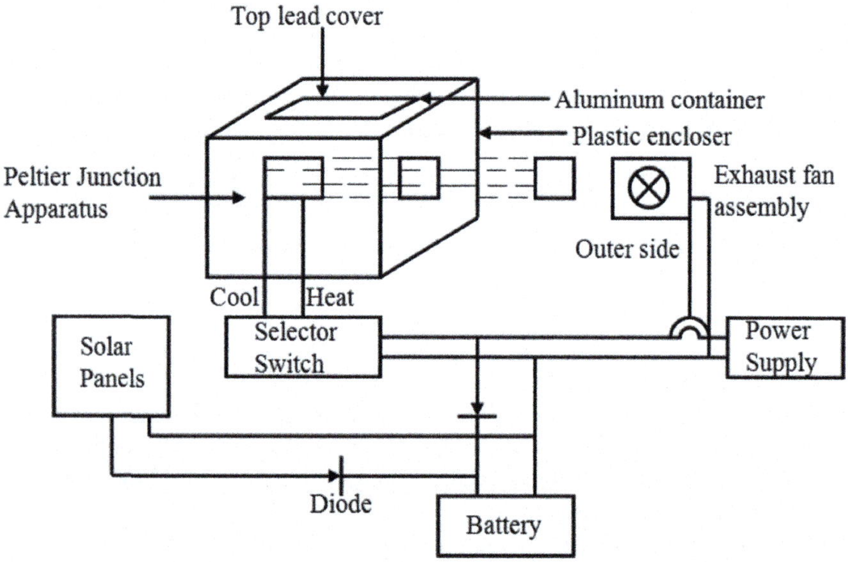

Fig. 9.13 A solar PV thermoelectric cooler

The efficiency of a solar *PV* system is

$$\eta_{\mathrm{PV}} = \frac{V_{oc} I_{sc} (\mathrm{FF})}{A_{\mathrm{PV}} G} \quad (9.113)$$

The power input required to thermoelectric cooler can be given as

$$P_{\text{in}} = \alpha I(T_H - T_C) + I^2R \tag{9.114}$$

The cooling power of the thermoelectric cooler can be given by Eq. (9.50). Then the overall energy efficiency and exergy efficiency of the solar *PV* thermoelectric cooler given as

$$\eta_{\text{PV-TEC}} = \frac{\alpha T_C I - \frac{I^2R}{2} - K(T_a - T_C)}{V_{oc}I_{sc}(FF)} \tag{9.115}$$

$$\Psi_{\text{PV-TEC}} = \frac{\left(\alpha T_C I - \frac{I^2R}{2} - K(T_a - T_C)\right)\left(\frac{T_H}{T_C} - 1\right)}{V_{oc}I_{sc}(\text{FF})} \tag{9.116}$$

It should be noted that the energy input to the solar PV thermoelectric cooler is 100% exergy since it the electrical energy.

The energy efficiency (COP) and the cooling power of solar PV thermoelectric cooler for a typical operating condition are shown in Fig. 9.14. It should be noted that the period of operation of the thermoelectric cooler in the morning and evening have higher COP because input power to the thermoelectric cooler from a solar PV panel is low because of lower solar radiation. The COP of the thermoelectric cooler is lower but the cooling power is higher during the mid-day because the power input to the thermoelectric cooler is higher as the solar radiation is higher. This is the major advantage of the solar cooling system, because the cooling is more required during the mid-day because of higher solar radiation and the solar PV-TEC system will work according to the requirements of the user (naturally!).

9.4 Thermoelectric Materials

The energy efficiency of the thermoelectric devices are strongly dependent on the figure of merit of thermoelectric materials. If FOM is high, the performance of the thermoelectric device is better. This led researchers to search for better thermoelectric materials.

In this section, we briefly discuss the FOM of the thermoelectric materials from the past to present.

The FOM of the thermoelectric device can be written (Eq. 9.117) by substituting electrical conductivity for electrical resistivity. The total thermal conductivity in a material is the sum of thermal conductivity by electrons k_{el} and by the lattice thermal conductivity k_{ph} (by phonons or lattice vibrations).

$$Z = \frac{\alpha^2\sigma}{k_{el} + k_{ph}} \tag{9.117}$$

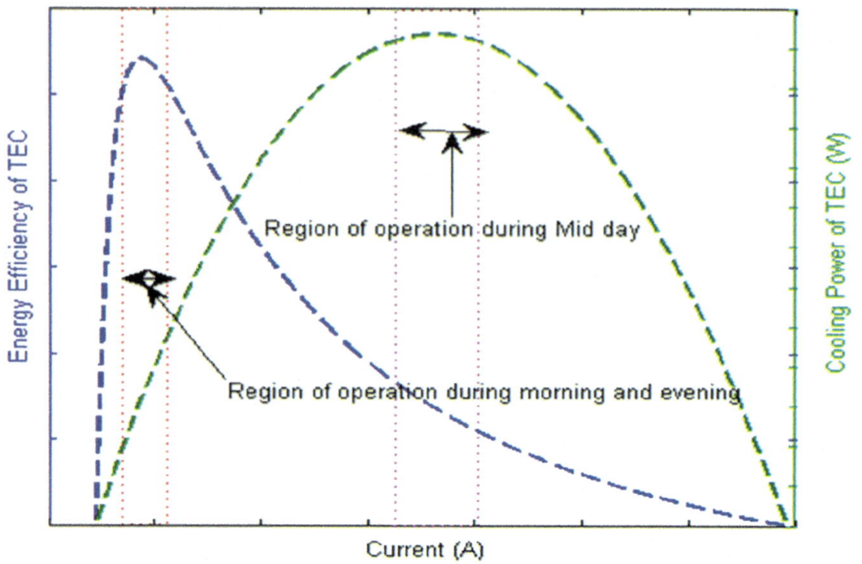

Fig. 9.14 Energy efficiency and cooling power of solar PV thermoelectric cooler

where the numerator is called the power factor of the thermoelectric material. Therefore, the objective of material researchers/scientists is either to make the power factor higher and/or reduce the thermal conductivity of thermoelectric material. The Seebeck coefficient of one energy level *p*-type thermoelectric material can be given as

$$\alpha = \frac{E - \mu}{qT} \tag{9.118}$$

where E is the energy levels, μ the chemical potential or Fermi energy level and q the electron charge. From Eq. (9.118), to get higher Seebeck coefficient, band gap energy ($E - \mu$) have to be higher, but to have high electrical conductivity ($E - \mu$) have to be minimum, e.g., metals. For metals, the bandgap energy is very low or nearer to zero, so the electrical and thermal conductivity is very high. But due to this, the metals have very low Seebeck coefficient and poor power factor. In the case of insulators, the bandgap is higher, but the electrical conductivity is lower, so the power factor is poor. In the case of semiconductors/semimetals, the bandgap is optimum (between metals and insulators) and this is the reason for semiconductors/semimetals to have relatively high FOM as compared to metals and insulators. The FOM of various thermoelectric materials has been presented in Fig. 9.15. Skutterudites are the naturally present thermoelectric materials and were first found in Skutterud, Norway (thus its name). The lattice thermal conductivity of these naturally occurring materials ($CoSb_3$) is higher at the room temperature, therefore, some chemical treatments will be required. The FOM of some of the Skutterudites is presented in Table 9.3.

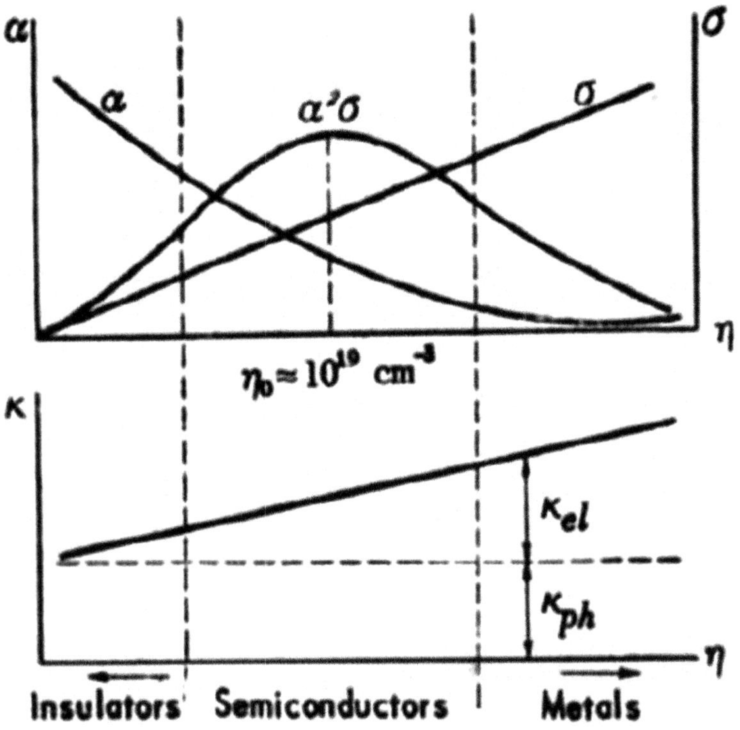

Fig. 9.15 Power factor and thermal conductivity of materials (Ioffe et al. 1959)

Clathrates are compounds in which molecules of one component are physically trapped within the crystal structure of another. The FOM of some of the Clathrate and alkali thermoelectric materials are given in Table 9.4. With the advancement in nanomaterials, the electronic structure of a material can be altered. The lattice thermal conductivity can be reduced if the length of the thermoelectric material is made lower than the average length of collision. This makes the ballistic transport of electrons, thereby contributing thermal conductivity only through the electrons. The FOM of some of the nano-thermoelectric materials is given in Table 9.5.

The maximum coefficient of performance of the thermoelectric cooler with present and futuristic values of the figure of merit has been compared with COP of vapour compression refrigeration (VCR) and vapour absorption refrigeration (VAB) systems. The present value of FOM for bulk bismuth telluride thermoelectric material is ~1. With this, the maximum COP of the thermoelectric cooler is comparable with VCR systems, but, keeping in mind that the cooling power of thermoelectric cooler is minimum at its maximum COP and cooling power of thermoelectric cooler is very much less when compared to the VCR systems. The VAB systems have COP in the order of 1, but it is only used for high cooling loads (Fig. 9.16).

Table 9.3 FOM of skeutterudite thermoelectric materials

Material	ZT	Temperature range (K)	References
$COSb_3$	0.3–1.4	300–1000	Sales et al. (1997)
$Yb_{0.19}CO_4Sb_{12}$	~1	600	Nolas et al. (2000)
$Ba_{0.30}Ni_{0.05}CO_{3.95}Sb_{12}$	~1.25	900	Tang et al. (2005)
$Ca_{0.18}Ni_{0.03}CO_{3.97}Sb_{12.4}$	~1	800	Puyet et al. (2005)
$Ba_xCO_4Sb_{12}$	~0.8	800	Dyck et al. (2002)
$Ca_xCO_4Sb_{12}$	~0.45	800	Puyet et al. (2004)
Bi_2Te_3-based inks	~1	600	Kim et al. (2018)

Table 9.4 FOM of Clathrate and alkali thermoelectric materials

Material	ZT	Temperature range (K)	References
$Sr_8Ga_{16}Ge_{30}$ and $Ba_8In_{16}Sn_{30}$	0.5–1.7	300–800	Nolas et al. (1998)
$Ba_{24}Ga_xGe_{100\text{-}x}$ ($x = 15$)	~1.25	943	Kim et al. (2006)
Bi_2Te_3	~1.35	300	Zhao et al. (2005)
PbTe	~1.5	773	Dughaish (2002)

Table 9.5 FOM of nano-thermoelectric materials

Material	ZT	Temperature range (K)	References
$Ag_{1-x}Pb_{m+y}Sb_{1+z}Te_{2+m}$ LAST-m with $m = 18$–22 (LAST—lead antimony silver telluride)	~1.7	700	Hsu et al. (2004)
$AgPb_mSn_nSbTe_{2m+n}$	~1.4	700	Hogan et al. (2007)
$NaPb_mSbTe_{2+m}$ SALT-m with $m = 20$ (SALT—sodium antimony lead telluride)	~1.6	675	Hoang et al. (2005)
$(Pb_{0.95}Sn_{0.05}Te)_{1-x}$ with $x =$ 0.08	~1.5	650	Androulakis et al. (2007)
PbTe/EuTe superlattice	~0.66	300	Dughaish (2002)
Silicon nano wires	~1	300	Hochbaum et al. (2008)

The advantage of the thermoelectric cooler is that it can be used for very low cooling powers with higher COP, where the VCR and VAR systems will not accommodate and serve the purpose, e.g., chip cooling, temperature control of electronics etc. It should also be noted that the COP of the thermoelectric cooler can be improved by the improved thermoelectric materials with higher FOM. If made possible, then there is no doubt that thermoelectric cooler will also become a major player in the HVAC industry.

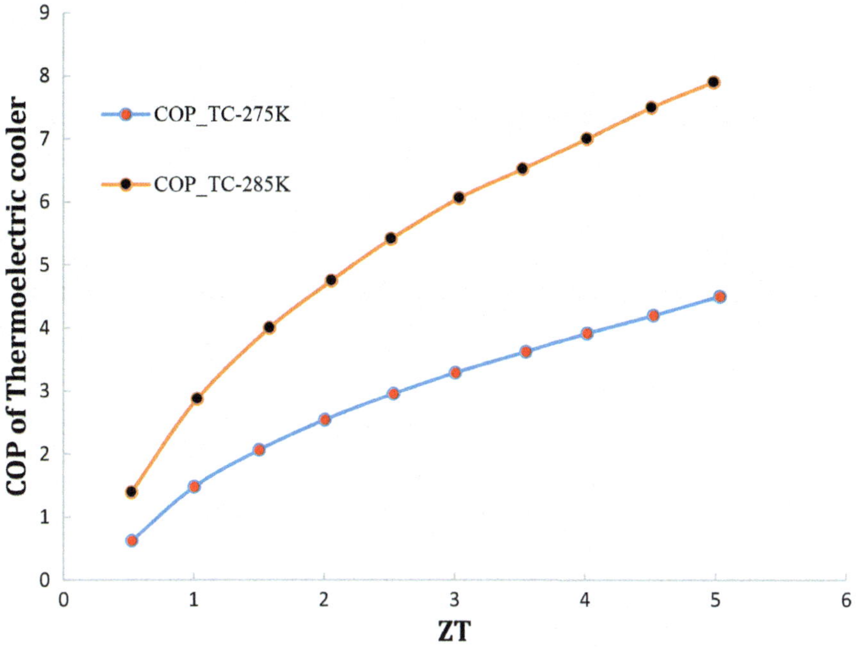

Fig. 9.16 Comparison of COP of TEC, VCR and VAR system

9.5 Advantages of Thermo-electric Cooler

(i) Compact in size.
(ii) The target temperature obtainable is too precise.
(iii) Thermoelectric coolers require small size heat exchanger systems.
(iv) Good choice for the electronic cooling systems.

9.6 Current Status and Application in Building

There are two types of thermoelectric coolers available, viz. domestic and portable. The major difference between these two is the availability of electrical power. The thermodynamic efficiency of the thermoelectric coolers is about 1% as compared to 14% of Stirling and VCR systems. Portable thermoelectric coolers have promising outdoor use, either using battery or powering by solar PV panels. According to recent studies, the COP of both portable and domestic thermoelectric refrigerators is usually less than 0.5 when operating at an inside/outside temperature difference of 20–25 °C.

Due to low COP nowadays, the applications of thermoelectric cooling has been restricted to in space missions, scientific and medical equipment, where COP is not as important as energy availability, reliability and quiet operating environment.

The small size and simplicity make TE the best choice to be applied for cooling application. One of the main applications that can be considered is the cooling of connected electronics, e.g., in telecommunication cabinets.

The current application of thermoelectric cooling can be categorised as:

(1) Cooling of small appliances: TE units are used in civil markets for the cooling of small enclosures such as domestic and portable refrigerator, portable ice box, cool beverages, cooler for picnic basket etc.
(2) Medical applications.
(3) For the heat dissipation in the industrial temperature control and electronics cooling.
(4) In automobile applications like mini-refrigerators, and automobile air conditioning applications.
(5) For generating freshwater, PV-TE hybrid integrated systems and thermal management in high power LEDs.
(6) Some researchers are working in the direction to make the thermoelectric domestic air conditioning systems with the hope that they may be competing with VCR systems in future.

For thermoelectric cooling applications, following general summaries can be drawn from literature:

(i) Thermoelectric coolers are used in the medical field, scientific applications, automobile heating and cooling applications etc.
(ii) Thermoelectric coolers are suitable in domestic and industrial applications.
(iii) Reported values of cooling COP of thermoelectric refrigerators are generally in the range 0.16–0.64.

The commercially available thermoelectric coolers are given in Table 9.6.

Comparison between the thermoelectric and vapour compression cooling system is depicted in the Table 9.7.

Table 9.6 Summary of a commercial type of thermoelectric refrigerators (Dyck et al. 2002; Zhao et al. 2017; Vining 2009)

Type of system	Company name	Maximum heat (Q) (W)	ΔT_{max} (°C)
CP1.4-127-045	Melcor Corp.	72b	75b
CP1.4-127-045L	Melcor Corp.	72	65
CP1.4-127-06L	Melcor Corp.	51.4	67
CP1.4-7-06L	Melcor Corp.	2.8	68
CP2-127-06	Melcor Corp.	120	68
CP2-127-06L	Melcor Corp.	120	67
HT8-12-40	Melcor Corp.	72	67
PT4-12-40	Melcor Corp.	32	67
TEC1-12706	HB Corp.	50	66

(continued)

Table 9.6 (continued)

Type of system	Company name	Maximum heat (Q) (W)	Δ T_{max} (°C)
TEC1-12708	HB Corp.	71	66
TEC1-6308	HB Corp.	37.4	67
6L	Marlow Ind.	50	66
DT12-4	Marlow Ind.	36	66
UT8-12-40-RTV	Ultra TEC Series	72	–
TB-127-1.4-1.2	Kryotherm Corp.	75	70
TB-127-1.4-1.5	Kryotherm Corp.	60	70
TB-199-2.0-0.9	Kryotherm Corp.	310	69
TECA 980-127	TECA Corp.	83.2	72
9500/127/085B	Ferrotec Corp.	80	72
HM3930	Acetec Co.	16.7	69
CP10-127-05	Laird Tech.	34.3	67

Table 9.7 Comparison between the thermoelectric and vapour compression cooling system

Chiller component	Refrigerant-based chiller	Thermoelectric-based chiller
Cooling engine	Refrigerant compressor	Thermo-electric device
Heating device	Heater	Thermo-electric device
Power	Electric motor	Variable output DC supply
Control method	On/off	Infinitely variable
Moving parts	Compressor fan valves	Fan pump
Working fluid	Refrigerant gas	

References

Androulakis, J., Lin, C. H., Kong, H. J., Uher, C., Wu, C. I., Hogan, T., Cook, B. A., Caillat, T., Paraskevopoulos, K. M. & Kanatzidis, M. G. (2007). Spinodal decomposition and nucleation and growth as a means to bulk nanostructured thermoelectrics: Enhanced performance in Pb_{1-x} Sn_x Te−PbS. *Journal of the American Chemical Society, 129*(31), 9780–9788.

Dughaish, Z. H. (2002). Lead telluride as a thermoelectric material for thermoelectric power generation. *Physica B: Condensed Matter*, *322*(1–2), 205–223.

Dyck, J. S., Chen, W., Uher, C., Chen, L., Tang, X. & Hirai, T. (2002). Thermoelectric properties of the *n*-type filled skutterudite $Ba_{0.3}$ Co_4 Sb_{12} doped with Ni. *Journal of Applied Physics, 91*(6), 3698–3705.

Goldsmid, H. J. (1964). Applications of Thermoelectric Refrigeration. *Thermoelectric Refrigeration*, 210–223. https://doi.org/10.1007/978-1-4899-5723-8_9

Hoang, K., Desai, K. & Mahanti, S. D. (2005). Charge ordering and self-assembled nanostructures in a fcc Coulomb lattice gas. *Physical Review B, 72*(6), 064102.

Hochbaum, A. I., Chen, R., Delgado, R. D., Liang, W., Garnett, E. C., Najarian, M., Majumdar, A. & Yang, P. (2008). Enhanced thermoelectric performance of rough silicon nanowires, *Nature, 451*, 163–167.

Hogan, T. P., Downey, A., Short, J., D'Angelo, J., Wu, C. I., Quarez, E., Androulakis, J., Poudeu, P. F., Sootsman, J. R., Chung, D. Y. & Kanatzidis, M. G. (2007). Nanostructured thermoelectric materials and high-efficiency power-generation modules. *Journal of Electronic Materials, 36*, 704–710.

Hsu, K. F., Loo, S., Guo, F., Chen, W., Dyck, J. S., Uher, C., ... & Kanatzidis, M. G. (2004). Cubic $AgPb_m$ $SbTe_{2+m}$: Bulk thermoelectric materials with high figure of merit. *Science, 303*(5659), 818–821.

Ioffe, A. F., Stil'Bans, L. S., Iordanishvili, E. K., Stavitskaya, T. S., Gelbtuch, A. & Vineyard, G. (1959). Semiconductor thermoelements and thermoelectric cooling. *Physics Today, 12*(5), 42.

Kim, J. H., Okamoto, N. L., Kishida, K., Tanaka, K. & Inui, H. (2006). High thermoelectric performance of type-III clathrate compounds of the Ba–Ge–Ga system. *Acta materialia, 54*(8), 2057–2062.

Kim, F., Kwon, B., Eom, Y., Lee, J. E., Park, S., Jo, S., ... & Son, J. S. (2018). 3D printing of shape-conformable thermoelectric materials using all-inorganic Bi2Te3-based inks. *Nature Energy, 3*(4), 301–309.

Nolas, G., Cohn, J. L., Slack, G. A. & Schujman, S. B. (1998). Semiconducting Ge clathrates: Promising candidates for thermoelectric applications. *Applied Physics Letters, 73*(2), 178–180.

Nolas, G. S., Kaeser, M., Littleton IV, R. T. & Tritt, T. M. (2000). High figure of merit in partially filled ytterbium skutterudite materials. *Applied Physics Letters, 77*(12), 1855–1857.

Puyet, M., Lenoir, B., Dauscher, A., Dehmas, M., Stiewe, C. & Müller, E. (2004). High temperature transport properties of partially filled Ca_x Co_4 Sb_{12} skutterudites. *Journal of Applied Physics, 95*(9), 4852–4855.

Puyet, M., Dauscher, A., Lenoir, B., Dehmas, M., Stiewe, C., Müller, E. & Hejtmanek, J. (2005). Beneficial effect of Ni substitution on the thermoelectric properties in partially filled Ca_y Co_{4-x} Ni_x Sb_{12} skutterudites. *Journal of Applied Physics, 97*(8), 083712.

Sales, B. C., Mandrus, D., Chakoumakos, B. C., Keppens, V. & Thompson, J. R. (1997). Filled skutterudite antimonides: Electron crystals and phonon glasses. *Physical Review B, 56*(23), 15081.

Tang, X., Zhang, Q., Chen, L., Goto, T. & Hirai, T. (2005). Synthesis and thermoelectric properties of *p*-type-and *n*-type-filled skutterudite R_y M_x Co_{4-x} Sb_{12} (R: Ce, Ba, Y; M: Fe, Ni). *Journal of Applied Physics, 97*(9), 093712.

Vining, C. B. (2009). An inconvenient truth about thermoelectrics. *Nature materials, 8*(2), 83–85

Zhao, X. B., Ji, X. H., Zhang, Y. H., Zhu, T. J., Tu, J. P. & Zhang, X. B. (2005). Bismuth telluride nanotubes and the effects on the thermoelectric properties of nanotube-containing nanocomposites. *Applied Physics Letters, 86*(6), 062111.

Zhao, W., Liu, Z., Wei, P., Zhang, Q., Zhu, W., Su, X., ... & Pei, Y. (2017). Magnetoelectric interaction and transport behaviours in magnetic nanocomposite thermoelectric materials. *Nature Nanotechnology, 12*(1), 55–60.

Chapter 10
Alternative Cooling System Options

Conventional cooling systems such as vapour compression, vapour absorption, vapour adsorption, desiccant cooling etc. have been described in previous chapters. In this chapter, some of the alternative solar cooling options are introduced and discussed.

10.1 Gas Compression Cycle Cooling System

Conventionally, air has also been used as a working fluid for producing refrigeration thereby, absorbing heat from the low-temperature source and discharging the same to high-temperature sink. Since there is no phase change associated during heat addition and heat rejection process, the heat-carrying capacity per unit mass is small as compared to that of the vapour compression cooling system. In an air refrigeration system, the working fluid enters at lower temperature and absorbs heat from the sink. Its temperature increases above the atmospheric temperature by compressing it in a compressor and it is further cooled in the heat exchanger up to the atmospheric temperature. Thereafter, it undergoes expansion in the expander lowering its temperature and enters into the refrigerator for absorbing heat and then the cycle repeats. The major advantages of these systems are lower weight per tonne of refrigeration, ease of availability and low cost of the fluid. However, the major disadvantages are low COP, the possibility of freezing of the moisture in the air and larger mass of the air required to be circulated in the system.

A non-conventional heat-driven gas refrigerator was first patented by Vuilleumier in 1918 and designed in 1951, which produces cooling at low temperature with power being supplied by a heat source at a high temperature and heat being rejected at an intermediate temperature. It is also a combined compressor expander machine driven by the heat to produce cooling and can utilize solar energy to produce refrigeration at not too low temperature. The steady-state operation of the idealized Vuilleumier

S. C. Kaushik et al., *Solar Cooling*, https://doi.org/10.1007/978-3-031-42410-6_10

cooling cycle was reported by Rule and Quale (1969) and predicted that cycle COP can exceed unity if a high-temperature heat source is used.

Figure 10.1 shows the gas cycle solar refrigerator (Chen 1978a, b) consisting of two cylinders, connecting space and two displacers which are driven by a small electric motor and can reciprocate in the cylinder with a relative phase difference angle. Both the displacer have a respective regenerator inside. A solar collector is attached to the hot regenerator side. The hot fluid from solar collector heats the top part of regenerator (T_H.). The displacer of hot regenerator can slide up and down. As it move down, the average gas temperature will be high as more gas occupies in the hot area. The rise in temperature leads to increase in pressure in the area. When the hot displacer moves up, very little gas is in the hot area, the average gas temperature will be low and so is the pressure. Thus, when the hot displacer moves up and down in the phase relationship with the motion of the cold displacer, the same pressure and volume variations are produced in the space of the cold cylinder C. This is similar to Stirling machine and, therefore, cooling effect is produced to provide useful refrigeration. Thermodynamically, the gas refrigeration cycle cooling system can be considered consisting of two Stirling engines with their cylinder pistons placed at an angle of 90° to each other. The gas refrigerator thus has a thermocompressor instead of a mechanical compressor and the ideal COP is given by Chen (1978a, b).

$$\text{COP} = \left(\frac{T_H - T_O}{T_H}\right) \cdot \left(\frac{T_C}{T_O - T_C}\right) \tag{10.1}$$

Equation (10.1) is identical to COP of an absorption refrigeration system.

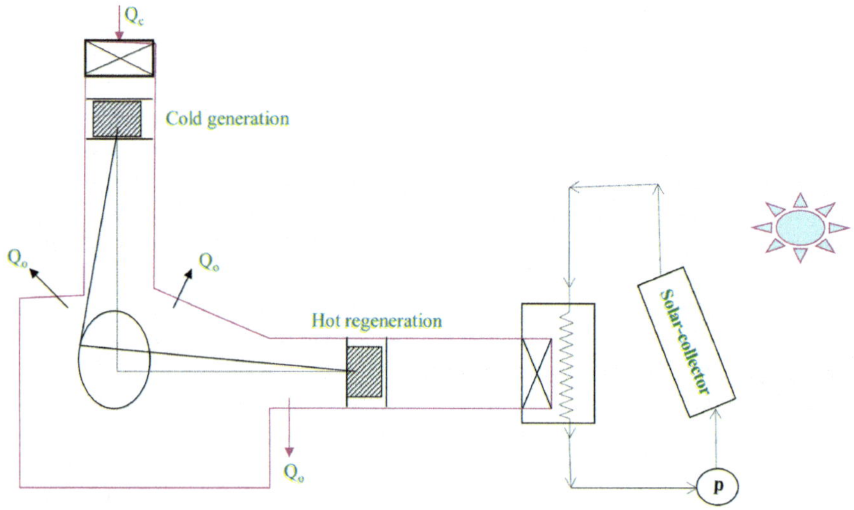

Fig. 10.1 Schematic of gas cycle refrigeration system

10.1.1 Thermodynamic Analysis

The gas cycle cooling unit is a combination of two Stirling machines, with one engine working between temperatures T_H and T_A, and the other a refrigerator working between temperatures T_C and T_O. Assumptions made in the thermodynamic modeling to assess the ideal performance are given as follows (Chen 1978a, b):

1. Refrigerant gas is considered as an ideal gas. Hence, it follows the ideal gas equation.
2. Gas temperatures such as T_H, T_A and T_C are maintained at constant values.
3. There are no thermodynamic losses in the regenerators.
4. Since the machine has no valves, solid and fluid friction is neglected. All volumes are nominal at the same pressure at any instant.
5. No heat conduction across the cylinder wall surface.

The variable cylinder volume is the sum of constant dead volumes and sinusoidal varying stroke volumes. The displacer in the hot volume leading the one in the cold volume by an angle ϕ, the instantaneous volume and temperature can be determined as follows.

The following abbreviations are used in the derivation.

V, volume; *T*, temperature; *P*, pressure; *H*, hot cylinder; *C*, cold cylinder; V_{tot}, total volume; *HR*, hot regenerator; *CR*, cold regenerator; *p*, connecting space; *DP*, dead volume in connecting space; *DH*, dead volume in the hot cylinder; *DC*, dead volume in the cold cylinder.

The total internal volume is given by

$$V_{\text{tot}} = V_C + V_H + V_A + V_{CR} + V_{HR} \tag{10.2}$$

Consider hot cylinder

$$V = V_H = V_{tot}\left[W_{DH} + \frac{W_H}{2}(1 - \cos\omega t + \phi)\right] \tag{10.3}$$

$$T = T_H$$

For cold cylinder, the corresponding equations are given by

$$V = V_{tot}\left[W_{DC} + \frac{W_C}{2}(1 - \cos\omega t)\right] = V_c \tag{10.4}$$

$$T = T_c$$

For connecting space

$$V = V_A = V_{tot}\left[W_{DP} + \frac{W_C}{2}(1 + \cos\omega t) + \frac{W_H}{2}(1 - \cos\omega t + \phi)\right] \quad (10.5)$$

Temperature, $T = T_A$.
For cold regenerator,

$$V = V_{CR} = V_{tot} W_{CR} \quad (10.6)$$

The temperature inside cold regenerator is approximated by the average value of T_A and T_C

$$T = T_{CR} = \left(\frac{T_A + T_C}{2}\right) \quad (10.7)$$

Similarly, for the hot regenerator

$$T = T_{HR} = \left(\frac{T_A + T_H}{2}\right) \quad (10.8)$$

Volume,

$$V = V_{HR} = V_{\text{tot}} W_{HR} \quad (10.9)$$

Next step is to find the heat flow into each cylinder and regenerator. Applying first law of thermodynamics to these isothermal open systems

$$Q_{HR} = \oint P dV_{HR}$$
$$Q_{CR} = \oint P dV_{CR}$$
$$Q_A = \oint P dV_A$$
$$Q_C = \oint P dV_C$$
$$Q_H = \oint P dV_H \quad (10.10)$$

Since

$$dV_{CR} \text{ and } dV_{HR} = 0$$

$$Q_{HR} = \oint P dV_{HR} = 0 \text{ and } Q_{CR} = \oint P dV_{CR} \quad (10.11)$$

Since it is assumed that the total mass of the working fluid is a constant and it is equal to sum of masses in five control volumes

$$M = M_C + M_H + M_{CR} + M_{HR} + M_A \tag{10.12}$$

According to ideal gas equation

$$PV = mRT$$

$$m = \frac{PV}{RT} \tag{10.13}$$

Applying Eq. (10.13) to five control volumes, Eq. (10.12) becomes

$$M = \frac{P}{R}\left[\frac{V_C}{T_C} + \frac{V_H}{T_H} + \frac{V_{CR}}{T_{CR}} + \frac{V_{HR}}{T_{HR}} + \frac{V_A}{T_A}\right] \tag{10.14}$$

$$= \frac{P}{RT_A}\left[\frac{V_C T_A}{T_C} + \frac{V_H T_A}{T_H} + \frac{V_{CR} T_A}{T_{CR}} + \frac{V_{HR} T_A}{T_{HR}} + V_A\right] \tag{10.15}$$

Applying the condition at $\omega t = 0$, $P = P_0$, $V_C = V_{C0}$, $V_H = V_{H0}$

$$M = \frac{P_0}{RT_A}\left[\frac{V_{C0} T_A}{T_C} + \frac{V_{H0} T_A}{T_H} + \frac{V_{CR} T_{CR}}{T_{CR}} + \frac{V_{HR} T_{HR}}{T_{HR}} + V_{A0}\right] \tag{10.16}$$

Now define work

$$W_{DPN} = \left[W_{DP} + \frac{W_{DC} T_A}{T_C} + \frac{W_{DH} T_A}{T_H} + \frac{W_{CR} T_{CR}}{T_{CR}} + \frac{W_{HR} T_{HR}}{T_{HR}}\right] \tag{10.17}$$

Applying the condition $\omega t = 0$, evaluate Eqs. (10.2)–(10.5)

$$K_C = 1 + \frac{T_A}{T_C},\ K_H = 1 + \frac{T_A}{T_H}$$

$$B_C = 1 - \frac{T_A}{T_C},\ B_H = 1 - \frac{T_A}{T_H} \tag{10.18}$$

Equation (10.16) becomes

$$P_0 = \frac{MRT_A}{V_{tot}[W_{DPN} + W_C + W_H(K_H + B_H \cos\phi)/2}$$

and

$$P = \frac{MRT_A}{V_{tot}[W_{DPN} + \{W_C K_C + W_H K_H + W_C B_C \cos \omega t + W_H B_H \cos(\omega t + \phi)\}/2]}$$

$$P = P_0 \frac{W_{DPN} + W_C + W_H(K_H + B_H \cos \phi)/2}{[W_{DPN} + \{W_C K_C + W_H K_H + W_C B_C \cos \omega t + W_H B_H \cos(\omega t + \phi)\}/2]} \tag{10.19}$$

Let

$$A = \frac{W_{DPN} + (W_C K_C + W_H K_H)/2}{[W_{DPN} + W_C + W_H(K_H + B_H \cos \phi)/2]}$$

$$B = \frac{W_C^2 G_c^2 + 2W_C W_H B_C B_H \cos \phi + W_H^2 B_H^2)^{1/2}}{2[W_{DPN} + W_C] + W_H(K_H + B_H \cos \phi)]}$$

$$\cos \theta = -\frac{W_C B_C + W_H B_H \cos \phi}{\left(W_C^2 G_c^2 + 2W_C W_H B_C B_H \cos \phi + W_H^2 B_H^2\right)^{1/2}\Big]}$$

$$\sin \theta = -\frac{W_H B_H \sin \phi}{\left(W_C^2 G_c^2 + 2W_C W_H B_C B_H \cos \phi + W_H^2 B_H^2\right)^{1/2}\Big]} \tag{10.20}$$

The pressure can be now expressed as

$$P = \frac{P_0}{A - B\cos(\omega t + \theta)} \tag{10.21}$$

The refrigeration effect produced and the heat input in the hot cylinder can be determined from the above equations.

$$\frac{Q_C}{Q_H} \tag{10.22}$$

$$\begin{aligned} Q_C &= \oint P dV_C \\ &= \int_0^{2\pi} \left[\frac{P_0}{A - B\cos(\omega t + \theta)}\right] \times \left[\frac{V_{tot} W_C \sin \omega t}{2}\right] d\omega t \\ &= P_0 V_{tot} W_C \pi \sin \theta \left[1 - \frac{A}{\left(A^2 - B^2\right)^{1/2}}\right] \frac{1}{B} \end{aligned} \tag{10.23}$$

Using the mathematical relationship

$$\cos \theta \sin \phi - \sin \theta \cos \phi = \frac{W_C B_C}{W_H B_H} \sin \theta \tag{10.24}$$

Heat transfer rate in the hot cylinder is given by

$$Q_H/Q_C = \frac{T_H}{T_C}\left(\frac{T_A - T_C}{T_H - T_A}\right)$$

$$Q_H = Q_C\frac{T_H}{T_C}\left(\frac{T_A - T_C}{T_H - T_A}\right) \tag{10.25}$$

$$COP = \frac{Q_C}{Q_H} \tag{10.26}$$

$$\text{COP} = \frac{Q_C}{Q_C\frac{T_H}{T_C}\left(\frac{T_A-T_C}{T_H-T_A}\right)}$$

$$\text{COP} = \frac{T_C(T_H - T_A)}{T_H(T_A - T_C)} \tag{10.27}$$

The cooling capacity for a given operating temperature is dependent on the dimension of the cylinder and displacer phase angle. Chen has obtained the optimum values of relative size and phase angle and studied the performance of the solar gas cycle refrigerator by taking into account the heat loss, regenerator effectiveness and the gas properties. It was concluded by Chen that the cooling system performance of the gas cycle system can be increased with higher collector fluid temperature and hence a concentrating solar collector is feasible for meeting the cooling requirements. This is an advantage over the conventional absorption refrigerator which is unable to take advantage of much higher temperature to increase overall COP as it is limited by thermodynamic constraints regardless of the possibly high temperature. Furthermore, there is no pressure difference across the displacer in the gas cycle refrigerator, the only mechanical/electrical power needed is the power required to overcome the friction and the motor load is very small. There is a good scope of research and development work to be undertaken on the gas cycle solar cooling system.

10.2 Jet Ejector Compression Cooling System

Initially, the principle of jet ejector compression (JEC) was used for water vapour compression cooling and is based on the principle of boiling the water at a lower temperature by maintaining low pressure at the water surface. This low pressure on the surface of the water was maintained by throttling the system through jet nozzles. In the process, high-pressure motive steam is supplied to the nozzle inlet and it is expanded. Water vapour from the evaporator chamber is entrained with the high-velocity steam jet and is further compressed in the thermocompressor. The KE of the working fluid is converted into static pressure and the mass is discharged to the condenser. The steam boiler retrieves back the condensate to complete one cycle of operation.

Jet ejector cooling is a thermally driven technology that has been used for cooling applications for many years due to their advantages of simplicity and no moving parts. A keen interest in the development of JEC is its ability to produce cooling when low heat source temperature is available.

The ejector system shown in Fig. 10.2 consists of two loops, namely, power and refrigeration. The system consists of components such as condenser, expansion device, evaporator, boiler, ejector and a mechanical pump. The boiler is driven by a low-grade energy source such as solar energy to evaporate the refrigerant (process 1-2). The high-pressure refrigerant then flows to the ejector part. The primary fluid coming out from the ejector will be at very high velocity due to the reduction in pressure while passing through the nozzle of the ejector. At the state point 3, pressure reduction causes generation of secondary fluid. Two fluids mix together in the mixing chamber and flow to the diffuser part, where the flow decelerates and pressure recovery occurs. This mixture is then subjected to condensation in the condenser by rejecting heat to the environment (Q_c). A portion of the condensate at point 5 is then pumped back to the boiler and the remainder of the liquid is expanded through an expansion device and enters the evaporator of the refrigeration loop at point 6. In the evaporator, the refrigerant evaporates producing a cooling effect (Q_e). The refrigerant vapour flows into the ejector at point 3, where it mixes with the primary fluid.

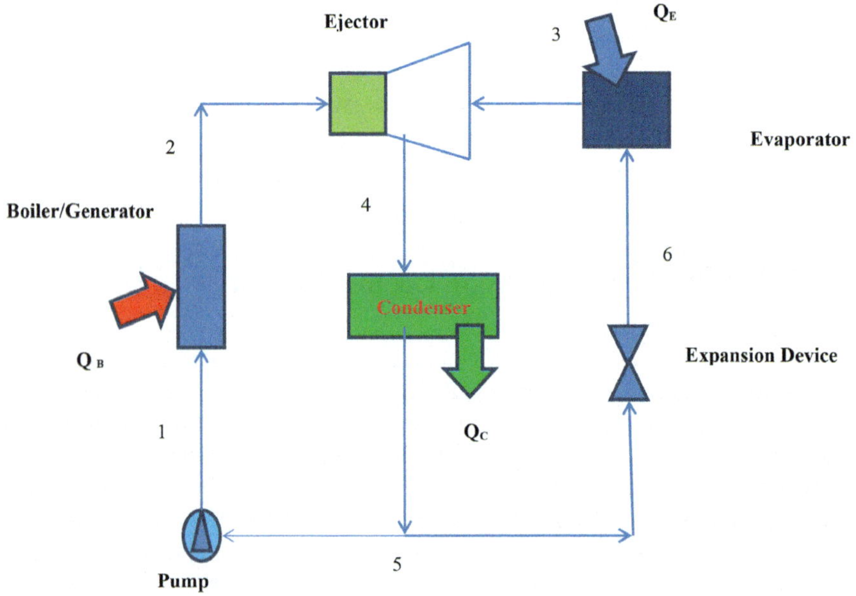

Fig. 10.2 Steam jet ejector cooling system

Now, the fluid is compressed in the diffuser part and sneak in the condenser at point 4. The mixed fluid condenses in the condenser and completes one cycle of operation.

10.2.1 Development of Jet Ejector Cooling System

The concept of a jet ejector cooling system device was introduced by Maurice Leblanc in 1910. The use of ejector system was popular for airconditioning applications until the development of chlorofluorocarbon refrigerants in the 1930s and their use in vapour compression cycle gained attention as they were more efficient as compared to the alternative thermally driven refrigeration systems. After the development of VCR systems, steam jet systems have been almost standstill as most of the research concentrated on the effort to improve the performance of VCR systems. Currently, various cooling systems based on the JEC have been refined with varying capacities from low up to 60,000 kW and more. The obstacles for the use of such systems in commercial aspects are their very low COP. The COP of such systems is reported to be less than 0.2, which is very low as compared to the other existing cooling systems. The JEC systems performance can be improved by integrating with conventional VCR or VAB systems. These systems are described in the Sect. 10.2.3. The most important applications and limitations of jet ejector cooling system are given below.

Applications

(a) Food processing factories: In this, ejector systems are recommended for process and product cooling.
(b) Tri-generation: Combination of heat, power and cooling systems. The ejector cooling systems produces the necessary cooling effect in the combination.
(c) Combined VCR, ejector and VAB systems to produce cooling: The better thermal integration of three units can yield a good coefficient of performance with the desired cooling effect. The combination of the cooling systems can be effective in the waste heat and other low grade energy heat sources applications.

Limitations

(a) Unavailability of sophisticated and readily available compact systems suitable for low, medium and high cooling load applications and the lack of its necessary data.
(b) Even though it can be easily powered by solar energy sources, it has very low COP (of the order of 0.2) as compared to vapour compression system.
(c) The COP also drops significantly at operation away from the design point.

10.2.2 Working Fluids

In general, following criteria should be taken for the selection of working fluids in JEC systems:

(1) Direct contact of refrigerant in food handling processes should be avoided.
(2) Chemically stable at operating temperature level. Non-toxic and non-flammable.
(3) It should have very high latent heat of vapourisation to obtain the large cooling effect.
(4) Fluid pressure should be within the available limits. High pressure leads to the need of big chambers and power requirement of the pump also increases.
(5) Condenser pressure and evaporator pressure should be within the acceptable range.
(6) Refrigerant should satisfy the ideal gas properties. The ideal refrigerants for ejector systems are: R-21, R-110, R-114, R-142b, R-152a, RC-318, R-717 and R-718.

10.2.3 Solar Driven Jet Ejector Compression Cooling System

A solar energy-driven jet ejector compression cooling system is shown in Fig. 10.3. Solar energy can be directly given to the boiler to produce high-pressure refrigerant vapour, it can entrain the low-pressure freon vapour leaving the evaporator. Thus, the compression is achieved within the jet pump by expanding the primary fluid which gets heat from the solar collector. The ejector pump is simply a convergent-divergent nozzle. The mechanical compressor in a Rankine cycle is replaced with an ejector in the jet ejector cooling systems. Basic components involve, a boiler, an ejector, a condenser and the liquid pump.

The main attractive feature of the JEC system is its ability to be driven by a low-temperature heat source such as solar energy. Solar JEC system can be fabricated in two ways as shown in Fig. 10.4. The system basically consists of a solar collector to harness solar energy and can be utilised to drive the system. The cooling unit consists of a condenser, an expansion device and an evaporator. According to the characteristics and investment cost, JEC systems are basically suited for airconditioning applications rather than refrigeration purposes. There are several designs of such solar JEC systems available, and two different configurations depending on the mode of operation are shown in Fig. 10.4a and b In the first mode (Fig. 10.4a) of its operation, the solar cycle acts as a boiler for the refrigeration system. In this case, the refrigerant is enforced through the absorber tube of collector. In this type, it is very difficult to control the flow rate of refrigerant. In addition to this, any leakage in the refrigerant will damage the system and affect its proper functioning. The design of pump according to the pressure and flow rate is also difficult. Hence, this configuration is not preferred.

In the next mode of operation (Fig. 10.4b) of solar jet ejector compression systems, a boiler is placed in-between the solar and refrigerator subsystem. Both systems are

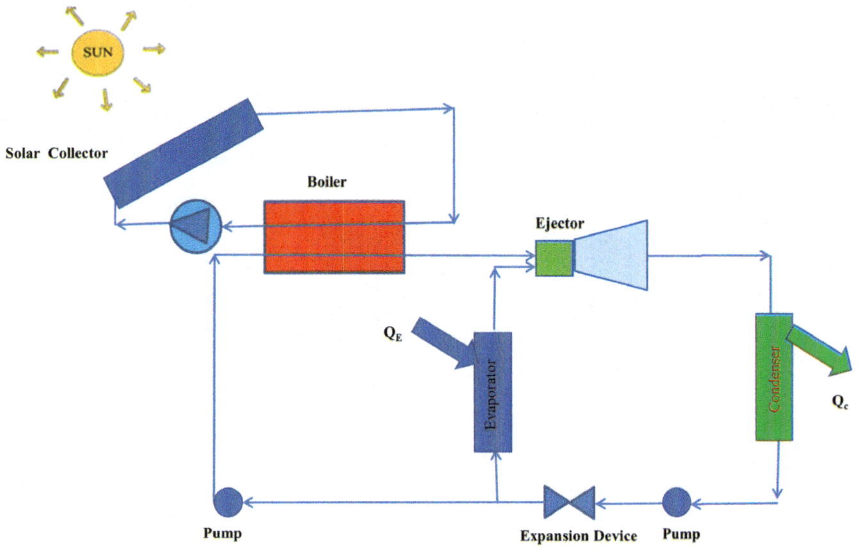

Fig. 10.3 Solar driven jet ejector compression cooling system

separated by a boiler heat exchanger, hence this mode of operation is called as separated mode. In this case, a heat transfer medium having high boiling point than the high temperature of the system is used to transfer heat from boiler heat exchanger to the refrigerant used.

The overall efficiency of the solar jet refrigeration cycle is expressed as

$$\mathrm{COP}_{overall} = \eta_{Solar} \times \mathrm{COP}_{jet\,refrigeration} \tag{10.28}$$

It is obvious that the overall efficiency depends on performance of the cooling system as well as the efficiency of solar collector integrated into the system.

10.2.4 Thermodynamic Analysis

Figure 10.5a and b shows the various thermodynamic processes within the ejector and the corresponding pressure variations along the length of the ejector. The refrigerant vapour is the high-pressure motive gas supplied from the high-pressure boiler (corresponding to the state point 1) and its total pressure is equal to the boiler pressure P_b. From there, the motive gas expands through the nozzle to a static pressure P_e (evaporator pressure). This low-pressure refrigerant vapour drawn in the ejector, then flows to the mixing chamber where it completely mixes till it reaches at the state point 3. The mixture of refrigerant vapour is then compressed in the diffuser section of the ejector pump and leaves the ejector part at the condenser pressure P_c. The

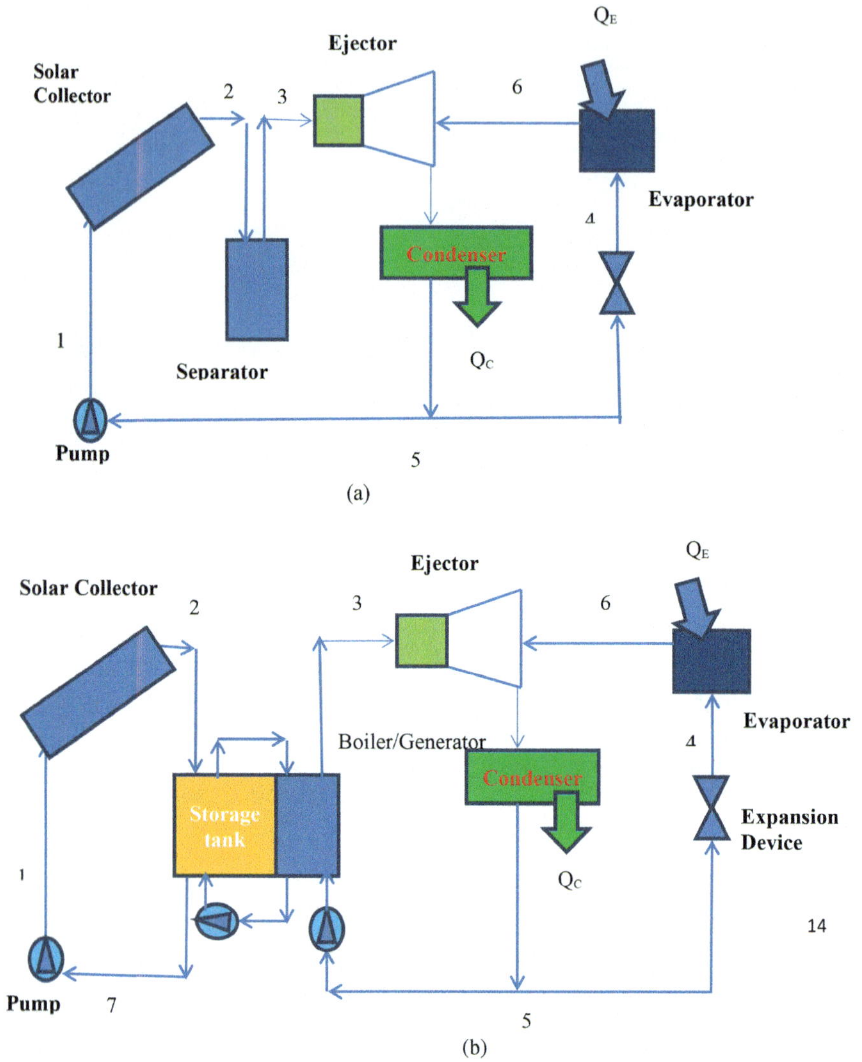

Fig. 10.4 A solar jet cooling system **a** combined and **b** separated

ejector efficiency is defined based on the entrainment ratio. The entrainment ratio is defined as the ratio of secondary and primary mass flow rate of the refrigerant vapour.

Efficiency of the ejector is thus defined by

$$\eta_e = \frac{(\text{Actual entrainment rate})}{(\text{Ideal entrainment rate})} \tag{10.29}$$

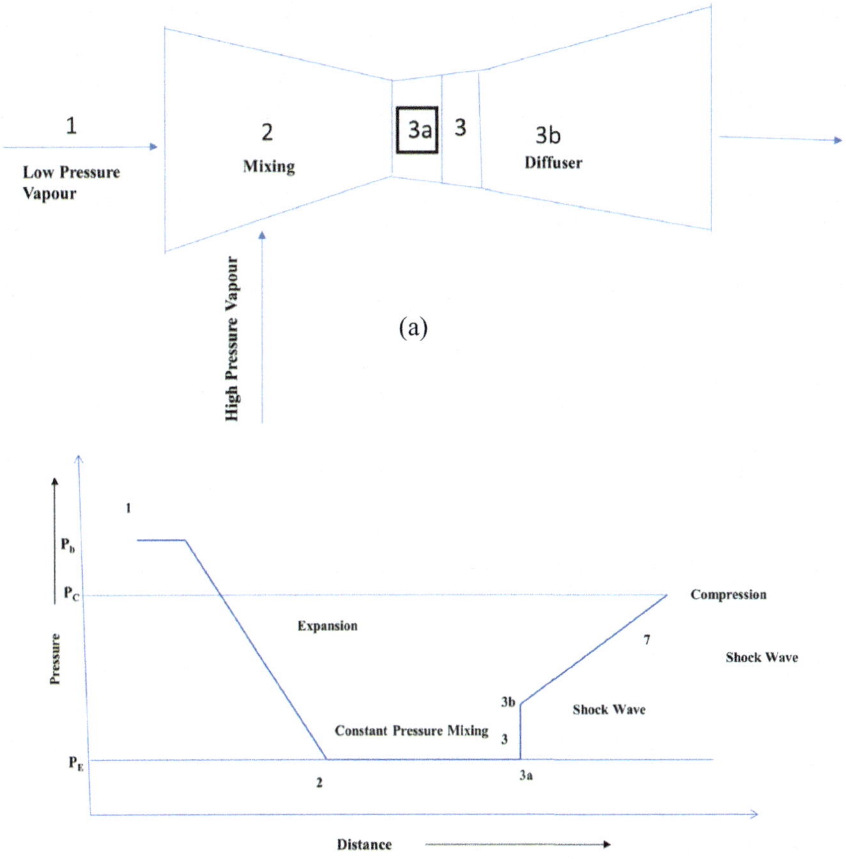

(b)

Fig. 10.5 **a** Thermodynamic process in an ejector system. **b** Variation of pressure along the length of ejector

The ideal entrainment ratio is defined for reversible expansion, compression and mixing process. If η_n is the efficiency of the nozzle, the jet velocity is given by

$$\sqrt{(2\eta_n(h_1 - h_2))} \tag{10.30}$$

The governing equations of mass, momentum and energy of each section in the ejector can also be used on the basis of the following assumptions:

(i) Adiabatic section (no heat transfer to the system).
(ii) Expansion of primary fluid is perfect, completely expanded from boiler pressure to the evaporator pressure in nozzle.
(iii) Pressure is constant between states 2 and 3 in the mixing section.

(iv) Wall friction is neglected.
(v) The potential and kinetic energy of the gas streams are neglected. The mass, momentum and energy balance equation for the mixing process is given by

$$m_1 + m_0 = m_2 = m_3 \tag{10.31a}$$

$$(m_1 + m_0)V_3a - m_1 V_2 - m_0 V_0 = 0 \tag{10.31b}$$

or

$$m_2 V_3a - m_1 V_2$$

Since V_0, the velocity of the suction vapour is negligible

$$m_1 h_1 + m_0 h_0 = m_3\left(h3a + \frac{V_3^2}{2}\right) \tag{10.31c}$$

For convenience, the momentum transfer leading to instantaneous pressure shock is not considered and then state point 3a is the same as state point 3b or say 3 only. The efficiency of entrainment η_e is considered through the mechanical energy balance, then we can define

$$\eta_e m_1 \frac{V_2^2}{2} = (m_1 + m_0)\frac{V_3^2}{2} = m_3\left(\frac{V_3^2}{2}\right) \tag{10.32}$$

The kinetic energy at state point 3 is converted into enthalpy in the diffuser section. Neglecting the velocity at the end of the diffuser, we have

$$(h_4 - h_3) = \frac{V_3^2}{2} = \frac{h_4' - h_3}{\eta d} = \frac{\Delta h_{1s}}{\eta d} \tag{10.33}$$

where Δh_{1s} is the isentropic enthalpy rise and $\eta_d = \frac{(h_4 - h_3)}{(h_4' - h_3)}$ is the diffuser efficiency.

Solving Eqs. (10.31a)–(10.33) for the mass ratio $\frac{m_1}{m_0}$, which is the mass ratio of the motive vapour to the suction vapour, we have

$$\frac{m_1}{m_0} = \frac{(h_4' - h_3)}{\left[(h_1 - h_2)\cdot \eta_n \cdot \eta_e \cdot \eta_d - (h_4' - h_3)\right]} \tag{10.34a}$$

and defining the overall ejector efficiency η as the product of $\eta_n \cdot \eta_e \cdot \eta_d = \eta$, we have

$$\frac{m_1}{m_0} = \frac{(h_4' - h_3)}{\left[\eta(h_1 - h_2) - (h_4' - h_3)\right]} \tag{10.34b}$$

For an ideal ejector, $\eta = 1$, $h_2 \sim h_3$, and $h'_4 - h_4$, we have

$$m_1 = m_0 \frac{(h_4 - h_3)}{(h_1 - h_4)} \tag{10.35}$$

Thus, we see that to keep the mass ratio low, the enthalpy h_1 and hence the temperature and pressure of the motive refrigerant vapour should be as high as possible. The COP of the non-ideal ejector cycle is defined as

$$\frac{m_0(\Delta h_r)}{m_1(\Delta h_b)} \tag{10.36}$$

where Δh_r is the refrigerating effect per unit mass flow of the fluid in the evaporator and Δh_b the heat energy supplied per unit mass flow of the fluid in the boiler. Thus, the COP and the required mass ratio of the motive vapour to the suction vapour can be evaluated as a function of boiler/evaporator and condenser temperature as well as the efficiency of the nozzle, diffuser and the ejector component. Another consideration for the design of the ejector system is the choice of the mixer chamber area for each operating condition. Further, design improvements for the ejector system includes regeneration and multi-staging to enhance system performance. In the regenerative ejector cycle, two heat exchangers (one on the flow loop between ejector and condenser and the other between the evaporator and the ejector) are added to preheat the subcooled liquid before going to the boiler by the heat contained in the ejector exhaust flow and to subcool the liquid flow before going to the expansion valve by the heat contained in the evaporator exhaust flow.

In case of multi-staging of ejector system with the same amount of cooling for each stage of an n-staged ejector refrigeration system the average COP is defined in terms of COP of each stage as

$$\mathrm{COP} = \frac{n}{\left[\frac{1}{\mathrm{COP}_1} + \frac{1}{\mathrm{COP}_1} + \cdots \frac{1}{\mathrm{COP}_n}\right]} = \frac{n}{\sum_{i=1}^{n}\left(\frac{1}{\mathrm{COP}_i}\right)} \tag{10.37}$$

Thus, an infinite stage ejector refrigeration system will theoretically reach the maximum COP. For a two-stage cooling ejector system, COP is improved by 7%. Hence, ejector compression cooling system is simple in construction, highly reliable which is not available with conventional Rankine cycle refrigeration systems. Such systems using freons have further extended their use for the production of sub-zero evaporator temperatures and utilization of low-grade heat thermal source like solar, geothermal etc.

10.2.5 Different Configurations of Solar Driven Jet Ejector Cooling System

The most important and difficult task in the design of a thermal system is to find a solution, which gives better efficiency for the less cost. The optimum system configuration is not easily available, it requires lots of simulations and design optimizations. Several researchers proposed new designs to improve the performance of solar-driven ejector cooling systems. Some of those are:

(a) Using multiejector cooling systems.
(b) Jet ejector compression system with a booster.
(c) Integration with other cooling systems such as VCR, VAB and VAD (hybrid) system.

Single-stage ejector cooling is simple in its construction and operation. The performance of solar cooling mainly depends on the availability of solar irradiation, cooling water temperature, air temperature and wind speed. The most important decisive factor in solar cooling system, particularly in the ejector system performance is the heat rejection temperature. In order to obtain the high performance, low heat rejection temperature is recommended (i.e., low sink temperature). It is difficult to keep the system running at optimum conditions of single stage JEC systems due to variation of working conditions. Like high operating temperature or low solar insolation leads to operation difficulties. One way to tackle this situation is to do the staging of ejectors. In the multiejector systems few ejectors are kept in parallel before the condenser as illustrated in Fig. 10.6. The overall entrainment ratio is the thick line. Condenser pressure decides the sequence of working of ejector. If the condenser pressure is above P_{c1}, ejector 1 operates. Similarly, when the condenser pressure is in between P_{c1} and P_{c2}, ejector 2 is in action and ejector 3 operates at a condenser pressure between P_{c2} and P_{c3}. Bejan proposed this arrangement in 1996.

The other possibility of improvement in the performance of the ejector system is to use booster or compressor. A booster is located in-between the condenser and evaporator for lifting the pressure of refrigerant coming out from the evaporator. Vapour refrigerant from the evaporator is first boosted to the intermediate pressure by using booster (process 9-8), the refrigerant from the intercooler then flows to the ejector. Figure 10.7 shows the schematic representation of an ejector system with a booster.

Research indicates that the use of booster and intercooler leads to improvement in the performance of the ejector cooling system (Fig. 10.8). Heating during winter, cooling in summer and hot water supply could be the load of the ejector power system. This machine is termed solar-powered compression enhanced ejector air conditioner. In this device, a boiler is placed between the solar and refrigeration subsystem. An intercooler is provided in-between the condenser and evaporator. The refrigerant coming out from the evaporator will be subjected to pressurisation in the booster. The vapour refrigerant then flows back to the ejector through intercooler. Liquid refrigerant coming out from the condenser is pumped back to the boiler,

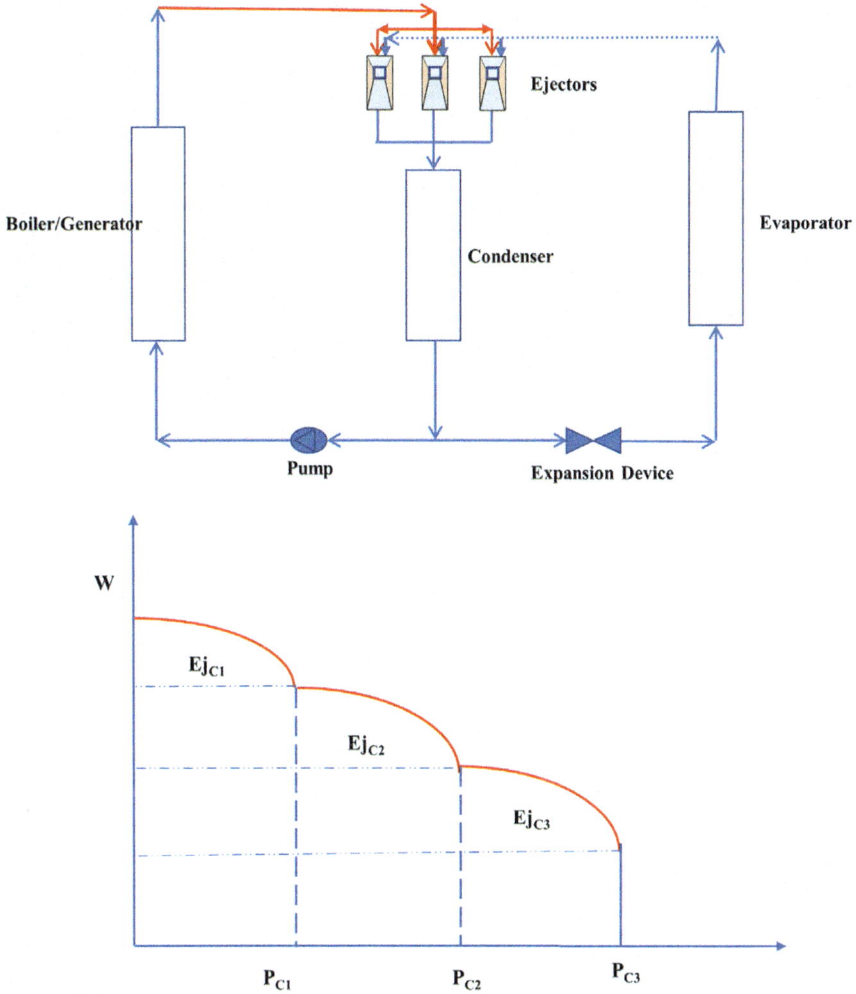

Fig. 10.6 Multiejector system

where it is heated by exchanging heat with the solar-heated fluid in the boiler heat exchanger. The vapour generated then flows back to the ejector unit. Energy in the form of mechanical or electricity is given to the booster. The system utilizes R114 as the working medium. For 4 °C evaporating temperature and 50 °C condensing temperature, the theoretical COP of the ejector system reported is 0.85 and COP of 0.5 for the overall cycle.

The other methods of improvement of the JEC systems performance is the integration with other commercial cooling systems, such as VCR, VAB and VAD systems.

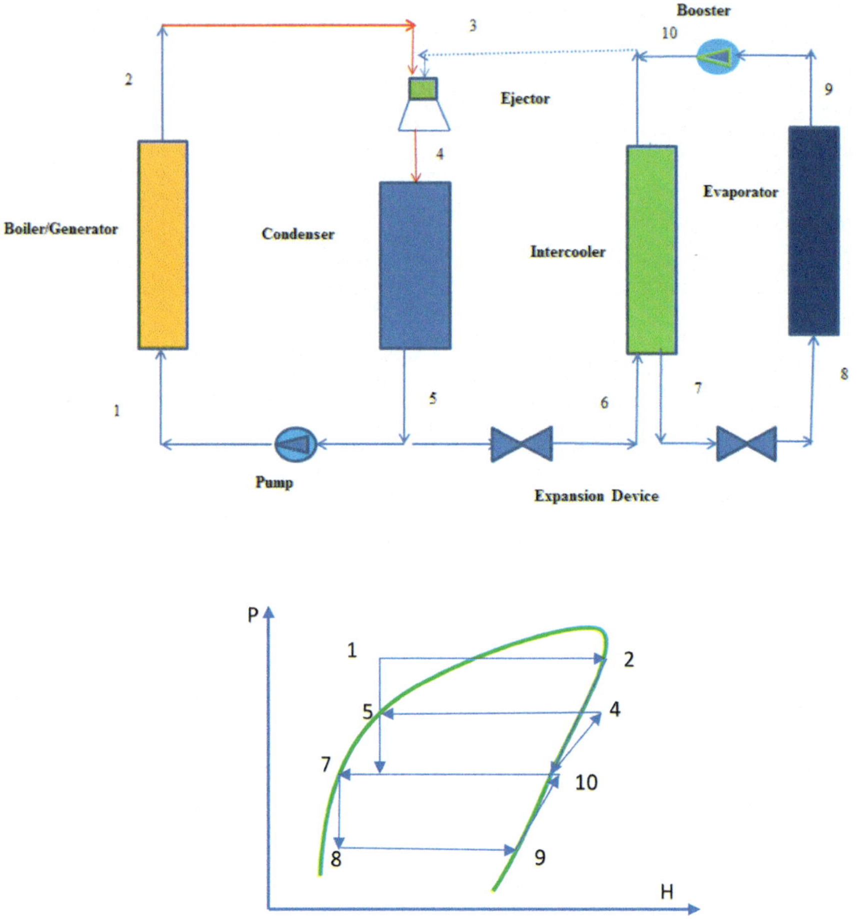

Fig. 10.7 A jet compression system with a booster

10.2.6 Jet Ejector-Compression Cooling System

A coupled steam jet ejector-vapour compression cooling system is illustrated in Fig. 10.9. Basically it has two parts. The first subsystem is a simple steam jet ejector system and second a conventional vapour compression cooling system. The system utilises two different working fluids. The ejector system utilizes water as the refrigerant and R104a is used in the VCR system. The condenser of VCR act as evaporator of the ejector system. The functions of remaining components in the system are similar to the conventional ejector cooling system and vapour compression cooling system (Wang et al. 2019).

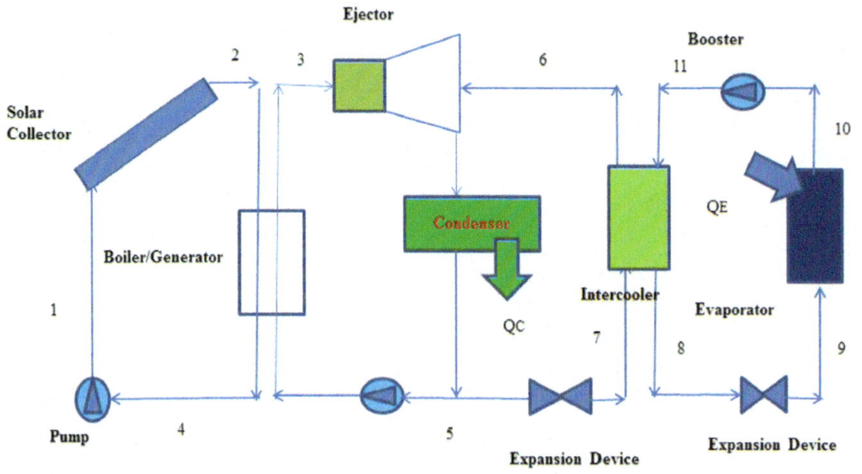

Fig. 10.8 Solar-powered jet ejector system with an intercooler and a booster

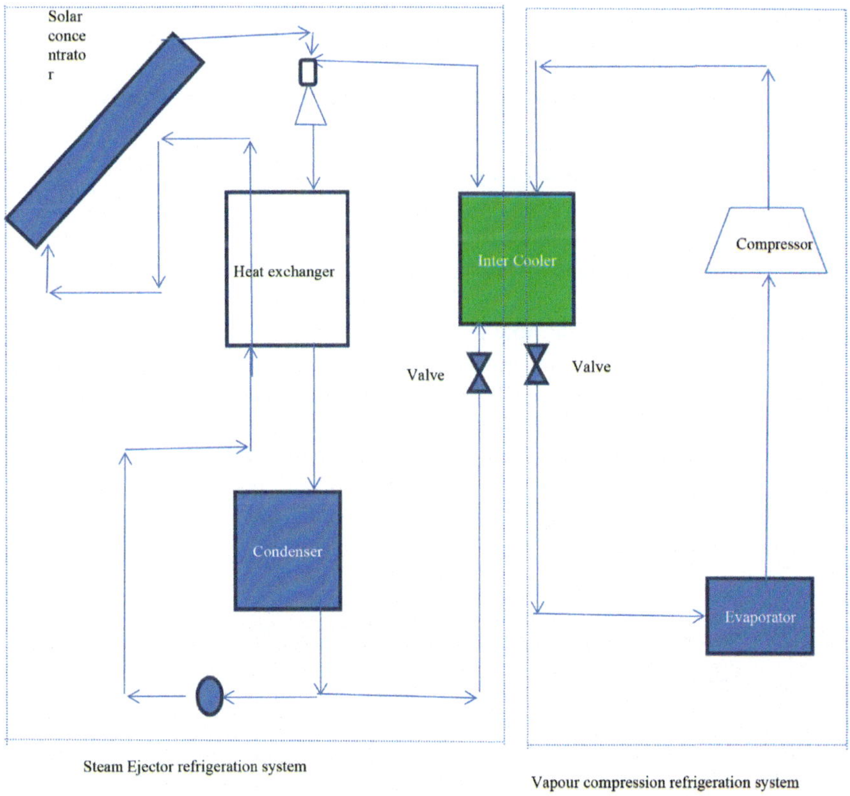

Fig. 10.9 A coupled steam jet ejector-vapour compression cooling system (Srikhirin et al. 2001)

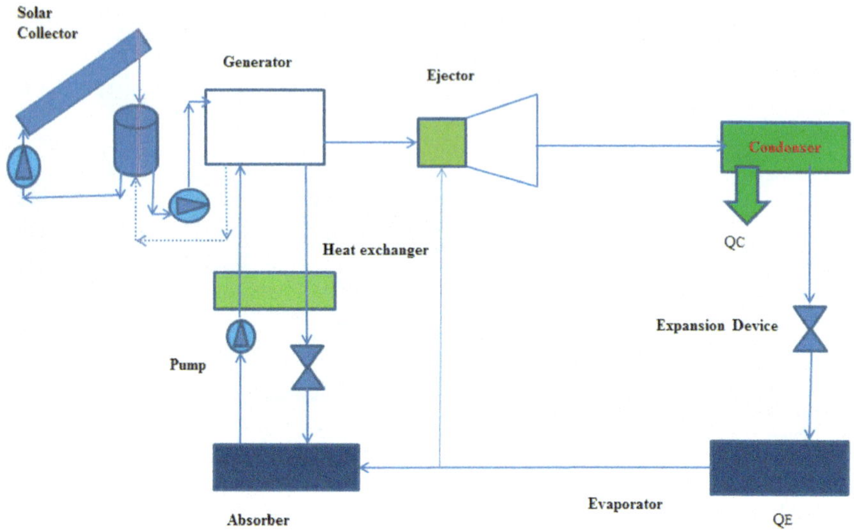

Fig. 10.10 A combined vapour absorption-ejector cooling system

10.2.7 Combined Absorption-Jet Ejector Cooling Systems

A coupled system (Fig. 10.10) clearly shows that the ejector is added in-between the condenser and generator part of ordinary single-effect vapour absorption cooling system. In addition to the ejector, it consists of components such as solar collector, absorber, a heat exchanger, condenser, an evaporator etc. as in the case of a conventional single-effect solar vapour absorption cooling system.

In a modified system, the low-pressure LiBr–water mixture leaving from the absorber is pumped to the generator. The generator is operated by using a high-temperature solar energy source, drives off the water from this solution and weak solution flows back through the solution heat exchanger into the absorber unit. In the ejector part, the secondary flow from the evaporator and the primary flow from the generator get mixed in the mixing chamber and passing through the diffuser of the ejector. The mixed stream after passing through the ejector, enters the condenser.

10.2.8 Solar Driven Vapour Adsorption-Jet Ejector Cooling System

A combined hybrid system consists of a heating system and a cooling system is illustrated in Fig. 10.11. The heating system consists of a solar adsorber, a storage tank and a valve. Components in the cooling system are an ejector, condenser, throttle

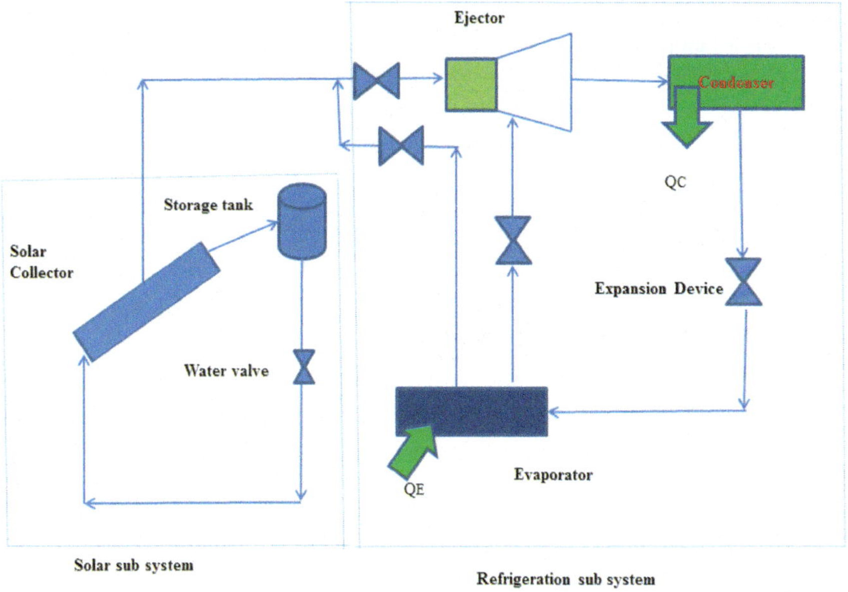

Fig. 10.11 A combined ejector-adsorption cooling system

valve and an evaporator. The solar adsorber is packed with enough quantity of adsorbent (zeolite or activated carbon or silica gel). Water is used as a refrigerant (Mosleh et al. 2019). Initially, valve V2 should be closed and V1 remains open. The solar adsorber heats due to solar energy with water desorbing from the adsorbent matrix once it attains the sufficient desorption temperature.

The water vapour at very high pressure and temperature desorbed from the adsorbent matrix enters the ejector. Refrigerant vapour gets accelerated in convergent-divergent part of the nozzle. At the exit, supersonic jet of refrigerant is mixed with the entrained refrigerant from the evaporator by opening valve V3. The mixture then enters the diffuser part to obtain its sufficient pressure and is then forced into the condenser. The condensed liquid after passing through the expansion device enters the evaporator part. In the afternoon, when the temperature reaches maximum desorption temperature, valves are exchanged so that connection to the evaporator (valve V2 opens and valve V1 closes). During this time, the cold water from the water tank will be circulated through solar collector by natural convection. Once the adsorber attains the sufficient temperature for adsorption, water refrigerant evaporated from evaporator will be adsorbed in the bed matrix. The comparison of various jet ejector compression cooling system is given in Table 10.1.

Table. 10.1 Performance comparison of various jet ejector compression cooling system

Configuration	COP
Single jet ejector compression system	0.2–0.3
Multijet ejector system	0.6
VEC system with booster	0.85
JEC-VCR system	1.04
JEC-VAB system	0.83–1.03
JEC-VAD system	0.6

10.3 Thermoacoustic Cooling

One of the most important applications of the concept of thermoacoustic principle is thermoacoustic cooling.

Thermoacoustics is the interaction between temperature, density and pressure variations of acoustic waves. The thermoacoustic devices can be easily driven by solar energy. Thermoacoustics systems are suitable for low grade energy sources like waste heat or solar energy. The mathematical foundation of thermoacoustics was laid by Rott (1980). The field was inspired by the work of Swift and his co-workers (Hayden and Swift 1997; Swift 1988; Wheatley et al. 1983).

A thermoacoustic refrigerator (TAR) is a refrigerator that uses sound waves in order to provide the cooling. In TAR, helium-argon mixture is used as working fluid. The loudspeaker performs the function of mechanical compressor used in VCR systems. They are more favourable due to the absence of moving mechanical parts, simpler control and lack of environmental issues.

Thermoacoustic cooling devices use sound to generate cooling power. The main components of a STC are, an acoustic source (loud speaker), attached to a resonator (tube) filled with a gas (such as air or any other inert gases). A stack contains several parallel plates and two heat exchangers are connected to the resonator as shown in Fig. 10.12.

Thermoacoustic coolers are classified based on the source of the acoustic energy input. Input energy can be derived out from thermoacoustic engines or acoustic drivers. The thermo acoustically driven thermoacoustic refrigerator (TADTAR) uses energy provided by a thermoacoustic engine. However, if the acoustic energy is provided by an acoustic driver e.g., a loudspeaker, it is termed as acoustically driven thermoacoustic refrigerator (Swift 2007).

The efficiency of acoustic refrigerator is almost 20% of the Carnot coefficient of performance (Garrett and Backhaus 2000). The efficiency of thermoacoustic refrigerators can be enhanced by the better design of resonator, improved stack, and loud speaker.

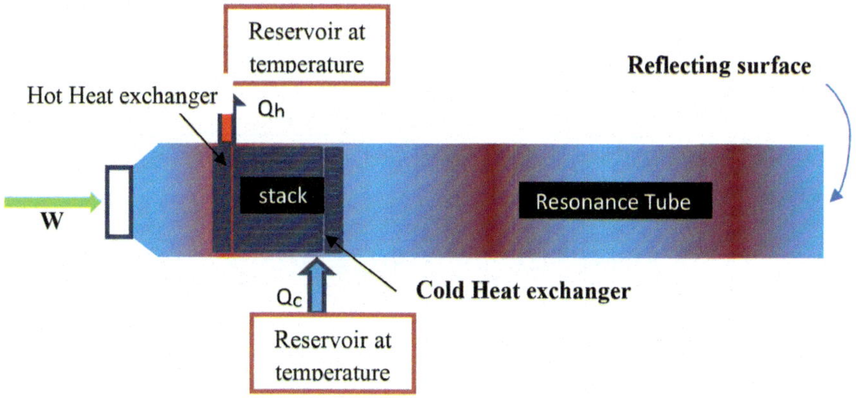

Fig. 10.12 Thermoacoustic cooler

10.3.1 Working Principle of Thermoacoustic Cooling System

Basic components of a thermoacoustic cooling system are:

(i) Acoustic driver
(ii) Stack or regenerator
(iii) Heat exchanger
(iv) Resonator.

The system consists of inert gases like helium or xenon filled at higher pressure and the sound waves are produced, which may be further amplified by the temperature difference. In contrast, the sound waves of the desired frequency are able to produce a temperature difference. When the loudspeaker blasts around 180 dB a standing wave resonates in the chamber. This standing wave reflects after colliding with other end of tube. When the incident and reflected waves overlap, they interface producing a single waveform. The standing waves vibrate in constant position and orientation around the stationery nodes. The position of the node is located where these two-component sound waves interface to create areas of zero net displacement. Antinode is the position in the area of maximum displacement and it is located half-way between the two nodes as in Fig. 10.13. The standing waves are thus useful since it requires small input power to create a large-amplitude wave. The most important parameter that governs the performance of the thermoacoustic refrigerator is its frequency.

The equation for the frequency of a wave travelling through a closed tube is given by (Swift 2007)

$$f = \frac{V}{4L} \tag{10.38}$$

where f is the frequency, V the velocity of wave, and L the length of the tube.

The COP can be defined as:

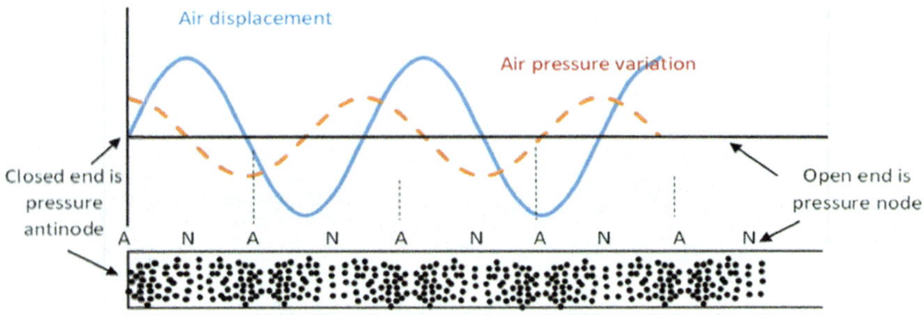

Fig. 10.13 Relationship between the phases of wave (Swift 2007)

$$\mathrm{COP} = \frac{\text{Heat extracted at lower temperature } (T_r)}{\text{Work done on the machine}} \tag{10.39}$$

The COP of the thermoacoustic refrigerator is defined as the ratio of cooling load to the electric power input E_{el}, into the acoustic driver, i.e.,

$$\mathrm{COP} = \frac{Q_{load}}{E_{el}} \tag{10.40}$$

The conversion of electrical power input E_{el} and acoustic power W_{tot} are connected by the term energy conversion the electroacoustic efficiency η_{elac} and can be defined as (Saechan and Jaworski 2018)

$$\eta_{elac} = \frac{W_{tot}}{E_{el}} \tag{10.41}$$

The main advantages and disadvantages of thermoacoustic systems are listed below.

Advantages

- Environmentally friendly. Since no toxic working fluid is used, helium is used as the working fluid.
- Less weight as compared with the vapour compression refrigeration system.
- Cost of the system is very low. Materials used don't need any special treatments.
- Very large cooling capacity.
- Range of temperature is large.
- No moving parts for the thermodynamic cycle, so very reliable and a long-life span is possible.

Disadvantages

- Efficiency of the thermoacoustic refrigerator is very low.
- High irreversibilities are associated with the thermoacoustic process. The irreversibilities due to viscous effects, the low heat transfer rate of the sound waves, thermal conduction losses etc. are leads to the decrease in the efficiency.
- The talent bottleneck is another problem with thermoacoustics, being a novel system there are very few people who expertise in the field.
- The application is also limited to very low power such as 10TR (35 kW approximately) per square meter of stack cross-sectional area at working fluids pressure below 20 atm.

  In view of the above limitations there is a need to improve further by:

  - Isolating the system to reduce the sound.
  - Use of conductive material for the hot section of a resonator.
  - The composition of stack material can be changed to gold, silver or copper.
  - Replacement of closed cap with a speaker.

  With the improvements, the system can be used effectively for various applications including but not limited to the following:

  - Storage of fruits.
  - Storage of medicines and vaccines.
  - Space craft applications.
  - Thermally acoustic orifice pulse tube cryocoolers designated as the coolahoop and can be used in high-speed electronics.
  - Natural gas liquefaction plants.

References

Chen, L. T. (1978a). Gas-cycle solar refrigeration system performance. *Energy Conversion, 18*(3), 179–187.

Chen, L. T. (1978b). A heat driven mobile refrigeration system analysis, *Energy Conversion, 18*, 179–187.

Garrett, L. & Backhaus, S. (2000). 'The power of sound,' *American Science, 88*, 516–525.

Hayden, M. E. & Swift, G. W. (1997). Thermoacoustic relaxation in a pin-array stack. *The Journal of the Acoustical Society of America, 102*(5), 2714–2722.

Mosleh, H. J., Hakkaki-Fard, A. & Daqiq Shirazi, M. (2019). A year-round dynamic simulation of a solar combined, ejector cooling, heating and power generation system. *Applied Thermal Engineering, 153*, 1–14.

Rott, N. (1980). Thermoacoustics. *Advances in Applied Mechanics, 20,* 135–175.

Rule, T. T. & Qvale, E. B. (1969). Steady-state operation of the idealized Vuilleumier refrigerator. In: *Advances in Cryogenic Engineering: Proceedings of the 1968 Cryogenic Engineering Conference Case Western Reserve University Cleveland, Ohio August 19–21, 1968* (pp. 343–352). Springer US.

Saechan, P. & Jaworski, A. J. (2018). Thermoacoustic cooler to meet medical storage needs of rural communities in developing countries. *Thermal Science and Engineering Progress, 7*, 164–175.

Srikhirin, P., Aphornratana, S. & Chungpaibulpatana, S. (2001). A review of absorption refrigeration technologies. *Renewable and sustainable energy reviews, 5*(4), 343–372.

Swift, G. W. (1988). Thermoacoustic engines. *the Journal of the Acoustical Society of America, 84*(4), 1145–1180.

Swift, G. (2007). Thermoacoustics. In: Rossing, T. (eds) Springer Handbook of Acoustics. Springer Handbooks. Springer, New York, NY.

Wang, J. X., Li, Y. Z., Li, J. X., Li, C., Zhang, Y. & Ning, X. W. (2019). A gas-atomized spray cooling system integrated with an ejector loop: Ejector modeling and thermal performance analysis. *Energy conversion and management, 180*, 106–118.

Wheatley, J., Hofler, T., Swift, G. W. & Migliori, A. (1983). An intrinsically irreversible thermoacoustic heat engine. *The Journal of the Acoustical Society of America, 74*(1), 153–170.

Chapter 11
Hybrid Cooling Systems

Different refrigeration systems and their working in individual mode and existing technologies for the better utilization of solar energy for the production of cooling are discussed in earlier chapters. It is found from the previous discussions that out of the available technologies, the conventional VCR technology is having advantages in terms of its high COP, flexibility, compactness etc. However, its application is also limited with respect to the energy and environment scenario. The absorption systems are used for high cooling capacity and large airconditioning applications due to their complexity and high operational costs. Steam jet ejector systems are existing since the early 1990s, but their use and deployment have been limited due to their low COP. The solar vapour adsorption refrigeration systems are advantageous since their refrigerants are environment friendly, chemically stable and the system can be powered by low grade heat like solar energy and waste heat. But the problems with VAD systems are their low COP and high operating cost. The non-conventional refrigeration like thermoelectric cooling systems, thermoacoustic systems etc. are having their own advantages and limitations as described in the earlier chapter.

In view of the environmental and economical issues, the hybrid systems could play a crucial role in refrigeration and airconditioning scenario in the coming future. The hybrid systems can provide higher cooling capacity and better coefficient of performance as compared to the systems operated in individual mode.

11.1 Hybrid Vapour Compression—Absorption Cooling Systems

A heat-driven vapour absorption cooling system can be coupled with the conventional vapour compression system to form a hybrid system as shown in Fig. 11.1. This permits an efficient utilisation of low grade energy that may be provided by low grade energy resources like waste heat or solar energy, resulting in reduction in high

S. C. Kaushik et al., *Solar Cooling*, https://doi.org/10.1007/978-3-031-42410-6_11

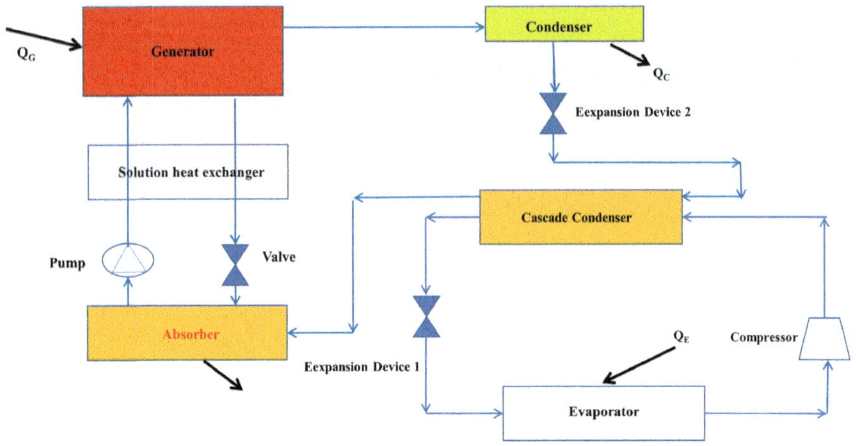

Fig. 11.1 Hybrid VCR-VAB cooling system

grade like electric energy consumption as compared to the standalone VCRS. This improves the efficiency of energy utilisation and also saves considerable amount of high-grade energy. The hybrid systems are more bulky, but its ability to use low-grade energy for operation makes them more favourable, and reduces the operating cost.

The schematic of a tyipcal hybrid VCR-VAB cooling system is shown in Fig. 11.1. The hybrid system includes compressor, condenser, expansion device and evaporator. The condenser of VCR system acts as the cascade condenser of the hybrid system. The vapour absorption cooling system consists of an absorber, generator, solution pump, heat exchanger and expansion device. The working pair for VAB system may be NH_3–H_2O or H_2O–LiBr and for VCR system, it can be R134a/R22. The system is explained for the working pairs of LiBr–H_2O in VAB and R22 for VCR system. The strong solution from absorber in VAB system is pumped into the generator through a solution heat exchanger. The desorbed water refrigerant vapour from the generator is subjected to condensation in the condenser of VAB system. The condensate is then expanded in the expansion device. The low-pressure water refrigerant enters into the cascade condenser, which is the evaporator of VAB system. Refrigerant vapour R22 of VCR is condensed by rejecting its heat to the refrigerant vapour (water) of VAB system in the cascaded condenser. This low-pressure water vapour then enters into the absorber unit of VAB system, where it is absorbed by the weak LiBr–H_2O solution to form a rich solution. The heat absorbed by the evaporator and the work done on the compressor of VCR system is equal to the heat absorbed by the cascade condenser of VAB system. The important design parameter of the system is the intermediate temperature between the two circuits, since it has effect on the COP of the system. The COP of VCR, VAB and hybrid system is calculated as following (Jain et al. 2013):

$$\mathrm{COP}_{\mathrm{VCR}} = \frac{Q_{\mathrm{evap}}}{W} \tag{11.1}$$

$$\mathrm{COP_{VAB}} = \frac{Q_{\mathrm{evap}}}{Q_g + W_p} \tag{11.2}$$

COP of hybrid system,

$$\mathrm{COP_{hsys}} = \frac{Q_{\mathrm{evap}}}{W + Q_g + W_p} \tag{11.3}$$

The other important parameter to be taken into consideration is the efficiency defect (δk) of kth component of the system and it is expressed in terms of input energy (Jain et al. 2013).

$$\delta k = \frac{IK}{W + Q_{\mathrm{g}}\theta_{\mathrm{Carnot}} + W_{\mathrm{p}}} \tag{11.4}$$

where the Carnot factor

$$\theta_{Carnot} = 1 - \frac{T_o}{T_g} \tag{11.5}$$

The term coefficient of the structural bond (CSB) has been applied for the hybrid system to measure the improvement of the performance of the system. The coefficient of structural bonds (CSB) is defined as (Kotas 1985).

$$\sigma_{ki} = \frac{[\partial I_t]/[\partial X_i]}{[\partial I_k]/[\partial X_i]} \tag{11.6}$$

where I_{t} is the total irreversibility given by

$$\begin{aligned}\mathrm{I_t} &= \mathrm{I_{generator}} + \mathrm{I_{absorber}} + \mathrm{I_{condenser}} + \mathrm{I_{cascade\ condenser}} + \mathrm{I_{compressor}} \\ &\quad + \mathrm{I_{evaporator}} + \mathrm{I_{expansion\ device\ 1}} + \mathrm{I_{expansion\ device\ 2}} + \mathrm{I_{pump}} \\ &\quad + \mathrm{I_{valve}} + \mathrm{I_{pipeline}} + \mathrm{I_{heat\ exchanger}}\end{aligned}$$

I_k is the irreversibility of kth component of the system under consideration and X_i the parameter of the system which produces the changes.

In addition to these, the other parameters to be considered are:

1. Heat utilisation factor (HUF)
2. Electrical energy utilization factor (EEUF)
3. Fuel energy savings ratio (FESR)

$$\mathrm{HUF} = \frac{Q_g}{Q_{\mathrm{evap}}} \tag{11.7}$$

The term EEUF, electrical energy utilisation factor relates total work input to the system (i.e., electrical energy supplied to the compressor and pump) and refrigeration effect produced (Hong et al. 2019; Jianbo et al. 2019).

$$\text{EEUF} = \frac{W + W_p}{Q_{\text{evap}}} \tag{11.8}$$

The total energy required for refrigeration effect produced is the fuel energy savings ratio.

$$\text{FESR} = \frac{Q_g + W + W_p - Q_{\text{evap}}}{Q_{\text{evap}}} \tag{11.9}$$

Simulation and thermodynamic modelling conducted by various researchers indicate that the average COP of the hybrid system is around 0.8. The power consumption of compressor in VCR system can be reduced by 62% due to a reduction in the mass flow rate of refrigerant and low temperature at the compressor exit in the hybrid system as compared to the standalone VCR system. It is also reported that the hybrid system reduces heat rejection in the condenser of VCR by 10%. The exergy analysis of such a system indicates that the irreversibility rate and efficiency defects are more in the compressor and generator of the unit. The generator contributes the highest irreversibility which may be attributed to the huge temperature difference between the source and working fluid.

Cimsit and Ozturk (2012) conducted a simulation study on a hybrid VCR-VAB cooling system. One of the important results obtained from their study is the variation of COP with generator temperature. As seen in Fig. 11.2, the coefficient of performance of the absorption cycle, COP_{abs} and cascade cycle, $COP_{cyclegen}$ increase with the increase in generator temperature. The COP of LiBr–H_2O has higher COP values than NH_3–H_2O system.

Different researchers studied different configurations of the combined VCR-VAB system. Pratihar et al. (2010) introduced a hybrid system in which the compressor is placed in-between the absorber and generator unit (Fig. 11.3).

In the hybrid system, the compressor is placed in-between the absorber unit and generator unit. An ideal mixture placed between the compressor and absorber. The ammonia refrigerant coming from the mixer is being absorbed by absorber of VAB unit. The separator is located in-between the desorber and compressor, which separates weak solution and the refrigerant vapour. Refrigerant vapour thus generated is compressed and discharged at the state point 2. During the processes 3–4, the weak solution is pumped back to the absorber at the state 5. The weak solution and strong solution leaving from the absorber at state 6 pass through a solution heat exchanger in reverse direction. As ammonia vapour is absorbed in the weak solution, strong solution is obtained at state 6. The strong solution is then subjected to throttling at point 7 and pressure decreases. The low-pressure strong solution entering at the state point 8 is desorbed in the desorber by utilising heat from the thermal energy

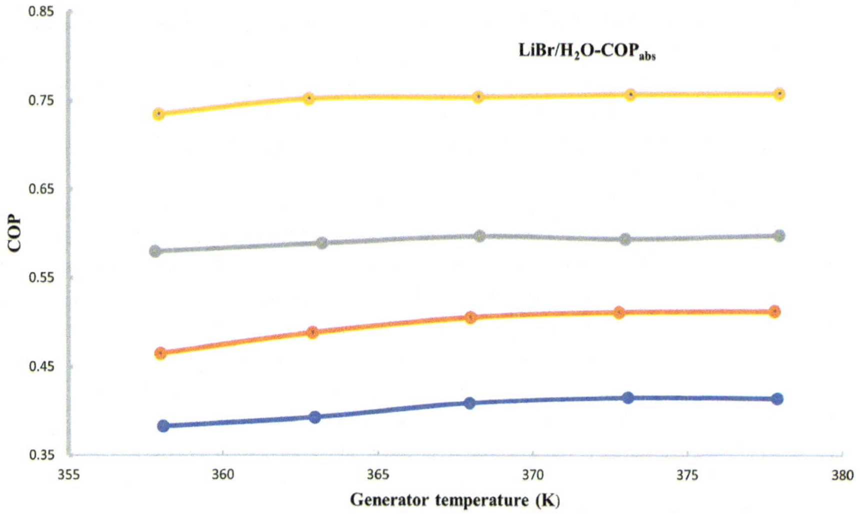

Fig. 11.2 COP versus generator temperature (Cimsit and Ozturk 2012)

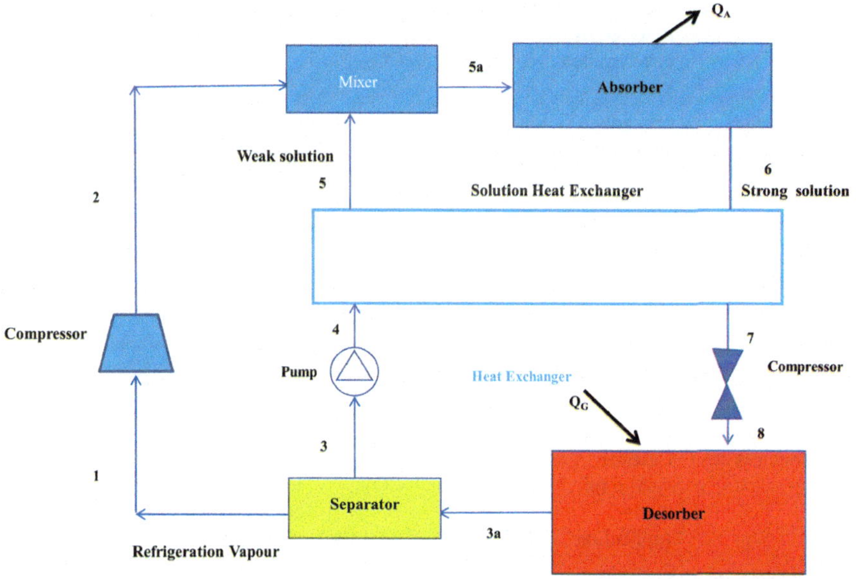

Fig. 11.3 Hybrid vapour compression—absorption cooling system (Pratihar et al. 2010)

source. The literature on various designs of the combined VCR-VAB system is given in Table 11.1.

Table 11.1 Various VCR-VAB hybrid system configurations (Jain et al. 2013)

Author	System	Refrigerants	Features of the hybrid cooling system
Cimsit and Ozturk	Cascade cycle	In absorption section—the working pairs H_2O–LiBr and NH_3–H_2O Vapour compression—R134a	Electrical energy consumption is less than the vapour compression system
Kairouani and Nehdi	Combined cascade system powered by geothermal energy	In absorption section—NH_3–H_2O Vapour compression—R717, R22 and R134A	COP of the combined system is 37–54% higher than the VCR cycle
Fernandez-Seara et al.	Cascade cycle powered by cogeneration system	In absorption section—NH_3–H_2O Vapour compression—CO_2	COP of compression section—2.602 with CO_2 and 2.463 for NH_3
Seyfouri and Ameri	Cascade system in which compressor is driven by a micro turbine	In vapour compression—R22 In vapour absorption—H_2O–LiBr	Authors considered four different configurations of integrated system. Analysis indicates an improvement of 114% in COP and 133% reduction in the energy consumption
Wang et al.	Cascade system driven by solar energy	In vapour compression—R134a In vapour absorption—H_2O–LiBr	COP of the system reported is 6.1 for solar intensity of 700 W/m^2. The power consumption reported to be reduced by 50% in comparison with conventional VCR refrigeration system
Garimella et al.	Cascade system for marine application	Vapour compression—CO_2 Vapour absorption—H_2O–LiBr	The electricity consumption was found to be reduced by 31% as compared with the conventional vapour compression system
Chinnappa et al.	Cascade system driven by solar energy	Vapour compression—R22 The vapour absorption—NH_3–H_2O	Power consumption reduced

11.2 Hybrid Vapour Compression—Adsorption Cooling Systems

Figure 11.4a shows schematic of the hybrid cycle, which comprises a conventional compressor and a thermal compressor unit, which is provided at the lower stage.

The conventional mechanical compression is supplemented by thermal compression using a string of adsorption compressors. The low-pressure vapour refrigerant from the evaporator was taken to the thermal compressor, where it is adsorbed by the adsorbent at near condensation temperature. The thermal compressor may be powered by solar energy. Then, the desorbed refrigerant vapour, which is nearly at the heat source temperature from intercooler, enters a mechanical compressor.

Thus, one cycle of operation is completed. The other possible configuration of hybrid system is shown in Fig. 11.4b. In this case, since the refrigerant vapour is

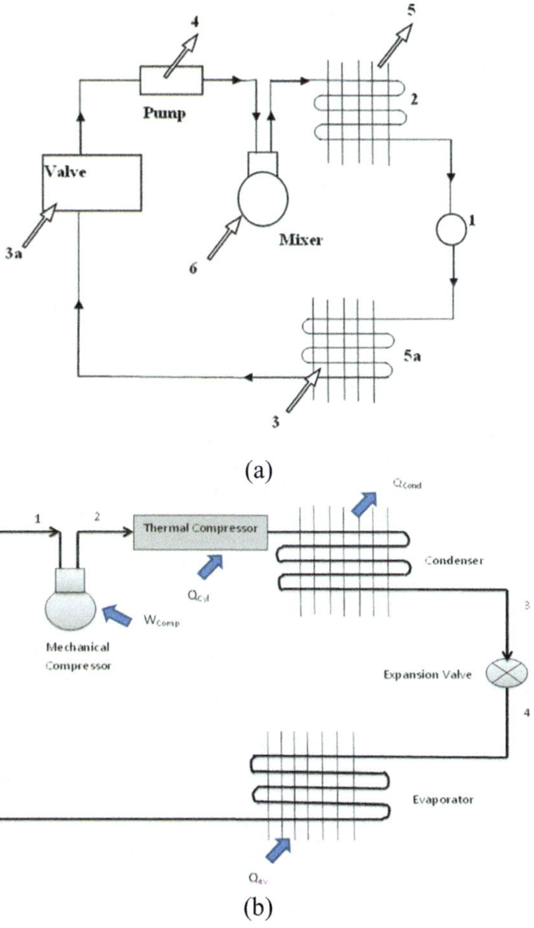

Fig. 11.4 Hybrid adsorption cooling systems with thermal compressor placed **a** before mechanical compression **b** after mechanical compression (Banker et al. 2011)

first subjected to mechanical compression, the use of intercooler can be avoided. The study showed that almost 40% energy saving was possible if the compression is performed in two stages of compression.

11.3 Combined Vapour Compression—Absorption-Jet Ejector Cooling System

A heat engine-driven cooling cycle based on the combination of VCR-VAB-ejector has been investigated in 2000 by Selahattin. The configuration of such a system is shown in Fig. 11.5.

The theoretical system proposed consists of a Carnot heat engine, Carnot VCR, Carnot VAB and Carnot ER. In this configuration, the condenser and absorber of the absorption unit discharge their heat to the evaporator of VCR and the generator of the ejector cooling system. The Carnot heat engine operates between heat

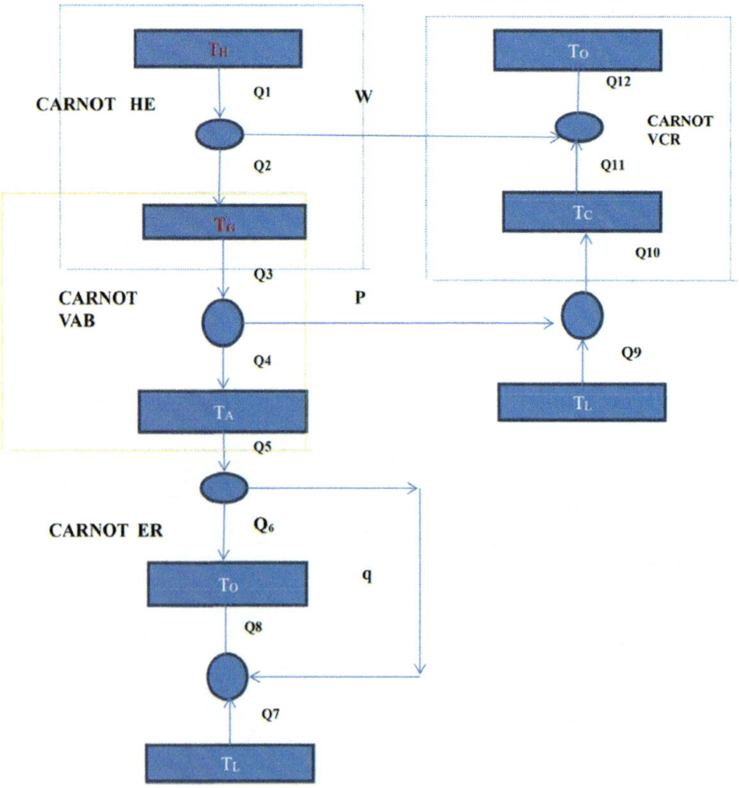

Fig. 11.5 Combined VCR-VAB-Ejector cooling system

source temperature T_H and generator temperature T_G of VAB system. Carnot heat engine drives both VCR and VAB units. The Carnot VCR unit operates between the condenser of VAB (heat source) at temperature T_C and the heat sink temperature T_O. Carnot VAB system drives the Carnot ejector unit and the system consists of a generator, absorber, condenser and an evaporator. The temperatures are T_G, T_A, T_C and T_L. The Carnot ejector cooling system consists of a generator, condenser and evaporator at temperatures of T_A, T_O and T_L, respectively.

According to Fig. 11.5, the overall COP of this reversible combined system is given by

$$\mathrm{COP} = \frac{T_L}{T_H}\left[\frac{(T_A - T_O)}{(T_O - T_L)} + \frac{(T_H - T_G)}{(T_O - T_C)}\right] \tag{11.10}$$

11.4 Solar Operated Heating, Cooling and Power Generation Systems (Tri Generation)

The combined heating, cooling and power (CHCP) systems are capable technologies to solve energy-related problems. The word CHCP is derived from the word cogeneration which is a proven technology with a history of more than 100 years. The cogeneration systems are able to produce combined production of electrical power output and heating from the same energy source. Exhaust heat from the conventional plants is recovered to meet the heating load. But in CHCP, the thermal or mechanical energy is further utilised to provide cooling. These systems are also called as trigeneration systems. In winter, many CHCP systems behave as a cogeneration unit, where there is no cooling demand. Figure 11.6 shows the general layout of CHCP system which consists of a power generating machine (PGM). The PGM set can also drive a vapour compression cooling system heat pump (HP). The thermal energy storage devices (TES) could also be integrated with the system.

Figure 11.7 is a typical CHCP system consisting of gas engine and an absorption chiller. The absorption chiller can be utilized to produce heating power in winter and cooling in summer. The engine is driven by natural gas and the mechanical energy produced is further converted into electricity by the coupled generator. The heat source for the absorption chiller unit is the exhaust gas and jacket water from the engine.

Following are the main components of a CHCP system:

(a) A prime mover
(b) Electric generator
(c) A heat recovery system
(d) Thermally activated equipment
(e) Management and control system.

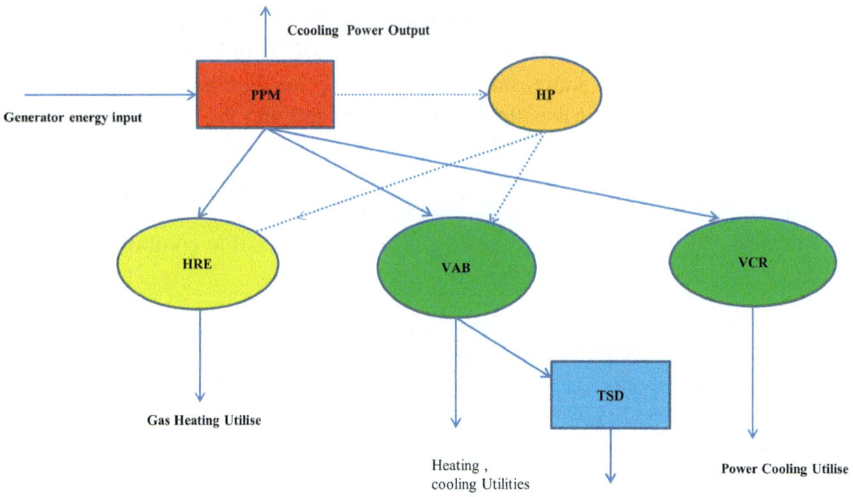

Fig. 11.6 General layout of a CHCP system

Fig. 11.7 Combined CHCP system

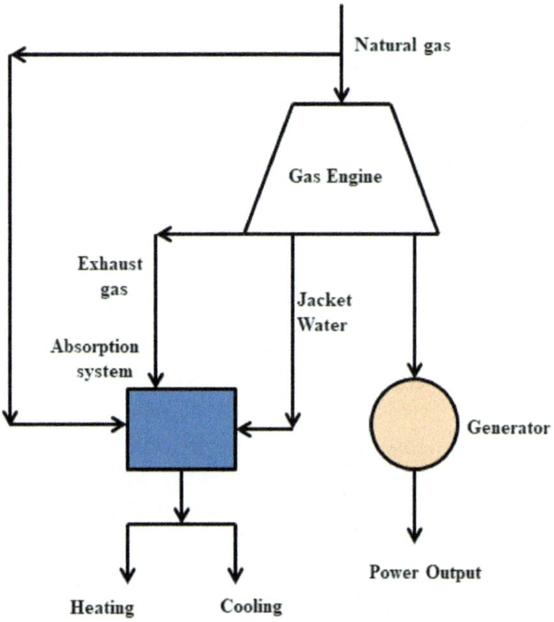

The prime mover may be a steam turbine, gas turbine, reciprocating IC engine, external combustion engine, micro turbine, Stirling engine or fuel cell. The thermal systems include VCR, VAB chillers, VAD systems, desiccant dehumidifiers and the other non-conventional cooling systems.

The advantages of the CHCP systems are as follows:

(a) Improved fuel energy utilisation
(b) Prime movers such as turbines and fuel cells have less emission as compared to conventional prime movers
(c) Combined cooling, heating and power systems are more reliable as compared to conventional systems.

11.4.1 CHCP Based on Vapour Adsorption Cooling System

Figure 11.8 shows hybrid solar heating, cooling and power generation system proposed by Zhai et al. (2009). The system consists of a parabolic trough solar collector, a helical screw expander and an adsorption refrigeration system and is divided into three main subsystems, namely, steam generation, power generation and waste heat utilisation subsystem.

The first subsystem that operates at high-temperature level includes a series of parabolic trough solar collectors and an assisted boiler. A boiler is integrated to work as an auxiliary unit. The other unit consists of a helical screw expander engine and a generator operators at medium temperature level. The parabolic collector collects solar energy and produces steam. The steam generated by parabolic trough collector can be used as the working fluid for the power cycle. A generator is coupled to the vane type expander. The energy content of high-pressure working medium is converted into mechanical power in expander, turning the shaft of generator and producing useful work. Two heat exchangers, a hot water tank, an adsorption chiller

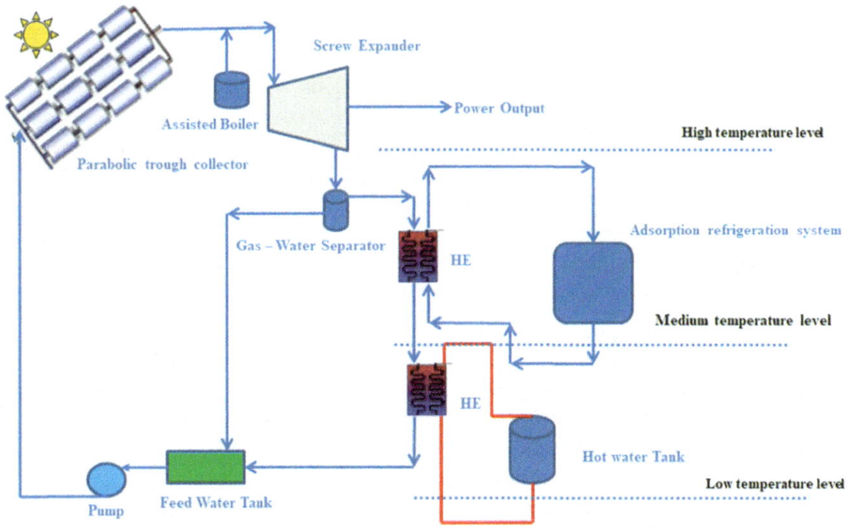

Fig. 11.8 Hybrid solar heating, cooling and power generation system (Zhai et al. 2009)

and a heating system constitute the third subsystem. A steam separator is integrated, which separates the exhaust steam from the helical screw expander and dry saturated steam is allowed into heat exchanger.

11.4.2 CHCP Based on Rankine-Vapour Absorption Cooling System

A combined cooling and power cycle working on the principle of vapour absorption and Rankine cycle is shown in Fig. 11.9. For this combined cycle, NH_3–H_2O is the preferred working pair because of its desirable thermodynamic properties in terms of the large heat capacity.

The system is divided into two subsystems, viz. turbine unit (power generation) and vapour absorption cooling system (cooling unit). The strong solution, enriched in ammonia coming from the absorber enters into the generation system.

The rectifier condenses out the water vapour entrapped in the NH_3 refrigerant before entering into the boiler. Refrigerant vapour coming from boiler gets superheated in superheater placed in the circuit. The superheated refrigerant is expanded in the turbine to produce work. Turbine acts as an expansion device for the refrigeration circuit. The expanded refrigerant has the potential for refrigeration. The main parameters to be considered while designing the system include absorber and system pressure, heat source and solution concentration. Heat sources can be solar energy, geothermal or waste heat sources. The analysis conducted by Tamm et al. (2004), indicates that the combined system have high second law efficiency than the conventional power cycles.

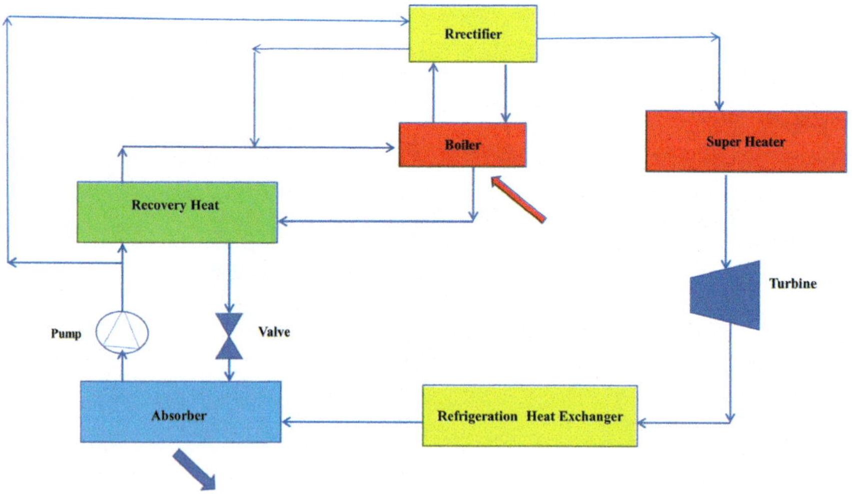

Fig. 11.9 A combined CHCP system (Tamm et al. 2004)

References

Banker, N. D., Prasad, M., Dutta, P. & Srinivasan, K. (2011). Experimental results of an activated carbon–HFC 134a adsorption cooling system for thermal management of electronics. *Applied Thermal Engineering, 31*(10), 1607–1612.

Cimsit, C. & Ozturk, I. T. (2012). Analysis of compression–absorption cascade refrigeration cycles. *Applied Thermal Engineering, 40*, 311–317.

Hong, S. J., Lee, S. M., Lee, C. H., Kim, I. G. & Park, C. W. (2019). Thermally-driven hybrid vapor absorption cycle: Simultaneous and flexible use of steam generation heat pump and refrigeration applications. *Energy Conversion and Management, 201*, 112100.

Jain, V., Kachhwaha, S. S. & Sachdeva, G. (2013). Thermodynamic performance analysis of a vapor compression–absorption cascaded refrigeration system. *Energy conversion and management, 75*, 685–700.

Jianbo, L., Shiming, X., Xiangqiang, K., Kai, L. & Fulin, C. (2019). Experimental study on absorption/compression hybrid refrigeration cycle. *Energy, 168*, 1237–1245.

Kotas, T. J. (1985). The Exergy Method of Thermal Plant Analysis. London, Great Britain.

Pratihar, A. K., Kaushik, S. C. & Agarwal, R. S. (2010). Simulation of an ammonia–water compression–absorption refrigeration system for water chilling application. *International Journal of Refrigeration, 33*(7), 1386–1394.

Tamm, G., Goswami, D. Y., Lu, S. & Hasan, A. A. (2004). Theoretical and experimental investigation of an ammonia–water power and refrigeration thermodynamic cycle. *Solar Energy, 76*(1–3), 217–228.

Zhai, H., Dai, Y. J., Wu, J. Y. & Wang, R. Z. (2009). Energy and exergy analyses on a novel hybrid solar heating, cooling and power generation system for remote areas. *Applied energy, 86*(9), 1395–1404.

Chapter 12
Economic Considerations in Solar Cooling Systems

12.1 Background

The value of a system utilising solar energy directly or indirectly must ultimately be judged on the basis of its economical viability. However, the cost of a cooling system should not be judged based on the cost of the solar energy systems but also on the cost related to environment and other harm the conventional refrigerants make to the environment in the form of ozone depletion potential (ODP) and global warming potential (GWP) besides, the emissions being emitted into the environment by the power generation unit. This chapter deals with some economic aspects and methods used for making economic evaluations of solar energy systems.

The main competitors of solar thermal cooling systems are conventional vapour compression cooling systems. Nowadays, it is commonly accepted that none of the solar-thermal powered cooling plants are cost competitive with the conventional vapour compression cooling plants, because the main capital cost in solar thermal cooling plants is the solar energy collection and conversion equipment cost (i.e., solar collector cost). In a solar-assisted cooling system, solar collector cost is between 50 and 80% of the overall system cost and is strongly dependent on the operating temperature of the solar-thermal driven chillers. Recent research to reduce solar collector price and technical improvements in solar collectors can reduce the solar thermal cooling systems cost. Solar cooling systems are environmental friendly and it reduces the greenhouse gas emissions. Certainly, these indirect benefits also have an economical value, but for now they do not have a direct effect on the solar thermal cooling plant. The advantage of the solar system is that once a system has been installed, a large fraction of the input energy is free whereas the energy required to operate a conventional system is a continuous recurring expense. The general criterion for a solar system to be economic is that the cost value of the energy saved by using the solar system added up over the life of the system should be greater than the extra costs incurred by installing the solar system. The value of the energy saved by the solar installation depends on both the quantity of energy saved and the price of

S. C. Kaushik et al., *Solar Cooling*, https://doi.org/10.1007/978-3-031-42410-6_12

the conventional energy supply that is being replaced by the system. The quantity of energy saved depends on the type and size of the system installed, the size of the load and the climate zone. For certain types of solar systems, methods are available, which enable the designer to estimate the fraction of the energy requirements of the load that can be supplied by the solar system. One of the problems in justifying economically a solar energy system is to estimate the future values of the energy savings produced by the system, which is a rather difficult task. As conventional energy prices rise, more solar applications would become economically viable and time would come when for solar energy to play a vital role as a future energy resource, there will be a strong industrial base to produce the solar equipment. Apart from domestic hot water and swimming pool heating application, passive house heating/cooling systems using solar energy are potential cost-effective propositions. Active solar cooling system needs to be made viable and acceptable in the market.

Although solar energy is available freely (without cost), solar powered cooling systems necessarily require more equipment than the conventional cooling systems and therefore have higher capital costs. In fact, solar powered cooling systems are hybrid in nature, capable of operating with solar energy which is not available continuously during a cooling season. Thus, the reduction in annual energy cost for cooling attainable with a solar system is the difference between the energy cost for a commonly used system and the cost of the supplementary energy source required for the solar system. The ultimate criterion for the implementation of a technically feasible solar system is its overall economic feasibility. The payback period approach is one of the common methods used to evaluate solar economics. In this method, the total cost of the system is divided by the cost value of the energy in the first year. This gives a figure indicating the time required for the energy savings to equalise the initial investment. However, in this method no allowance is made for the expected increase in conventional energy prices, which will substantially improve the cost-effectiveness of the solar system. The payback period thus directly relates the extra capital cost of solar equipment to the value of energy saved under the assumed operating conditions. This approach is relatively simple and provides a basic indication of economic feasibility, at the present stage of the development of solar systems. Presumably, with further development, solar cooling (and heating systems installed in buildings may be considered as part of the real estate) property investment, which normally appreciates in value with time. In such cases, payback and other capital amortisation approaches based on energy-saving would have limited applicability. Assuming that the competing solar and conventional systems would both adequately satisfy the required function and that the solar system would be more expensive. An appropriate amortisation criterion for the cost premium could eventually be based on the value of fuel savings versus the difference between the mortgage rate and probable appreciation rate. However, as with many applications of solar energy, it is very difficult to quantify any economic assessment of the cost of solar-powered refrigeration and cooling systems. Some examples may be given here for the indirect cost estimates of the sub-system and the complete system from both research/demonstration projects and monitored systems. Those active researches or monitoring work are less optimistic about the future prospects of the active solar cooling system. Because of their being closely involved, nevertheless, solar-based cooling systems are preferred over largely imported conventional cooling equipment. As there is increase in the energy use for space cooling and need ice-making in remote areas, solar cooling and refrigeration would play an increasingly important role.

Two terms to be considered for the economic study of solar cooling systems are investment cost and operational cost.

12.2 Investment Cost

Investment cost is the cost of solar collectors, pump, storage tank, chiller, controllers, heat exchangers etc. related to solar cooling system installation. The installation cost should also be included in the investment cost. Hence, the investment costs of the solar installation can be shown as (Duffie and Beckman 1974)

$$C_S = C_A \times A_c + C_E \tag{12.1}$$

where C_S is the total cost of solar energy equipment, C_A the cost of a solar collector per unit area (/m^2), A_c the collector area (m^2) and C_E the cost of other equipments.

The investment can be reduced by availing number of promotional incentives offered by the Central Government. Govt. offers subsidies for the solar collector equipments. Most of the state governments exempt solar energy systems from the sales tax.

12.3 Operational Cost

Operational costs are the running cost throughout lifetime of solar cooling system. This includes the amount used for energy operating fans, pumps and controllers. In addition to the parasitic energy consumption mentioned above, this also includes the cost of maintenance cost, taxes etc.

Annual operational cost is estimated by the equation given below (Bejan et al. 1996):

$$\begin{aligned}\text{Yearly cost} = {} & \text{Fuel expense} + \text{Maintenance and insurance cost} \\ & + \text{Parasitic energy cost} + \text{Cost of consumables} \\ & + \text{Property taxes} - \text{Income tax savings}\end{aligned} \tag{12.2}$$

Usually fuel expenses, repayment on loan and reduction in the taxes are the major terms in Eq. (12.2), and this equation is applicable in most of the situations. However, in some specific cases, the terms in the form of costs of benefits have to be added to the right hand side of Eq. (12.2).

In the case of an energy system using only conventional energy source then Eq. (12.2) has to be modified. In this case, there would not be no repayment installment and no tax reduction, since it is unlikely to be taken as loan. Hence, the equation becomes:

Annual cost of equivalent conventional energy systems

$$\frac{\text{Cost of the fuel}}{\text{Electricity Consumed}} \tag{12.3}$$

12.4 Economic Analysis

In order to evaluate the economic viability of the system (Sughatme and Nayak 2017) we are always interested in calculating the earnings on long term basis with the solar cooling system. It is clear, from the Eq. (12.2) of net cost of a system, that a solar energy system will help in saving the commonly used fuel like electricity. Tax deduction will also be a savings in this regard. The annual repayments such as the electricity charges, insurance, taxes and repayment of the loans etc. to be considered for the economic analysis.

12.4.1 Annual Savings (AS)

The annual savings is generally dependent on the savings in electricity bill. Annual savings is given by, AS = Savings in electricity/fuel + Tax deductions − Repayment of loan − Operating cost.

The sum of annual savings over a certain number of years *(n)* plus the cost of an equivalent conventional system minus initial down payment made at the time of installation of the solar system is the cumulative savings (CS).

$$\sum_{j=1}^{n} (AS)j - (DP) \tag{12.4}$$

where AS is the annual savings and DP means the initial down payment.

12.4.2 Life Cycle Savings (LCS)

The cumulative savings calculated over the life time of the system plus the resale value of the system at the end of its life time

$$\sum_{j=1}^{nt} (AS)j - (RV) + (CEQ) - (DP) \tag{12.5}$$

where *RV* is the resale value of the solar system and CEQ represents conventional system cost.

It is obvious that the solar cooling systems are economically viable if the value of LCS is positive.

12.5 Terms Used in Economic Analysis

The following terms are generally considered for the economical analysis.

Salvage Value
Salvage value of the item represents how much it will be worth once the component is outlived its usefulness.

Depreciation
Depreciation is the decrease in assets value with time. The monetary value of the assets decreases with time due to the use, obsolescence, wear and tear etc. The assets such as the equipment, machinery, currency depreciates over a specific period of time. Most important are straight-line method, double declining balance, and sum of years.

Number of techniques are commonly used for the comparison and optimisation of solar energy systems. It is to be ascertained that no specific guidelines are published to select a proper methodology to optimise a solar cooling type system. The life-cycle cost (LCC), annual life-cycle cost (ALCC), payback time (PB) and return on investment (ROI) are commonly used method. The term return of interest (ROI) represents the discount rate for CS and LCS is zero. The term payback period is used to judge the economic feasibility of a solar cooling systems.

Life Cycle Cost (LCC)

The life cycle cost comprises all the cost including investment and operational costs, replacement cost and maintenance cost associated with the cooling system over its lifetime are summed.

Life cycle cost of an equipment is calculated based on the following components:

$$\begin{aligned} \text{LCC} = {} & \text{Capital cost} + \text{Replacement cost} + \text{Maintenance cost} \\ & + \text{Energy cost} - \text{Salvage} \end{aligned} \tag{12.6}$$

In the Eq. (12.6), the term, capital cost is the present worth, all other terms such as the replacement cost, maintenance cost and energy cost are need to be converted to present worth. The salvage value of equipment represents the money obtained while disposing at the end of its useful life.

There are certain situations in which future worth of money has to convert into the present worth money.

The equation used to calculate the present worth is given by

$$Pr = F/(1 + i)^n \tag{12.7}$$

where F is the future worth of money, P the present worth of money and $(1 + i)^n$ the future worth factor (Eq. 12.8).

Future value of money is expressed by (Sukhatme and Nayak 2017)

$$F = \mathrm{Pr}(1 + i)^n \tag{12.8}$$

where F is the future value, Pr the present value, i the interest rate and n the time period.

Inflation

Inflation represents the necessary money to obtain the same product before the inflated price was present. Let C represent the money for buying a component and f the inflation rate. Assuming a uniform inflation rate over the years, the amount we need to pay for the same product can be determined from $C(1 + f)^n$. The present worth is given by

$$P_f = C(1 + f)^n/(1 + i)^n \tag{12.9}$$

12.6 Example: Cost Analysis of Solar VAD Cycle

The major components of this system considered here for calculating life cycle cost are:

1. Solar collector
2. Adsorbent bed
3. Chiller unit and the water tank
4. Transfer lines
5. Other accessories.

The cost of components are given in Table 12.1.

Now let us assume interest rate $i = 10\%$. Then the life cycle cost per kW is calculated as follows:

Table 12.1 Cost of various components of solar adsorption cooling system

Item	Cost in Rs/KW cooling	Life period
Solar collectors	25,000	20 years
Adsorbent bed and water tank	13,900	10
Chiller unit	5500	10 years
Accessories, tools	1000	5 years

$$\begin{aligned}
\text{Capital cost} &= \text{Cost of (Solar collectors} + \text{Adsorbent bed} \\
&\quad + \text{Chiller unit} + \text{Accessories)} \\
&= 25{,}000 + 13{,}900 + 5500 + 1000 \\
&= \text{Rs. } 45{,}400 \\
\text{Replacement cost} &= 13{,}900/(1+0.1)^{10} + 5500/(1+0.1)^{10} \\
&\quad + 1000/(1+0.1)^{5} + 1000/(1+0.1)^{10} \\
&\quad + 1000/(1+0.1)^{12} \\
&= \text{Rs. } 8725.5 \\
\text{Maintenance cost} &= 1\% \text{ of total capital cost per year} \\
&= \text{Rs. } 454 \\
\text{Therefore, life cycle cost per kW} &= 45{,}400 + 8725.4 + 454 \\
&= 54{,}579
\end{aligned}$$

12.7 Evaluation of Payback Period

Payback period is the time needed for cumulative savings on fuel or electricity or the income from power fed to the grid to be equal to the total initial investment on the solar system.

Probably all the solar refrigerators and air conditioners have been of prototype models. In order to have a cost assessment of solar cooling systems, it is also desirable to have knowledge of the payback period associated with them. By the definition, the payback period is the amount of time that it takes to recover your costs in a project.

Simple payback period is given by

$$\text{Payback period} = \frac{\text{Intial Investment}}{\text{Net Annual savings}} \tag{12.10}$$

The equations for payback period (Sukhatme and Nayak 2017) are given below. In the case of standalone system, the savings in fuel/electricity in the first year are given by *cfFE*.

Let the quantity be denoted by P. In the case of add-on-system let the income derived from the sale of electricity fed into the grid in the first year be denoted by the same symbol P.

Thus, for both systems without discounting the savings or income in the kth year is given by

$$(1 + i_f)^{k-1} P \tag{12.11}$$

If the payback period, n_{p} years, then

or

$$\sum_{k=1}^{np} (1 = i_f)^{k-1} P = R \tag{12.12}$$

or

$$\frac{P}{i_f}\left[(1 + i_f)^k - 1\right] = R$$

$$n_p = \frac{\ln\left[\frac{i_f R}{P} + 1\right]}{\ln\left[1 + i_f\right]} \tag{12.13}$$

without discounting,

$$\sum_{k=1}^{np} \frac{(1 + i_f)^{k-1}}{(1 + d)^k} P = R \tag{12.14}$$

or

$$np = \frac{\ln\left[1 - \frac{R(d - i_f)}{P}\right]}{\ln\left[\frac{1 + i_f}{1 + d}\right]} \quad \text{for } i_f \neq d \tag{12.15}$$

and

$$np = \frac{R(1 + i_f)}{P} \quad \text{for } i_f \equiv d \tag{12.16}$$

Sukhatme and Nayak (2017) presented an estimate of the payback period for a solar cooling system of 10 ton cooling capacity. For such a unit, a conventional VCR system would require about 12 kW. About two-thirds of the electrical energy would be saved in a solar system. Hence, only 5 kW would be needed for the auxiliaries like pumps etc. However, the initial cost would be much higher because of the cost of collectors, and because of the extra cost of vapour absorption cooling system as compared to a vapour compression system.

In a solar cooling system, the necessary collector area needed for the cooling is given by

$$A = \frac{Q_C}{Q_I \times \eta_C \times \eta_S \times COP} \tag{12.17}$$

where A is the area of collector, Q_c the cooling capacity, Q_I the total solar energy incident on the collectors, η_C the average collector efficiency, η_S the average efficiency of storage and heat transport and COP the coefficient of performance.

In order to calculate the collector area, we assume that

(1) COP = 0.5
(2) Cooling system works 16 h a day
(3) Average insolation incident on collector is 4600 kcal/m^2 a day
(4) Daily collector efficiency is 32%,

Thus, from Eq. (12.17)

$$4600 \times A \times 0.32 = \frac{30{,}000 \times 16}{0.5}$$

$$\mathrm{A} = 650\,\mathrm{m}^2$$

$$10\,\mathrm{ton} = 10 \times 3.52\,\mathrm{kW} = 35.2 \times 3600 \times 24 \frac{\mathrm{kJ}}{\mathrm{day}}$$

$$= \frac{35.2 \times 3600}{4.2} \frac{\mathrm{kcal}}{\mathrm{day}}$$

The installed cost of the collectors for this area is estimated as Rs. 700,000 while the extra cost of the adsorption system will be about Rs. 50,000. Thus, the net extra cost of having solar cold storage would be Rs. 750,000.

$$\text{The annual savings in electricity} = 10 \times 365 \times 16 = 58{,}400\,\mathrm{kWh}$$

Assuming that the average electricity price over the next 10 years will be Rs. 1.25/kWh, we find that the annual savings in electricity are equivalent to Rs. 58,400 × 1.25 = Rs. 73,000.

The annual cost of operating the collector system may be assumed to be 2% of the first cost, i.e., Rs. 14,000/-.

Thus, the net annual savings = Rs. 73,000 − 14,000 = Rs. 59,000.

The pay back for the extra initial cost = Rs. 750,000/59,000 = 12.7 years.

This payback period is quite high and for this reason, solar cold storage cannot be considered as a good proposition at the moment. However, it is likely that the cost of electricity will rise considerably in the next 10 or 12 years, while the cost of collectors will not increase at the same rate because the volume of production will increase. If this happens the payback period may reduce to a value of 5–6 years. Probably, the wisest investment at the moment is to obtain thermal energy by biomass and biogas for operating a vapour adsorption system.

Nowadays, more countries are endeavouring to exploit renewable energy than ever before. Therefore, the world community decided to utilise the renewable energy sources for refrigeration and cooling applications. The solar energy have been of considerable interest among the renewable energy resources because of their promising advantages. Today, there are about 1000 solar thermal cooling systems installed worldwide. Solar thermal cooling technologies are being used all over the world for industrial and home cooling purposes. The various solar refrigeration and cooling technologies available are compared in terms of performance and initial cost are presented in Table 12.2 and Fig. 12.1.

Although COP of thermoelectric and solar VCR systems is high, from Fig. 12.1, it is clear that these systems are more expensive. In solar thermomechanical and solar photovoltaic systems, vapour compression system is cost-effective than the solar photovoltaic systems.

VCR: Vapour compression refrigeration system, VAB (SE): Vapour absorption (single effect), VAB (DE): Vapour absorption (double effect), VAD: Vapour adsorption refrigeration system, ES: Ejector system, TE: Thermo electric refrigeration system, VAB (DE): vapour absorption (double effect), VAD: vapour adsorption refrigeration system, ES: ejector system, TE: thermoelectric refrigeration system.

Although an ejector system costs less than the other sorption cooling systems, it has very low COP and its unit cost is very high. Absorption and adsorption cooling technologies are comparable in terms of performance. But presently, an adsorption system is more expensive than an absorption system.

For this comparison, solar electric systems are assumed to be equipped with 10% efficient solar photovoltaic panels with a unit price of €5/Wp. These panels are able to convert 1000 W/m_2 into 100 W of electricity. The solar panels are integrated to the solar cooling devices transforms this electrical power to the cooling power based on its specified COPs., e.g., consider a VCR system of 1000 W capacity, solar energy

Table 12.2 Comparison of cycle COP and COST for various cooling options

Cooling system	Cycle COP	Cost €/kW cooling
VCR	3	2500
Absorption		
Single effect	0.8	1100
Double effect	1.2	1300
Ejector system	0.3	1700
Thermo electric	2	3300
Stirling	1.7	3800
Thermoacoustic	2	3600
Magnetic cooling	3	3500
Vapour adsorption	0.6	1400

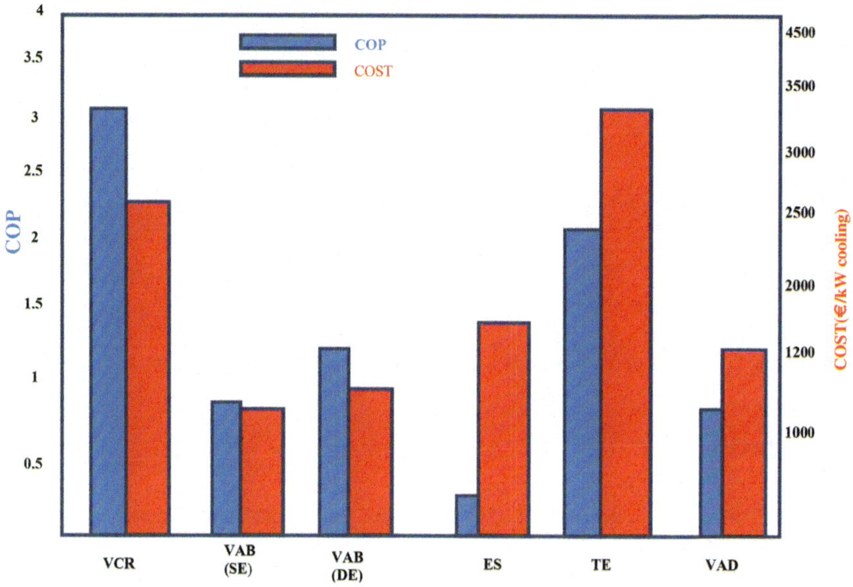

Fig. 12.1 Comparison of COP and cost of various solar cooling systems

converted into 100 W equivalent electrical energy by PV module. Assume the COP of the system as 3 then the cooling power produced is 300 W. For 300 W of cooling power the investment is electrical input, 100 W x€5/Wp = €500. So, for 1 kW cooling the investment for PV module is 1000 × 1.67 = €1700/-(approx.).

Cost of the chiller and engine unit is €800 (approx.). Hence the approximate investment (without installation cost) cost of solar VCR system is €2500. The research indicates that the low-temperature sorption systems are useful. This is because first, the cost of a solar collector system tends to increase with working temperature more rapidly than the COP of a sorption machine does. Secondly, VAB systems require a very high heat source temperature for its operation. Current solar absorption cooling systems is not likely to deliver much financial benefit. In these systems, the annual cost of solar systems was always higher than the conventional electric powered system of which the largest portion is taken up by the solar collectors. In order to reduce the initial cost of vapour absorption cooling systems, an absorption chiller should be made to work with less or cheaper solar collectors. That is, either the chillers COP should be increased, or its driving temperature should be lowered. The coefficient of performance of absorption cooling systems is better than that of adsorption systems. The adsorption system having the advantages of handling higher temperature issues. Moreover, solar hybrid cooling systems can provide higher capacity and better coefficients of performance by eliminating some of the problems encountered with individual working pairs.

On the basis of the cost and cooling capacity requirements, the status of suitability of various solar cooling options are as indicated in Table 12.3.

Table 12.3 Status of solar cooling options

Option	Single 1 tonne	Family 3 tonne	Multifamily capacity 10–20 tonnes	Commercial capacity 20–75 tonnes
Absorption	N	D	D	D
Adsorption	C	D	C	N
Desiccant heat pump	C	D	D	D
Rankine engine	N	D	D	D
Photovoltaic	C	D	C	C

Symbol are: D = under development, N = Not suitable C = Potential candidate

References

Bejan, A., Tsatsaronis, G. & Moran, M. (1996). Thermal Design and Optimization John Wiley and Sons. *Inc. New York*.

Duffie, J. A. & Beckman, W. A. (1974). Solar Energy Thermal Processes Wiley Interscience. *New York*.

Sukhatme, S. P. & Nayak, J. K. (2017). Solar Energy. McGraw Hill Education.

Index

S. C. Kaushik et al., *Solar Cooling*, https://doi.org/10.1007/978-3-031-42410-6

The manufacturer's authorised representative in the EU is Springer Nature Customer Service Centre GmbH, Europaplatz 3, 69115 Heidelberg, Germany. If you have any concerns regarding our products, please contact ProductSafety@springernature.com

Printed and bound by CPI Group (UK) Ltd, Croydon, CR0 4YY

01/10/2026

02225907-0013